The List of the 20th-Century Chinese Architectural Heritage [Vol. 2]

中国20世纪建筑遗产名录 [第二卷]

丛书主编　中国文物学会　　中国建筑学会
本卷编著　北京市建筑设计研究院有限公司
　　　　　中国文物学会20世纪建筑遗产委员会
　　　　　中国建筑学会建筑师分会

Series Editor　Chinese Society of Cultural Relics
　　　　　　　The Architectural Society of China

Volume Editor　Beijing Institute of Architectural Design (Group) Co., Ltd
　　　　　　　Chinese Society of Cultural Relics Committee on 20th-Century Architectural Heritage
　　　　　　　Institute of Chinese Architects, ASC

天津大学出版社
TIANJIN UNIVERSITY PRESS

《中国20世纪建筑遗产名录（第二卷）》编委会

丛书主编

中国文物学会　中国建筑学会

本卷编著

北京市建筑设计研究院有限公司　中国文物学会20世纪建筑遗产委员会　中国建筑学会建筑师分会

参编单位

北京市建筑设计研究院有限公司　中国建筑设计研究院有限公司　华东建筑集团股份有限公司　天津市建筑设计研究院有限公司
中国建筑西北设计研究院有限公司　深圳市建筑设计研究总院有限公司　北方工程设计研究院有限公司
河北省建筑设计研究院有限责任公司　广东省建筑设计研究院有限公司　中南建筑设计院股份有限公司
香港华艺设计顾问（深圳）有限公司　天津华汇工程建筑设计有限公司　重庆市设计院有限公司　重庆市文化遗产研究院
湖北省古建筑保护中心　广州市设计院　等

顾　问

吴良镛　谢辰生　关肇邺　傅熹年　彭一刚　陈志华　张锦秋　程泰宁　江欢成　何镜堂　郑时龄　钟训正　费　麟
刘景樑　黄星元　魏敦山　王小东　王瑞珠　唐玉恩　刘叙杰　邹德侬　李拱辰　孙国城　布正伟　顾孟潮　郭　旃

主　任

单霁翔　修　龙　马国馨

副主任

伍　江　崔　愷　孟建民　王建国　常　青　庄惟敏　徐全胜　张复合　黄　元　路　红　李　琦　龙卫国　熊中元　周　恺
马华山　何智亚　李秉奇　张　宇　周　岚　张立方　沈　迪　文　兵　宋　源　刘若梅　孙兆杰　郭卫兵　徐千里

名誉主编

马国馨　单霁翔　修　龙

委　员（姓氏笔画为序）

马震聪　王时伟　孔宇航　付清远　叶依谦　孙宗列　刘伯英　刘克成　刘若梅　刘志华　刘临安　刘　谞　刘燕辉　刘曙光　吕　舟
朱光亚　永昕群　李兴钢　汤羽扬　邵韦平　邱　跃　张之平　张玉坤　张　兵　张　杰　张　松　张　颀　张　谨　李华东　李　季
李　沉　李海霞　杨昌鸣　杨立新　杨剑华　吴　晓　陈同滨　陈伯超　陈爱兰　陈　雄　陈　薇　陈日飙　陈　飞　陈　雳　陈　雄
陈　纲　宋雪峰　青木信夫　周　岚　周学鹰　金　磊　范　欣　罗　隽　侯卫东　胡　越　胡　斌　胡　燕　柳　肃　赵万民
赵元超　赵燕菁　倪　阳　桂学文　贾　珺　奚江琳　徐苏斌　徐　锋　殷力欣　郭　玲　郭　旃　袁东山　曹晓东　崔　彤　崔　勇
梅洪元　龚　良　董　明　彭长歆　韩林飞　韩振平　舒　莺　舒　平　覃　力　赖德霖　谭玉峰　薛　明　薛绍睿　戴　俭　戴　路

主　编

徐全胜　宋　源　熊中元

策　划

金　磊

副主编

金　磊（常务）　殷力欣　张　松　彭长歆　李　沉　罗　隽

执行主编

金　磊　韩振平　苗　淼　朱有恒

执行编辑

李　沉　殷力欣　苗　淼　朱有恒　董晨曦　刘东昭　舒　莺　林　娜　周高亮　张丽丽　郭　颖　季也清
金维忻　陈　雳　王　展　王　碑　刘博超

美术编辑

朱有恒　董晨曦

图片提供（由中国文物学会20世纪遗产委员会组织）

侯凯源　万玉藻　杨超英　刘锦标　傅　兴　李　沉　殷力欣　马国馨　朱有恒　杨雁云　刘东昭　张广源　金　磊　陈　鹤
彭长歆　周高亮　傅忠庆　舒　莺　叶诗锋　路　霞　刘晓乐　等（部分图片来自网络）

新媒体支持

慧智观察微信公众号　新潮澎湃微信公众号

翻　译

中译语通科技股份有限公司
《中国建筑文化遗产》编辑部　朱有恒　苗　淼

特别鸣谢

国家图书馆 李东晔、秦皇岛市建筑设计研究院 倪明、北戴河文物保管所 闫宗学、长春伪满皇宫博物馆 杨宇、广东省文物考古研究所 崔俊、太原师范学院 尉宏波、黑龙江省文化和旅游厅 盖立新、黑龙江省东北烈士纪念馆 张彬、黑龙江省大庆市文化旅游局 颜莉、大庆市博物馆副馆长 颜祥林、哈尔滨文旅集团有限公司 张元、齐齐哈尔市文保中心 霍晓东、天津市保护风貌建筑办公室 李琦琳　等

谨以此书献给为中国 20 世纪建筑创造经典且不懈传承遗产的人们

This book is dedicated to the people who have made the classic works and inherited the heritage of Chinese architecture in the 20th century

PREFACE 序

单霁翔

中国文物学会会长
故宫博物院院长
中国文物学会 20 世纪建筑遗产委员会会长
President of the Chinese Society of Cultural Relics
Director of the Palace Museum(former)
Chairman of the CSCR-C20C

（本序系单霁翔为《中国20世纪建筑遗产名录[第一卷]》所作）

2016年7月，中国又有两个遗产项目——"广西左江花山岩画文化景观"和"湖北神农架"，分别进入联合国世界文化与自然遗产之列，至此中国的世界遗产项目已达50项。同时，我也十分欣喜地看到，世界现代主义设计大师、著名建筑家勒·柯布西耶分别建在7个国家的17个项目终于入选世界文化遗产名录。这不仅是对申报国家坚持申报的褒奖，更是对以柯布西耶为代表的20世纪"旗手"般的建筑大师的作品与思想的尊重与奖赏，更是向全世界建筑师发出了信号：要像大师那样去创作、去思考，每个建筑师手中的建筑作品都有成为令人仰慕的文化遗产的机会。由申报"世遗"的作品，我想到了中国文物学会20世纪建筑遗产委员会所做的工作，翻阅了《中国20世纪建筑遗产名录（第一卷）》的清样之后，尤为兴奋。因为经过多年的努力，中国不仅有了20世纪建筑遗产保护的专家团队，还评选出98个"首批中国20世纪建筑遗产"项目。我由衷地为《中国20世纪建筑遗产名录（第一卷）》一书的出版而高兴，并对此表示衷心的祝贺。

In July, 2016, two more Chinese heritage properties, "Rock Paintings of the Huashan Mountain in Zuojiang, Guangxi Zhuang Autonomous Region" and "Shennongjia Nature Reserve in Hubei Province", were put on the lists of UNESCO World Cultural and Natural Heritage respectively. Up to now, there are 50 world heritage sites in China. I am also delighted to see that 17 items in 7 countries designed by Le Corbusier, the world-renowned master of modern architecture, had eventually been selected into the World Cultural Heritage List. This is not only a reward for the states' effort to have the properties inscribed, but, more importantly, a tribute to iconic architects of the 20th century represented by Le Corbusier for their works and thinkings. It has sent a signal to the architects all over the world that they should treat their creative works like a master, because every architecture they made has the chance to become world heritage. The application for the World Heritage of architectural works reminds me of the work of the 20th-Century Architectural Heritage Committee under the Chinese Society of Cultural Relics (CSCR). I feel even more excited after I went through the final proof of *The List of the 20th-Century Chinese Architectural Heritage (Vol. 1)*. Because after many years of effort, China now not only has an expert team committed to the conservation of the 20th-century architectural heritage, but also has identified the first list—altogether 98 properties—of the 20th-century architectural heritage. I am truly elated at, and would like to congratulate on, the publication of *The List of the 20th-Century Chinese Architectural Heritage (Vol. 1)*.

"20世纪遗产"，顾名思义是根据时间阶段划分的文化遗产的集合，包括了20世纪历史进程中产生的不同类型的遗产。20世纪是人类文明进程中变化最快的时代，对中国来说，20世纪具有特殊的意义，在20世纪的一百年间，我国完成了从传统农业文明到现代工业文明的历史性跨越。没有哪个历史时期能够像20世纪这样，慷慨地为人类提供如此丰富、生动的文化遗产；面对如此波澜壮阔的时代，才能让文化遗产在百年历程中呈现得最为理性、直观和广博。以20世纪所提供的观察世界的全新视角，反思和记录20世纪社会发展进步的文明轨迹，发掘和确定中华民族百年艰辛探索的历史坐标，对于今天和未来都具有十分重要的意义。我很高兴首次"中国20世纪建筑遗产名录"的公布得到了中国建筑学会的大力支持，并且我们能够携手推进这项功在当代、利在千秋的伟大事业。

"The 20th-century heritage", obviously, is an aggregate of cultural heritage properties of that specific period of time. It includes heritage of different types generated in the 20th century. The 20th century was an era that saw the fastest changes in the course of human civilization. For China, it was particularly significant. The 100 years saw China achieve a historic leap from traditional agricultural civilization to modern industrial civilization. No other historic era has so generously left such abundant, vivid cultural heritage as the 20th century. Only with the greatness of the era in mind can we highlight the cultural heritage in the most reasonable, objective and broad fashion. The time dimension of the 20th century and offers a brand-new perspective of observing the world to reflect upon and record the social development and progress in the 20th century and explore and determine the historical coordinates of the Chinese nation's strenuous quest over the past century. It is extremely significant to the present and the future. I am also pleased to see that the debut of the List of the 20th-Century Chinese Architectural Heritage has won vigorous support from the Architectural Society of China. We can join hands to advance this great cause contributing to not only this generation, but many more to come.

早在2004年8月，马国馨院士领导的中国建筑学会建筑师分会就向国际建协等学术机构提交了一份20世纪中国建筑遗产的清单，关注点是对那些存在损毁危险或需要立即得到保护的建筑予以重视。这些建筑既包括20世纪上半叶的燕京大学、上海外滩建筑群等建筑遗产，也包括20世纪下半叶的重庆人民大礼堂（1954年）、北京儿童医院（1954年）、北京电报大楼（1958年）、厦门集美学村（1934—1968年）等22处现代建筑，蕴含着大量20世纪的珍贵历史信息。2008年4月，在无锡召开的中国文化遗产保护无锡论坛通过了《20世纪遗产保护无锡建议》，同时，国家文物局发布了《关于加强 20 世纪建筑遗产保护工作的通知》。2012 年 7 月，中国文物学会、天津大学等单位在天津举办"首届中国 20 世纪建筑遗产保护与利用论坛"，会上通过了涉及六方面意向的《中国 20 世纪建筑遗产保护的天津共识》。2014 年 4 月 29 日，在来自全国各地的长期以来为保护中国文化遗产而积极呼吁的各位保护专家，为延续中华文明的辉煌历史而辛勤创造的各位建筑大师的见证下，酝酿已久的中国文物学会 20 世纪建筑遗产委员会（Chinese Society of Cultural Relics Committee on the 20th-Century Architectural Heritage，CSCR-C20C）正式成立，自此，中国 20 世纪建筑遗产保护工作有了自己的专家团队。

As early as August, 2004, the Institute of Chinese Architects, subordinated to the Architectural Society of China, under the leadership of Academician Ma Guoxin, submitted a list of the 20th-century Chinese architectural heritage to academic institutions including the International Union of Architects, with the purpose of calling attention to the conservation of those buildings facing the danger of destruction or demanding immediate protection. These buildings include not only the properties of architectural heritage from the the first half of the 20th century, such as Yenching University and the architectural complex of the Bund, but also 22 modern buildings like Chongqing People's Auditorium (1954), Beijing Children's Hospital (1954), Beijing Telegraph Building (1958), Xiamen Jimei School Village (1934-1968), which contain abundant precious information of the 20th century. In April, 2008, Wuxi Forum on the Conservation of China's Cultural Heritage, passed the *Wuxi Proposals on the Conservation of the 20th-Century Heritage*. In the meantime, the State Administration of Cultural Heritage promulgated the *Notice on Strengthening the Conservation of the 20th-Century Architectural Heritage*. In July, 2012, the Chinese Society of Cultural Relics and Tianjin University jointly, etc. held the First Forum on the Conservation and Exploitation of the 20th-Century Chinese Architectural Heritage. The forum passed the *Tianjin Consensus on the Conservation of China's 20th-Century Architectural Heritage*, which was involving 6 intentions. On April 29, 2014, the experts from all over China who had actively called out for the conservation of Chinese cultural heritage officially launched the Chinese Society of Cultural Relics Committee on the 20th-Century Architectural Heritage (CSCR-C20C) after a long time of preparation, as witnessed by many master architects who had diligently created works to continue the splendid history of Chinese civilization. Since then, the conservation of the 20th-century Chinese architectural heritage has been in the hands of an expert team.

自 2014 年 5 月，经过中国文物学会 20 世纪建筑遗产委员会 97 位顾问专家委员们的权威推荐、严谨把关，在北京市方正公证处的监督下，历经严密的初评、终评流程，最终于 2015 年 8 月 27 日产生了"首批中国 20 世纪建筑遗产项目"，共计 98 项。作为"首批中国 20 世纪建筑遗产项目"评选认定活动的重要成果之一，《中国 20 世纪建筑遗产名录（第一卷）》以图文并茂的形式全面展示了中国 20 世纪建筑遗产项目的风采，不仅为更多的业界人士及公众领略 20 世纪建筑遗产的魅力与价值提供了重要渠道，更向世界昭示了中国 20 世纪不仅有丰富的建筑作品，也有对世界建筑界有启迪意义的建筑师及其设计思想。为此，本书挖掘了主持创作"首批中国 20 世纪建筑遗产"项目的建筑师并将其作品和思想呈现在书中。这种"见物见人"的做法是文化遗产保护、传承、发展的好经验，应大力推广。

Starting from May, 2014, recommended strictly by 97 expert consultants of the CSCR Committee on the 20th-Century Architectural Heritage, supervised by Beijing Fangzheng Notary Public Office, through rigorous preliminary and final review, the First List of the 20th-Century Chinese Architectural Heritage including a total of 98 properties was published. As one of the important achievements of the selection of the First List of the 20th-Century Chinese Architectural Heritage, *The List of the 20th-Century Chinese Architectural Heritage (Vol.1)* fully showcases the properties of the 20th-century Chinese architectural heritage with excellent illustrations and texts. Not only does it provide an important channel for the industry and the public to appreciate the charm and value of the 20th-century architectural heritage, but, more importantly, it shows the world that China in the 20th century has not only created a multitude of architectural works, but also left architects who, with their design philosophies, are sources of inspiration to world architecture. The book has searched for the architects in charge of the properties on the First List of the 20th-Century Chinese Architectural Heritage and put them in the book. This practice which pays attention to both architecture and their architects offers good experience for the conservation, inheritance and development of cultural heritage. Thus it should be vigorously promoted.

认真研究优秀的20世纪建筑遗产，思考它们与当时社会、经济、文化乃至工程技术之间的互动关系，从中汲取丰富的营养，成为当代和未来理性思考的智慧源泉。文化遗产是有生命的，这个生命充满了故事。20世纪遗产更是承载着鲜活的故事，随着时间的流逝，故事成为历史，历史变为文化，长久地留存在人们的心中。如果说改变与创新需要智慧，那我更认为对中国20世纪建筑遗产保护事业要有敬畏之心，要有跨界思维，要有文化遗产服务当代社会的新策略。中国文物学会将一直支持为20世纪中国建筑设计思想"留痕"的工作，将与中国建筑学会合作，热情期待这项旨在保护中国城市文脉、建设人文城市之举持续地开展下去。

Studying carefully the outstanding the 20th-century architectural heritage, reflecting upon their interaction with society, economy, culture and construction technology and absorbing rich nutrition from therein have become a source of wisdom in contemporary and future rational thinking. Cultural heritage has its life—a life filled with stories. The 20th-century heritage embodies lively stories. With the passage of time, stories will become history and history will become culture that takes roots in people's heart. If change and innovation demand wisdom, I believe that we should hold in awe the cause of conserving the 20th-century architectural heritage. We should have interdisciplinary perspectives and new strategies enabling cultural heritage to serve contemporary society. The Chinese Society of Cultural Relics will continue to support the passing-down of Chinese thinkings on architectural design. It will work together with the Architectural Society of China to carry on the initiative to conserve the culture of Chinese cities and build cities with culture.

特此为序。

Such is the preface.

2016年7月

July, 2016

PREFACE 序

马国馨

中国工程院院士
全国工程勘察设计大师
中国文物学会 20 世纪建筑遗产委员会会长
北京市建筑设计研究院有限公司总建筑师
Academician with the Chinese Academy of Engineering
Chinese National Master of Engineering Survey and Design
Chairman of the CSCR-C20C
Chief Architect of BIAD

（本序系马国馨为《中国 20 世纪建筑遗产名录 [第一卷]》所作）

近来，新闻媒体发表了习近平总书记关于文化遗产保护的一系列重要论述和批示，引起了全社会的热切关注。习总书记在北京考察工作时说："历史文化是城市的灵魂，要像爱惜自己的生命一样保护好城市历史文化遗产。北京是世界著名古都，丰富的历史文化遗产是一张金名片，传承保护好这份宝贵的历史文化遗产是首都的职责。要本着对历史负责、对人民负责的精神，传承历史文脉，处理好城市改造开发和历史文化遗产保护利用的关系，切实做好在保护中发展、在发展中保护。"在考察新农村建设时，他指出，要"注意乡土味道，保留乡村风貌，留得住青山绿水，记得住乡愁"。他还强调："让文物说话，把历史智慧告诉人们，激发我们的民族自豪感和自信心，坚定全体人民振兴中华、实现中国梦的信心和决心。"

Recently, a series of important arguments and instructions by General Secretary Xi Jinping have been published in the media and caught the keen attention of the entire society. When making a inspection in Beijing, Xi said: "History and culture are the soul of a city. We should cherish the historical and cultural heritage of our cities as our own lives. Beijing is a world-famous ancient capital. Its abundant historical and cultural heritage is its golden business card. Passing on and conserving the precious heritage are the responsibility of the city. We should fulfil the responsibility for history and for the people by passing on the heritage, properly handling the relationship between urban reconstruction and heritage's conservation and utilization. Development and conservation should go hand in hand." When he inspected new rural construction, he pointed out that "we should preserve the country style and the landscape of the countryside, keep mountains green and water clean, and retain our nostalgia." He also stressed: "Let cultural relics speak for itself and tell historical wisdom to the people, kindling pride for and confidence in our nation and steeling the people's determination for the rejuvenation of the nation and the fulfilment of the Chinese dream."

正是在这一系列重要指示精神的指引下，2014年4月成立的中国文物学会20世纪建筑遗产委员会明确了委员会成立的重要使命，确定了自己的定位，也看到了我们所面临的挑战，把人们酝酿已久的20世纪建筑遗产保护正式列入议程，为第一批20世纪建筑遗产的遴选和认定做了大量深入细微的工作。在认真研究了中国20世纪建筑遗产的认定标准，经反复讨论取得共识的基础上，经过全国各地97名建筑、文博、文化学者和专家的提名、投票，提出了将98项近现代建筑作为第一批认定的20世纪建筑遗产加以公布。这是一个十分有意义的开端，相信以此项工作为契机，将进一步推进全国20世纪建筑遗产保护工作，这也是学习贯彻习总书记关于文化遗产保护的一系列指示、勇于承担历史责任、主动实践的重要举措。

It was precisely under the guidance of this series of important instructions that the Chinese Society of Cultural Relics Committee on the 20th-Century Architectural Heritage, founded in April, 2014, identified its important mission and determined its orientation. It has also enabled us to see the challenges and officially launched the conservation of the 20th-century architectural heritage, which has been planned for a long time. A lot of work has been done conscientiously to select and identify the First List of the 20th-Century Architectural Heritage. After studying the standards for identifying the 20th-century architectural heritage in China, on the basis of reaching a consensus after considerable discussion, a panel of 97 experts including architects, museologists and cultural scholars nominated and voted to inscribe 98 properties into the First List of the 20th-Century Architectural Heritage, which have been published. This is a meaningful, important beginning. We believe that it will be an opportunity for us to further advance the conservation of the 20th-century architectural heritage nationwide, as well as an important measure for us to learn and implement General Secretary Xi's instructions on cultural heritage conservation, shoulder historical responsibility with courage and take concrete actions.

20世纪是一个充满了重大变革、跌宕起伏的时代，在这百年间，中国的各个领域，包括政治、经济、社会、科学、文化等都发生了巨大的变化。在人类文明不断进步的过程中，人们的价值观、历史观、审美观发生了许多变化。无论是清末，还是国民政府时期以及新中国成立以后，人们所创造的建筑遗产都准确地反映了那个时代，反映了国家和民族的复兴之路，反映了社会的进步和曲折，反映了建筑技术和建筑材料的推陈出新，反映了对创造多样的建筑形式和类型的追求。随着时间的不断流逝，我们身边的建筑物成为了时代的重要历史见证，其历史价值和文化价值逐渐为人们所认识。尤其是随着城市化建设的热潮涌现，城市范围不断扩大，人口不断增长，城市发展和保护的矛盾也日渐突出。因此，20世纪建筑遗产的认定和保护已成为一个极为紧迫的严肃课题。

The 20th century was a era full of important changes, ups and downs. The 100 years had seen China undergo tremendous changes in all fields—politics, economy, society, science and culture, etc. As human civilization continues to progress, a lot of changes have taken place in people's values, history and aesthetics. Whether in the late Qing Dynasty, or in the period of the Republic of China, or after the founding of PRC in 1949, architectural heritage that created by human beings, can accurately represent and reflect the time, the path of rejuvenation along which our nation has travelled, the progress our society has made and the vicissitudes it has been through, the advancements in architectural technology and materials, and the efforts made to diversify architecture in form and type. Therefore, with the passage of time, the architecture around us is an important witness to the time and history. Its historical and cultural value has gradually been recognized, particularly amid our current rush to urbanization whilst cities are constantly expanding and the population is continuously growing. The contradiction between urban development and conservation is also becoming prominent day by day. Therefore, the identification and conservation of the 20th-century architectural heritage have become a pressing question compelling us to answer.

在 2016 年 6—7 月，两件事情引起了学界有识之士的密切关注。

During June and July, 2016, two incidents caught close attention in academic circles.

2016 年 6 月 21 日，北京中关村一座具有 63 年历史的科研建筑被拆除了。这栋不那么起眼的被称为"共和国科学第一楼""中国原子弹的起点"的五层建筑，为了新建国家纳米科学中心而在机械的轰鸣声中化为一堆残垣断壁。在该建筑被拆除的前一天，《科技时报》在头版以《京城之大，能容得下小小的原子能楼吗？》为标题进行了呼吁。同时配发了整版文章《共和国科学第一楼的尘封往事》，详细介绍了这栋"颜值"不高的"大将"如何成为了我国第一个综合性科学研究机构，又如何衍生出许多国家科学研究机构。当年中科院成立研究所时，近代物理研究所名列名单之首，吴有训、钱三强先后任所长，1958 年时更名为"原子能研究所"，"原子能楼"由此而得名。科研人员白手起家，一切从零开始，为我国的原子能研究事业奠定了坚实的基础。更重要的是，中国核物理科学的重要奠基人和开拓者都集聚于此，从这栋楼里走出了六位"两弹一星功勋奖章"的获得者，两位国家最高科技奖的获得者，三十多位科学院和工程院的院士。许多科学家曾多方呼吁保留此建筑，但最后还是在"手续齐全"的说辞下被拆除。据称有关方面也研究过将老楼的南墙嵌入新楼保留老楼中的加速器这一设备，但恐怕研究保护方案的难度和审批周期都有问题，将"影响新建项目的进度"。人们在质疑，以现代人们的智慧，难道就想不出一个两全其美的办法，既保留历史的记忆，又珍惜人们的情感吗？

On June 21, 2016, a scientific research building with a history of 63 years, situated at Zhongguancun, Beijing, was demolished. The inconspicuous five-story building had been known as "No.1 Building of Science of the People's Republic of China" and "the Starting Point of China's A-Bomb Project". It was reduced to debris in the roar of machinery to make space for the National Center for Nanoscience and Technology. One day before the demolition, the *Science and Technology Times* appealed against the demolition in its front page report titled, "Can't a City as Large as Beijing Accommodate the Small Atomic Energy Building?" It also published a full-page article titled "The Forgotten Past of the No.1 Building of Science of the People's Republic of China", which introduced in a detailed manner how this not-so-tall building became China's first comprehensive scientific research facility and derived many of China's scientific research institutions. Among the institutes established by the Chinese Academy of Sciences, the Institute of Modern Physical Science ranked the first, with Wu Youxun and Qian Sanqiang as directors in succession. In 1958, it was renamed Atomic Energy Research Institute. Hence the building was called "Atomic Energy Building". Researchers started from scratch to lay a solid foundation for

China's research in atomic energy. More importantly, major founders and trailblazers of Chinese nuclear physical science had gathered here, including six winners of the "Two Bombs, One Satellite" Meritorious Medal, two winners of the highest national science and technology award and more than thirty academicians of Chinese Academy of Sciences and Chinese Academy of Engineering. Many scientists had appealed to all authorities concerned to keep the building. However, it was demolished as "all required papers being available". It is said that some interested parties studied the possibility of embedding the south wall of the old building into the new building and keeping the accelerators in the old building. But it did not work. The most important reason might be that not only was a plausible conservation plan hard to work out, but it would also take longer time to go through the approval procedure, which would "affect the progress of the new project". People have doubted: By contemporary wisdom, couldn't we think of a perfect way to resolve this conflict so that we could keep the memories of the past and care for people's feelings?

与此形成对比的是过了不到一个月，7月10日—20日在土耳其伊斯坦布尔召开的第40届世界遗产大会将法国建筑师、艺术家勒·柯布西耶（1897—1965年）的17项建于法国、德国、瑞士、比利时、日本、印度、阿根廷的各类建筑选为世界文化遗产，其中包括早期的法国萨伏伊别墅（1930年）和晚期的印度昌迪加尔（1962年），类型有教堂、公寓、办公建筑、美术馆、城市规划等。委员会称，这些革命性的建筑作品"实现了建筑技术的现代化，满足了社会及人们的需求，影响了全世界，为现代建筑奠定了基础"。他的这些项目在2009年和2011年的两次申请中未成功。与此相类似的是美国这次也申报了美国本土建筑大师赖特的十项建筑作品，虽然这些项目被咨询机构评估为"推迟录入"，但如果联系到1987年联合国教科文组织将巴西建筑师迈耶设计建造的巴西新首都巴西利亚（1960年建成）列入世界文化遗产名录，1994年世界遗产委员会就提出关注"20世纪遗产"的倾向性政策，此后霍尔塔（比利时）、密斯·凡·德·罗（美国）、伍重（丹麦）、巴拉干（墨西哥）等建筑师设计的20世纪知名建筑陆续被列入世界遗产名录，虽然"共和国科学第一楼"和柯布西耶等人的作品在历史文化价值上可以有不同的解读，但仍可以看出中国在20世纪建筑遗产保护上与当前的世界潮流存在着时间和空间上的巨大差距，需要我们努力工作，迎头赶上。

In contrast, less than one month later, during July 10-20, at the 40th World Heritage Conference convened in Istanbul, Turkey, 17 works of different types designed by French architect and artist Le Corbusier (1897-1965) and lcoated in France, Germany, Switzerland, Belgium, Japan, India, and Argentina were inscribed on the World Heritage List. Among them are Villa Savoye (1930) in France, one of his early works, and Chandigarh (1962) in India, one of his later works. The architectures include churches, apartment buildings, office buildings, galleries, urban planning projects, etc. The Committee stated that these revolutionary works of architecture "have realized the modernity of architectural technology, met the needs of society and people, had worldwide influence and laid the foundation for modern architecture". Application for the inscription had been made twice—in 2009 and 2011 respectively—but failed. Similarly, the United States also tried to get ten architectural works designed by American architect Frank Lloyd Wright to be inscribed, though the assessment of the application is "delayed" by the consultancy concerned. Considering that the UNESCO inscribed Brasilia, new capital of Brazil constructed from the design of Oscar Niemeyer (completed in 1960), into the List of World

Cultural Heritage, the World Heritage Committee introduced policies calling attention to the 20th-century heritage in 1994. And the 20th-century renowned buildings designed by architects such as Victor Horta (Beligum), Ludwig Mies van der Rohe (US), Jorn Utzon (Denmark) and Luis Barragán (Mexico) had been inscribed on the List of World Cultural Heritage in succession, though the No.1 Science Building and the works by Le Corbusier are open to different interpretations in terms of historic and cultural value, it can be seen that China is lagging behind the trends of the world in the conservation of the 20th-century architectural heritage. We need to work hard to catch up.

正如习近平总书记所说："不忘本才能开辟未来，善于继承才能更好创新。"建筑遗产，包括20世纪建筑遗产是我国文化遗产的重要组成部分，也是人类文明的重要组成部分。对建筑遗产的历史价值、文化价值、社会价值的认同是中华民族强大凝聚力的具体体现，也是我们自觉、自信的体现。当然，我们也面临着科学界定，及时的法律保障，成熟的技术和经验，保护及合理利用等一系列问题。如前所述的"共和国科学第一楼"的拆除，从法律上说得过去，拆除理由看上去也还充分，该建筑未列入文保单位的名单，但《文物保护法》中专门列出了"尚未核定公布为文物保护单位的不可移动文物"，这说明对遗产价值的认识尤其是潜在价值的认识是有过程的，许多尚未列入文保单位或世界遗产的项目很可能就是日后的文保单位或世界遗产，这正是对我们智慧的考验。如果以习总书记所说的"见证历史，以史鉴今，启迪后人""把历史智慧告诉人们，激发我们的民族自豪感和自信心，坚定全体人民振兴中华，实现中国梦的信心和决心"来对照，不是十分值得我们深思吗？

Just as General Secretary Xi said, "Only if we do not forget about the past can we usher in a better future; only if we are good at passing on heritage can we fare better in innovation." Architectural heritage, that of the 20th century included, is an important component of China's cultural heritage and human civilization. Our awareness of the historical, cultural and social values of architectural heritage is a representation of the cohesiveness of the Chinese nation, as well as our self-consciousness and confidence. Of course, we are faced with a series of questions with regard to scientific definition, timely legal protection, sophisticated technology and experience and the relationship between conservation and proper utilization. The demolition of the No.1 Science Building of the People's Republic of China was not against the law; the reason for its demolition seemed fairly tenable as it was not the building named a historic site under protection. However, the Cultural Relics Protection Law has mentioned specifically "immovable cultural relics that have not been declared historic sites are under protection". This shows that it takes time to see the value, especially potential value, of heritage. Some properties that have not been placed under protection are likely to be protected. Before that our wisdom is put to trial. If we can "witness history, examine the present with history as a mirror and enlighten the posterity" as the General Secretary had put, if we can "tell historical wisdom to the people, kindling pride for and confidence in our nation and steeling the people for the rejuvenation of the people and the fulfilment of the Chinese dream", aren't there many things well worth contemplating?

建筑遗产作为不可移动的物质遗产，有着不同的使用寿命、结构寿命、商业寿命、遗址历史文化寿命。在迅猛的城市化大潮中，我们往往急功近利、目光短浅，造成了不可挽回的损失和遗憾。在中国文物学会和中国建筑学会的指导下，中国20世纪建筑遗产委员会将以本次遗产名录的认定为起点，和全国的有志之士一起，把关系到我们的城市、城市中的人物和事件的历史记忆更好地传承下去，让这些丰富的人文内涵充实我们的民族记忆，继承伟大的民族精神。

Architectural heritage, as immovable material heritage, varies in usable life, structural life, business life and cultural life. In the hustle and bustle of this urbanization, we tend to be obsessed with profits and gains and short-sighted. As a result, some irretrievable damages and regrets have been caused. Directed by the Chinese Society of Cultural Heritage and the Chinese Society of Architecture, the Chinese 20th-Century Architectural Heritage Committee will start with the identification of the heritage list, work together with those across China who have the will to carry on the memories of personages and events related to our cities, so that their cultural connotation can enrich the memory of our nation and inherit the great national spirit.

2016 年 7 月

July, 2016

PREFACE 序

修　龙

中国建筑学会理事长
中国建筑科技集团股份有限公司董事长
Director of the Architectural Society of China
Board Chairman of China Construction Technology Consulting Co. Ltd

（本序系修龙为《中国20世纪建筑遗产名录[第一卷]》所作）

记得2016年2月2日，农历丙申年到来前夕，我曾在故宫博物院与中国文物学会会长单霁翔先生会面，当时就中国建筑学会与中国文物学会通力合作、共同促进建筑文化遗产保护事业等议题进行了深入广泛的探讨，尤其对中国20世纪建筑遗产保护方面展开深入合作达成共识。半年后的今天，《中国20世纪建筑遗产名录（第一卷）》一书的清样呈现在我的面前。认真品读后，我认为这是一部凝结了中国文物学会和中国建筑学会共同对20世纪建筑遗产保护研究成果的心血之作，更是双方学会在中国20世纪建筑遗产保护事业中携手同行的合作标志。

I remembered that it was on February 2nd, 2016, right before the Year of Bing Shen according to Chinese lunar calendar, when I had a meeting with Mr. Shan Jixiang, president of the Chinese Society of Cultural Relics (CSCR) at the Palace Museum. We had an extensive and in-depth discussion on the collaboration between the Architectural Society of China (ASC) and CSCR and the joint effort to promote the conservation of architectural heritage; and then reached consensus upon deepening our cooperation in the conservation of the 20th-century Chinese architectural heritage. Today, half a year later, lying right before my eyes is the final proof of *The List of the 20th-Century Chinese Architectural Heritage (Vol. 1)*. After perusal of it, I believe that this is the fruition of the hard work for ASC and CSCR in the conservation of the 20th-century Chinese architectural heritage, as well as a symbol of the collaboration between the two societies in the cause of the 20th-century Chinese architectural heritage conservation.

中国建筑学会成立于 1954 年，至今已走过 62 年的历程，从成立之初就一直重视传统建筑文化的传承与弘扬，更沿袭着梁思成、刘敦桢等建筑文化大家的文化遗产观。中国建筑学会的宗旨就是推进中国建筑文化的发展与繁荣，而中国 20 世纪建筑遗产作为中国建筑文化不可缺少的璀璨瑰宝，对它的保护与利用从根本上就是与中国建筑学会工作理念完全契合的。事实上，中国建筑学会在历年的奖项评选工作中也一直关注近现代建筑项目的挖掘，最典型的是 2009 年举行的"新中国成立 60 周年建筑创作大奖"评选活动。活动回顾了新中国成立 60 年以来不同时期的优秀建筑作品，专家们认真评选出 300 项获奖作品，其中包括人民大会堂、民族文化宫、人民英雄纪念碑、中国美术馆、重庆人民大礼堂等新中国建筑项目，而它们也恰恰出现在"首届中国 20 世纪建筑遗产项目"名单中，这充分验证了这些承载着城市文脉与文化记忆的建筑作品是经得起时间考验的，不愧为业内专家及社会大众心中永恒的经典之作。

The Architectural Society of China, founded in 1954, has been around for 62 years. Since its foundation, importance has been attached to carrying on and forwarding Chinese architectural traditions and inheriting the outlook on cultural heritage from architecture masters such as Liang Sicheng and Liu Dunzhen. The mission of ASC is to promote the development and prosperity of Chinese architectural culture. The 20th-century Chinese architectural heritage is an indispensable part of Chinese architectural culture, and its conservation and utilization are essentially congruous with the mission of ASC. As a matter of fact, in the award selection of the past few years, ASC has always paid attention to modern architecture projects. The most illustrating example is the selection of the Architectural Creation Award at the 60th Anniversary of the Founding of the People's Republic of China held in 2009. The event reviewed the outstanding works of architecture created in different periods after the foundation of the People's Republic of China. The experts carefully picked 300 winning buildings including the Great Hall of the People, the Cultural Palace of Nationalities, the Monument to the People's Heroes, the National Art Museum of China and Chongqing People's Auditorium. They also appeared in the First List of the 20th-Century Chinese Architectural Heritage. This fully proves that the buildings embodying the culture and history of the cities can withstand the trials of time and are rightfully the time-honored classics in the eyes of both the experts and the public.

2015 年 12 月，在中国建筑学会第十二届五次理事会暨九次常务理事会上，我在发言中提出"大建筑观"的构想，引发了与会专家的一致共鸣。我认为，"大建筑观"的树立，必须建立在梳理著名建筑学家的学术思想及贡献基础上。在思考中国建筑文化问题时，建筑先贤们往往从建筑全局出发，从建筑的过去、现在及未来整体走向出发，不会去人为地分割现代建筑与传统建筑。因此，他们留下的作品无论是传统的还是现代的，都堪称20世纪建筑遗产，都是当今中国建筑师应学习的经典项目。我们要号召广大建筑师学习传统匠人对自然的依赖与敬畏。同时，中国传统建筑文化要有现代性，要研究并分析传统建筑的价值应用，要发现何处是传统建筑的继承点，要辨识传统与现代之间的冲突点，要找到理论与实践的困惑点。中国建筑师希望创造自己的现代建筑，希望在保持中国建筑文化并吸纳西方建筑文化精华的基础上，实现中国现当代建筑的理想。

In December, 2015, at the 5th meeting of the 12th Council and the 9th Executive Council of the Architectural Society of China, I set forth the concept of "great outlook on architecture", which evoked the empathy of the attending experts. I believe that the concept of "great outlook on architecture" must be based on a clearer picture of the theories and contributions of the famous Chinese architects—when approaching Chinese architectural culture, architectural scholars of the past tend to follow the line from the past, to the present and to the future, instead of artificially dividing it into modern and traditional architecture. Therefore, their works, whether traditional architecture or modern architecture, can be safely said to be the 20th-century architectural heritage. They are all classic projects from which the contemporary Chinese architects should learn. We thus call upon architects to learn from the traditional craftsmen who depended upon and venerated nature. Moreover, traditional Chinese architectural culture should be endowed with modernity. The use value of traditional Chinese architecture should be studied and analyzed; the features worthy of being passed on should be found out; the conflict between tradition and modernity should be distinguished; and the puzzles in theory and practice should be identified. Chinese architects hope to create their own modern architecture and wish to create Chinese modern and contemporary architecture on the basis of retaining traditional Chinese architectural culture and assimilating the fine parts of the Western architectural culture.

"建筑遗产"本身蕴含着文物保护与建筑设计的双重含义。中国建筑学会关注建筑遗产保护，尤其是对20世纪建筑遗产的保护与利用事业责无旁贷。无论是中国文物学会与中国建筑学会联合向社会公布"首届中国20世纪建筑遗产名录"，还是联名倡言《中国20世纪建筑遗产保护与利用建议书》，这些工作仅仅是双方学会合作的开端。未来，中国建筑学会将借助其深厚的专家资源与专业平台，与中国文物学会携手共同投入到中国20世纪建筑遗产保护与利用事业，让珍贵的中国20世纪建筑遗产既融入人们的生活中，也为繁荣当代中国建筑创作找到可遵循的文化之根。

Architectural heritage itself entails a dual meaning of cultural heritage protection and architectural design. The Architectual Society of China pays close attention to the conservation of architectural heritage, particularly they regard the cause of the conservation and utilization of the 20th-century architectural heritage as their unshirkable responsibity. The joint publication of the First List of the 20th-Century Chinese

Architectural Heritage and the *Proposals on the Conservation and Utilization of the 20th-Century Chinese Architectural Heritage* are merely the beginning of the two societies' collaboration. In the future, the Architectural Society of China will rely upon its pool of scholars and its professional platforms to work together for the cause of conserving and utilizing the 20th-century Chinese architectural heritage, so that the precious the 20th-century architectural heritage may find its way into people's life and provide the roots of culture which gives nutrients to contemporary Chinese architectural creation.

再次祝《中国20世纪建筑遗产名录》(第一卷)出版,是为序。

Congratulations again to the publication of *The List of the 20th-Century Chinese Architectural Heritage (Vol. 1)*. Such is the preface.

2016年8月

August, 2016

IN FRONT
写在前面

传承 20 世纪建筑遗产乃北京建院的使命

Passing on the Architectural Heritage of the 20th Century: The Mission of BIAD

徐全胜

中国建筑学会副理事长
中国建筑学会建筑师分会理事长
北京市建筑设计研究院有限公司党委书记、董事长、总建筑师

Deputy President of the Architectural Society of China
President of the Architectural Society of China Architect Branch
Secretary of the Party Committee and Chairman and Chief Architect of Beijing Institute of Architectural Design

作为国有大型民用建筑设计研究机构，北京市建筑设计研究院有限公司（简称北京建院）始终秉承为新中国建设的使命，在服务首都北京的一系列城市建设中不断创造佳绩。以设计的现代延伸，向前辈建筑师及他们所创造出的经典建筑致敬，这不仅是我们北京建院人的崇尚，更是我们从历史走来、向未来迈进的延续与传承。在此特别祝贺《中国 20 世纪建筑遗产项目名录 [第二卷]》即将出版。

As a large-scale state-owned civil architectural design and research institution, Beijing Institute of Architectural Design Co., Ltd. (BIAD) has always adhered to its mission of building for new China and continuously made accomplishments in a series of urban construction projects serving the capital city of Beijing. With the modern iterations of classic designs, we pay tribute to earlier architects and the classic works created by them. They are not only what we admire, but also the continuations from history with which we advance toward the future. Here we'd like to give special congratulations on the forthcoming publication of *The List of the 20th-Century Chinese Architectural Heritage [Vol.2]*.

2019 年，北京市建筑设计研究院有限公司与新中国共同迎来 70 周年庆典。2018 年我院在国家博物馆举办"都·城 我们与这座城市——北京建院首都建筑作品展"。展示了这个有历史情怀，有责任担当的国有设计机构奉献国家、服务人民的成果，"都与城"最好地诠释出北京建院文化

背景下的全程贡献；2018年12月18日在时任故宫博物院院长单霁翔的指导下，由北京建院与中国文物学会20世纪建筑遗产委员会共同编制的《中国20世纪建筑遗产大典（北京卷）》首发式在故宫博物院举办。该书系新中国建筑史上第一次以省（市）为单元展示20世纪遗产项目，此外它"见物亦见人"，梳理挖掘了为百年北京建筑做出非凡贡献的数十位建筑师、工程师，从而也成为北京20世纪经典建筑作品选的"人物志"。2019年，北京建院为庆祝成立70周年，还隆重推出《北京市建筑设计研究院有限公司·五十年代"八大总"》一书。享誉全国的杨宽麟、杨锡镠、顾鹏程、朱兆雪、张镈、张开济、华揽洪、赵冬日"八大总"，不仅在设计理念与建筑技术上成为新中国初期北京建院乃至中国建筑界的引领型人物，他们中的大多数还有多项作品载入"中国20世纪建筑遗产"的史册。

In 2019, BIAD and New China celebrated the 70th anniversary together. In 2018, our institute held the exhibition, "Capital and City: Beijing Is with Us: BIAD's Architectural Works in the Capital", at the National Museum of China. The exhibition showcases the achievements of this state-owned design instituation with a commitment to passing on historical heritage and a sense of responsibility for serving the nation and people and interprets in the best way possible its contributions throughout the whole process. On December 18, 2018, directed by Shan Jixiang, then-curator of the Palace Museum, *The 20th-Century Chinese Architectural Heritage Classics: Beijing Volume* jointly compiled by BIAD and the 20th Century Architectural Heritage Committee of the Chinese Cultural Heritage Society was launched at the Palace Museum. This book is the first time in the history of New China's architecture to showcase the properties of the 20th-century heritage by provincial division. In addition, it "showcases both architecture and architects", sorting out dozens of architects and engineers who have made extraordinary contributions to Beijing's architecture in the past century and thus adding biographical data to Beijing's the 20th-century classic architectural works. In 2019, to celebrate the 70th anniversary of its founding, BIAD also launched the book *Eight Chief Architects of BIAD in the 1950s*. The eight nationally renowned chief architects and engineers, namely, Yang Kuanlin, Yang Xiliu, Gu Pengcheng, Zhu Zhaoxue, Zhang Bo, Zhang Kaiji, Hua Lanhong, and Zhao Dongri, not only became leading figures in design philosophy and technology in architecture in BIAD and in the sphere of Chinese architecture in the early stage of New China, but most of them also have multiple works recorded as China's 20th-Century architectural heritage.

在2020年北京建院为北京国际设计周举办"北京城市建筑双年展2020先导展"的十余个主题中，我们与中国文物学会20世纪建筑遗产委员会共同推出"致敬中国百年建筑经典——北京20世纪建筑遗产"特展。该展的两大"亮点"如下。其一，通过四个批次共计396项中国20世纪建筑遗产项目的区域分析，展示了约占全国总项目22%（共计88项）的北京属地设计机构的作品，其中北京建院有49项，占56%。翔实的数据说明我院在过往项目上的兢兢业业与坚守始终的文化情怀。其二，荣获20世纪建筑遗产项目推介名录，不仅是设计机构与建筑师的荣光，也反映了中国建筑界与文博正在向"世界遗产名录"看齐的态势。中共十九届五中全会提出建设"文化强国"之目标，这就要求作为文化遗产重要类型的20世纪建筑遗产要在传承与利用上有新贡献与新作为。

Among more than ten themes of the 2020 Preceding Exhibition of Beijing Urban Architecture Biennale organized by BIAD for Beijing International Design Week in 2020, we, jointly with the 20th Century Architectural Heritage Committee of the Chinese Society of Cultural Heritage, launched a special exhibition titled "Beijing's 20th-Century Architectural Heritage: A Tribute to China's Century-Old Architectural Classics". The two highlights of the exhibition are as follows. Firstly, through regional analysis of a total of 396 Chinese 20th-Century architectural heritage projects in four batches, a total of 88 works by design agencies based in Beijing were put on display, which accounted for about 22% of the country's total. Of them, 49 works were from BIAD, accounting for 56%. The detailed data shows BIAD's conscientiousness and commitment at its past projects. Secondly, being inscribed in the Recommendation List of the 20th-Century Architectural Heritage is not only the glory of design agencies and architects, but also reflects the trend that the Chinese architectural circles and cultural museums are aligning with the World Heritage List. The Fifth Plenary Session of the 19th Communist Party of China Central Committee set forth the goal of building a cultural power, which requires that new contributions and achievements should be made in passing on and utilizing the 20th-century architectural heritage, a main genre of cultural heritage.

2019年9月19日，中国文物学会20世纪建筑遗产委员会秘书处办公地落户北京建院，这表明了北京建院对20世纪建筑遗产的重视程度；表明作为一个有担当的国有大型设计机构在完成设计任务时，不忘城市历史底蕴和文化传承，展示了其对标世界一流建筑遗产保护理念及做法的态度，让更多反映北京历史发展的"活化石"绽放当代的建筑文化活力。有鉴于此，北京建院将以更加自觉的遗产保护传承姿态，推动自身乃至全国的20世纪建筑遗产保护事业，留下对于优秀历史建筑、当代建筑更富中国表达力的作品与思想。

On September 19th, 2019, the secretariat of the 20th Century Architectural Heritage Committee of the Chinese Society of Cultural Relics settled in BIAD, which shows that how much importance BIAD attaches to the 20th-century architectural heritage; and that as a responsible large-scale state-owned design agency, while doing designs, it has not forgotten the city's historical implication and cultural inheritance and shown its willingness to align with the concept and practice of world-class architectural heritage protection, so that more architectural "living fossils" reflecting the history of Beijing can have fresh vigor in the contemporary era. In view of this, BIAD will preserve the heritage more consciously, promote the cause of protecting its own and all the Chinese 20th-century architectural heritage, and give a richer Chinese treasure of outstanding historical and contemporary buildings.

2020 年 11 月

Novemember, 2020

Contents
目录

中国 20 世纪建筑遗产传承与创新研究纲要初步（代前言）
Preliminary Study Outline of Inheritance and Innovation of the 20th-Century Chinese Architectural Heritage

第一篇
Unit One

第二批中国 20 世纪建筑遗产项目
The Second List of the 20th-Century Chinese Architectural Heritage

20世纪中国建筑百年，其作品与设计思想已成为必不可缺的展示与传承方式，中国建筑越来越成为20世纪世界建筑大家庭中的风格与"流派"。本书呈现的 100 项作品，或多或少可梳理出中国现当代建筑的发展脉络，从而展示一个时代的整体风貌，它们分别从历史、科技、艺术与文化等不同侧面成为认知中国 20 世纪建筑的图鉴。

As architectural works and their design are necessary means to communicate and substantiate the tone of the times, the 20th-Century Chinese architecture, from legacy to innovation, from antiquity to modern, has evolved and defined its own genre in the world's architecture family in a hundred year span. This book includes Chinese 100 monumental architectural works as an architectural atlas to trace out the contemporary Chinese architectural strand of the 20th-Century from the perspectives of history, technology and culture.

国立中央博物院（旧址）	P046	The Old Site of National Central Museum
鼓浪屿近现代建筑群	P050	Gulangyu Modern Architectural Complexes
开平碉楼	P054	Kaiping Diaolou
黄埔军校旧址	P058	The Former Site of the Whampoa Military Academy
中国人民革命军事博物馆	P064	Military Museum of the Chinese People's Revolution
故宫博物院宝蕴楼	P068	Baoyun Building of the Palace Museum
金陵女子大学旧址	P074	The Former Site of Ginling College
全国农业展览馆	P080	National Agricultural Exhibition Center
北平图书馆旧址	P084	The Former Site of Peiping Library
青岛八大关近代建筑	P088	Morden Architectural Complex of the Qingdao Eight Great Passes
云南陆军讲武堂旧址	P096	The Former Site of Yunnan Military Academy
民族饭店	P100	Minzu Hotel
国立中央研究院旧址	P104	The Former Site of National Academia Sinica

国民大会堂旧址	P108	The Former Site of the Grand Hall of the Nationals
三坊七巷和朱紫坊建筑群	P112	Three Lanes and Seven Alleys and the Zhuzi Lane Architectural Complex
中东铁路附属建筑群	P114	Building Complex Attached to the China Eastern Railway
广州沙面建筑群	P120	Guangzhou Shamian Building Complex
马尾船政	P124	Mawei Shipbuilding
南京中山陵音乐台	P128	Nanjing Sun Yat-Sen Mausoleum Music Station
重庆大学近代建筑群	P132	Modern Architectural Complex of Chongqing University
北京工人体育馆	P136	Beijing Workers' Gymnasium
梁启超故居和梁启超纪念馆（饮冰室）	P140	Liang Qichao's Former Residence and Liang Qichao Memorial Hall (Yin-Bing Chamber)
石景山钢铁厂	P144	Shijingshan Iron and Steel Plant
中国银行南京分行旧址	P146	The Former Site of the Bank of China, Nanjing Branch
中央体育场旧址	P150	The Former Site of the Central Stadium
西安事变旧址	P156	The Former Site of the Xi'an Incident
保定陆军军官学校	P160	Baoding Military Academy
大邑刘氏庄园	P164	Liu's Manor in Dayi
湖南大学早期建筑群	P168	The Early Building Complex of Hunan University
北戴河近现代建筑群	P174	Beidaihe Modern Architectural Complex
国民党"一大"旧址（包括革命广场）	P180	The Former Site of the First National Congress of the Kuomintang (including the Revoluation Square)
北京国会旧址	P184	The Former Site of the Beijing Congress
京张铁路南口段至八达岭段	P186	The Beijing-Zhangjiakou Railway from the Nankou Section to the Badaling Section
齐鲁大学近现代建筑群	P188	Modern Architectural Complex in Cheeloo University
东交民巷使馆建筑群	P192	Embassy Architectural Complex in Dongjiaomin Lane
庐山会议旧址及庐山别墅建筑群	P196	The Former Site of Lushan Conference and Lushan Villa Complex
马迭尔宾馆	P200	Madie'er Hotel
上海邮政总局	P206	Shanghai General Post office
四行仓库	P210	Four-Bank Warehouse
望海楼教堂	P214	Wanghailou Church

长春电影制片厂早期建筑	P218	Early Buildings in Changchun Film Studio
798 近现代建筑群	P222	798 Modern Architectural Complex
大庆油田工业建筑群	P226	Daqing Oilfield Industrial Architectural Complex
国殇墓园	P230	National Martyrs' Memorial Cemetery
蒋氏故居	P234	Chiang Kai-Shek's Former Residence
天津利顺德饭店旧址	P238	The Former Site of Lishunde Hotel in Tianjin
京师女子师范学堂	P244	Jingshi Women's Normal School
南开学校旧址	P246	Former Site of Nankai School
广州白云宾馆	P250	Guangzhou Baiyun Hotel
旅顺火车站	P254	Lvshun Railway Station
上海中山故居	P258	Sun Yat-Sen's Former Residence in Shanghai
西柏坡中共中央旧址	P262	The Former Site of CPC Central Committee in Xibaipo
百万庄住宅区	P266	Baiwanzhuang Residential Area
马勒住宅	P268	Moller Residence
盛宣怀住宅	P272	Sheng Xuanhuai's Residence
四川大学早期建筑群	P276	The Early Buildings of Sichuan University
中央银行、农民银行暨美丰银行旧址	P284	The Former Site of Central Bank, Farmers' Bank and American-Oriental Bank of Fukien
井冈山革命遗址	P286	Mount Jinggangshan Revolutionary Sites
青岛火车站	P290	Qingdao Railway Station
伪满皇宫及日伪军政机构旧址	P294	The Former Sites of Puppet Manchurian Palace and Japanese Puppet Military Institutions
中国共产党代表团办事处旧址（梅园新村）	P302	The Former Office Sites of the Delegation of the Communist Party of China (Meiyuan New Village)
百乐门舞厅	P306	Paramount Hall
哈尔滨颐园街一号欧式建筑	P308	European Building at No.1 Yiyuan Street, Harbin
宣武门天主堂	P312	Cathedral of the Immaculate Conception
郑州二七罢工纪念塔和纪念堂	P316	Zhengzhou February 7th Strike Memorial Tower and Memorial Hall
北海近代建筑	P320	Beihai Modern Buildings
首都国际机场航站楼群	P324	Terminal Buildings of Capital International Airport
天津市解放北路近代建筑群	P330	Modern Building Complex on Jiefang North Road, Tianjin
西安易俗社	P336	Xi'an Yisu Community Theater
816 工程遗址	P340	816 Project Site
大雁塔风景区三唐工程	P346	Three Tang Projects in Big Wild Goose Pagoda Scenic Area

茅台酒酿酒工业遗产群	P352	Maotai Liquor Industry Heritage Group
茂新面粉厂旧址	P356	Former Site of Maoxin Flour Mill
于田艾提卡清真寺	P358	Aitika Mosque of Yutian County
大连中山广场近代建筑群	P362	Modern Building Complex in Zhongshan Square, Dalian
旅顺监狱旧址	P366	The Former Site of Lvshun Prison
唐山抗震纪念碑	P370	Tangshan Earthquake Monument
中苏友谊纪念塔	P376	Sino-Soviet Friendship Memorial Tower
金陵兵工厂	P380	Jinling Arsenal
天津广东会馆	P386	Guangdong Guild Hall in Tianjin
南通大生纱厂	P390	Nantong Dasheng Yarn Factory
张学良旧居	P394	Zhang Xueliang's Former Residence
中华民国临时参议院旧址	P398	The Former Site of Provisional Senate of the Republic of China
汉冶萍煤铁厂矿旧址	P400	The Former Site of Hanyeping Coal and Iron Plant
甲午海战馆	P404	Jiawu Naval Battle Memorial Hall
国润茶业祁门红茶旧厂房	P408	Old Factory Buildings of Qimen Black Tea of Guorun Tea Industry
本溪湖工业遗产群	P414	Lake Benxi Industrial Heritage Group
哈尔滨防洪纪念塔	P420	Harbin Flood Control Memorial Tower
哈尔滨犹太人活动旧址群	P424	The Former Sites of Jewish Activities in Harbin
武汉金城银行（现市少年儿童图书馆）	P430	Wuhan Jincheng Bank (Currently the Municipal Children's Library)
民国中央陆军军官学校（南京）	P432	Central Army Academy of the Republic of China, Nanjing
沈阳中山广场建筑群	P434	Buildings of Zhongshan Square in Shenyang
抗日胜利芷江洽降旧址	P438	The Former Site of Japanese Surrender Negotiation in Zhijiang
山西大学堂旧址	P442	The Former Site of Shanxi University
北京大学地质学馆旧址	P446	The Former Site of the School of Geology, Peking University
杭州西湖国宾馆	P448	Hangzhou West Lake State Guesthouse
杭州黄龙饭店	P452	Hangzhou Huanglong Hotel
淮海战役烈士纪念塔	P456	Huaihai Campaign Martyrs Memorial Tower
罗斯福图书馆暨中央图书馆旧址	P460	The Former Site of the Roosevelt Library and the Central Library
洛阳拖拉机厂早期建筑	P462	The Early Buildings of Luoyang Tractor Plant

P466 中国 20 世纪建筑遗产保护进程及展望 / 张松
The Evolution and Prospect of Preserving the 20th-Century Architectural Heritage in China / Zhang Song

P478 从建筑学视角看第二批中国 20 世纪建筑遗产项目中的现代建筑 / 戴路 李怡
Obsening the Modern Architecture in the Second Batch of Chinese 20th Century Architectural Heritage Projects from the Perspective of Architecture / Dai Lu, Li Yi

P484 中国 20 世纪建筑遗产的保护价值评价体系建构 / 林娜 张向炜 刘军
The Construction of the Evaluation System of the Protection Value of Chinese Architectural Heritage in the 20th Century / Lin Na, Zhang Xiangwei, Liu Jun

P492 苏联建筑风格及其影响下的中国 20 世纪建筑遗产 / 陈雳 张翰文
Soviet Architectural Style and 20th Century Architectural Heritage in China under its Influence / Chen Li, Zhang Hanwen

第二篇
Unit Two

中国 20 世纪建筑师
The 20th-Century Chinese Architects

中国 20 世纪建筑作品，融合建筑师、工程师及艺术家的创作是时代的需要，是普及公众建筑文化视野的需要，因为中国 20 世纪建筑教育太缺少"个人"建筑史。所以，从敬畏建筑先贤设计思想出发，从传承优秀的文化遗产出发，对他们的学术研究与追忆，都将是无比珍贵的设计精神遗产。
Including insightful introduction to architects, engineers and artists, the atlas attempts to offer the public the background elements of the great creations that would not exist without these masterminds. The academic studies and historical memory of these forerunners, in addition to our admiration of their ideas and the profound Chinese architectural, cultural legacy, will serve as the precious, contemporary and spiritual heritage.

P500　库尔特·罗克格 / Curt Rothkegel
P502　李守先 / Li Shouxian
P503　思九生（Robert Ernest Stewardson）
P505　沈理源 / Shen Liyuan
P507　柳士英 / Liu Shiying
P508　刁泰乾 / Diao Taiqian
P509　杨锡镠 / Yang Xiliu
P512　奚福泉 / Xi Fuquan
P513　陆谦受 / Lu Qianshou
P515　徐敬直 / Xu Jingzhi
P516　米哈伊尔·安德烈耶维奇·巴吉赤（Mikhail Andreevich Bagich）
P517　孙秉源 / Sun Bingyuan
P518　胡庆昌 / Hu Qingchang
P520　王时煦 / Wang Shixu
P522　严星华 / Yan Xinghua
P524　欧阳骖 / Ouyang Can
P526　李光耀 / Li Guangyao
P527　熊明 / Xiong Ming
P530　彭一刚 / Peng Yigang
P532　程泰宁 / Cheng Taining
P535　李拱辰 / Li Gongchen
P538　孙国城 / Sun Guocheng

P540　20 世纪中国建筑遗产大事要记（2016 年 10 月—2020 年 12 月）
Highlights of China's Architectual Heritage in the 20th Century (October 2016—December 2020)

P564　中国 20 世纪建筑遗产认定标准 2014
The Identification Standard for the 20th-Century Chinese Architectural Heritage, 2014

P568　第二批中国 20 世纪建筑遗产名录
The Second List of the 20th-Century Chinese Architectural Heritage

P572　中国文物学会 20 世纪建筑遗产委员会顾问及专家委员名单
The List of Consultants and Experts of CSCR Committee on the Twentieth-Century Architectural Heritage

P576　参考文献
References

中国 20 世纪建筑遗产传承与创新研究纲要初步（代前言）

Preliminary Study Outline of Inheritance and Innovation of the 20th-Century Chinese Architectural Heritage

金 磊

引言：这本名录是专门为 2017 年 12 月获第二批中国 20 世纪建筑遗产项目所编，但本文则是为展示近年来中国文物学会 20 世纪建筑遗产委员会的研究进程并提供相关思路的。截至 2019 年 12 月，中国文物学会、中国建筑学会议公布了四批共计 396 项中国 20 世纪建筑遗产项目。通过长达五年对中国 20 世纪建筑遗产项目与建筑师的追踪式研究、调查与分析，笔者在《中国建筑文化遗产》总第 24 辑撰写《跨越 21 世纪的〈北京宪章〉20 年——〈世界遗产名录〉20 世纪建筑遗产项目研究与借鉴》一文。文章不仅学习了两院院士吴良镛教授 1999 年发布的《北京宪章》精神，还以国际建筑大师项目入选《世界遗产名录》分析了在当今世界建筑界 20 世纪建筑遗产与经典地标备受关注的原因。本文试图对将 20 世纪建筑遗产新类型引入中国建筑与文博界有所推动，因为它的确是世界文化遗产界不能忽视的重要方面。

Introduction: This book is specially compiled for the second list of the 20th-century architectural heritage in China released in December 2017, but this article is to present the research process of the Chinese Society of Cultural Relics Comittee on the 20th-Century Architectural Heritage in recent years and to provide relevant ideas. By December, 2019, the Chinese Society of Cultural Relics and the Architectural Society of China had announced 396 Chinese architectural heritage projects in the 20th century in four batches. After five years of follow-up research, investigation and analysis of Chinese architectural heritage projects and architects in the 20th century, the author published the article "Two Decades of 'The Beijing Charter' Across the 20th- and 21st-Centuries: Research and Reference of the 20th Century Architectural Heritage Projects in the 'World Heritage List'" in the 24th Series of *Chinese Architectural Cultural Heritage*. This article not only addresses the spirit of "The Beijing Charter" published in 1999 by Professor Wu Liangyong, an academician of the Chinese Academy of Sciences and Chinese Academy of Engineering, but also analyzes the reasons why the architectural heritage and classic landmarks in the 20th century have attracted much attention in today's world architecture community, based on the international architecture masters' works in the "World Heritage List". This article attempts to promote the introduction of the 20th century architectural heritage into Chinese architectural and cultural circles, because the world cultural heritage circles indeed cannot afford to ignore the importance of the architectural heritages in 20th century.

2018 年 12 月 18 日在故宫博物院举办《中国 20 世纪建筑遗产大典（北京卷）》首发式上，时任故宫博物院院长单霁翔做了《中国 20 世纪建筑遗产传承与利用》的演讲，他不仅分门别类勾勒出中国 20 世纪建筑遗产项目的特点，还向数十计的百岁中国 20 世纪建筑师致敬，首发式上 8 位专家还针对 20 世纪建筑遗产保护与传承展开交流。其要点是：既然北京在 20 世纪建筑遗产项目上占优势，为什么北京不率先推动 20 世纪建筑遗产保护工作呢？古都北京，当代世界城市理应为中国现代建筑走向世界作出更大贡献。自 2019 年 1 月 4 日至 9 月 12 日，本人应北京人民广播电台《城市文化》栏目邀请，结合北京 20 世纪建筑遗产、长安街的北京建筑、新中国 70 年北京建筑等话题，做了 6 次讲座。我讲述了 20 世纪建筑遗产的概念，同时强调阅读北京的精髓在于读懂北京建筑，只有读懂，才会有情感，才可谓有敬畏之情与保护之意。应感谢的是，在北京市规划和自然资源委员会（简称规自委）与北京城市规划学会支持下，在北京市

注册规划师继续教育课题中,增加了20世纪建筑遗产内容。2019年3月29日我以《20世纪建筑遗产保护理念与问题研究兼议北京20世纪建筑遗产项目背后的人和事》为题作了讲授。2019年6月,北京市公布了第一批429项历史建筑名单,市规自委要求听取社会各界反馈,我们于6月27日正式提交反馈建议,主要内容是北京要突破局限于历史建筑说法,要用20世纪建筑遗产新概念,尤其要创造性保护20世纪建筑遗产。

On December 18, 2018, the launching ceremony of "The 20th-Century Chinese Architectural Heritage Classics (Beijing Volume)" was held in the Palace Museum. Shan Jixiang, then-curator of the Palace Museum, delivered a speech titled "Inheritance and Utilization of China's 20th-Century Architectural Heritage", not only outlining and classifying the characteristics of Chinese 20th-century architectural heritage projects, but also paying a tribute to dozens of centenarian Chinese architects in the 20th century. At the launching ceremony, eight experts also exchanged views on the protection and inheritance of 20th-century architectural heritage. Their main point was that: since Beijing had an advantage in the 20th century architectural heritage projects, why didn't Beijing take the

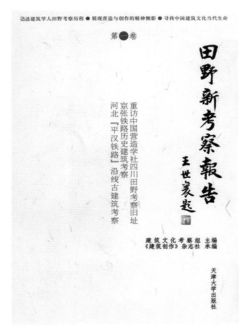

lead in promoting the protection of this category of heritage? Beijing, as an ancient capital and a modern metropolis, should make greater contributions to the introduction of Chinese modern architecture into the world. From January 4 to September 12, 2019, upon the invitation of the "Urban Culture" column of the Beijing People's Broadcasting Station, I rendered six lectures on topics such as Beijing's architectural heritage in the 20th century, Beijing architecture in Chang'an Street, and Beijing architecture in the seven decades since the founding of the People's Republic of China. I talked about the concept of architectural heritage in the 20th century, stressing that to understand Beijing we must understand Beijing's architecture. Only by understanding Beijing architecture will we have passion for it and be inspired to respect and protect it. Thanks to the support of the Beijing Municipal Commission of Planning and Natural Resources and the Urban Planning Society of Beijing, the content of the 20th-century architectural heritage has been added to the subject of continuing education for registered planners in Beijing. On March 29th, 2019, I delivered a lecture titled "Research on the Concept and Problems of the 20th-Century Architectural Heritage Protection and Discussion on the People and Events Behind Beijing the 20th-Century Architectural Heritage Projects". In June 2019, Beijing published the first list of 429 historical buildings. The Municipal Planning Commission sought opinions from various sectors of society. We gave feedback on June 27, mainly suggesting that Beijing should break through the concept of historical buildings, use the new concept of the 20th-century architectural heritage, and in particular, make creative efforts to protect the 20th-century architectural heritage.

一、中国 20 世纪遗产与百年包豪斯及"赛先生"之精神

对中国 20 世纪建筑遗产的梳理，要不忘中国现代建筑先驱的文脉坚守。2019 年 7 月 22 日，"归成——毕业于美国宾夕法尼亚大学的第一代中国建筑师"（简称：归成）主题展览在清华大学艺术博物馆举办。展览集中展示，1918—1935 年在宾夕法尼亚大学建筑系求学的 20 多位中国留学生的学习历程，以揭示他们在建筑学理论与设计实践的贡献。在中国第一代建筑学家中，梁思成、杨廷宝、范文照、童寯、陈植、林徽因等都毕业于这所大学美术学院的建筑系或艺术系，他们成为中国 20 世纪建筑事业的奠基人。1931 年，《中国建筑》（创刊号）有立足高远的发刊词："建筑之良窳，可以觇国度之野。"发刊词的作者赵深（1998—1978 年），早在 20 世纪 30 年代便学成回沪，他与同为留美归来的陈植、童寯共同组建了华盖建筑师事务所，成为中国现代建筑史上最早一批典范之作的发轫团队。中国开始有了现代城市景观的雏形且隐含着东方建筑美学与功能的意图与趣味。"归成"展部分复原了那些历史进程，

建筑编辑家杨永生笔下的"建筑四杰",除刘敦桢系留日外,梁思成、杨廷宝、童寯皆为宾夕法尼亚的同门才俊。空口无凭,仅从留美学生当年的课程作业"水彩画"即可见一斑,这其中有罗马式立柱和拱门洞、西式的静物如玫瑰花瓶,也有唐代佛像、唐三彩等,都那么心思端凝,斯文周正。据陈植回忆,他们常在交图前夕彻宵绘图或渲染,梁思成的渲染"水墨清澈,偶用水彩,则色彩雅淡,明净脱俗",这是奠定建筑先贤成长的基石。"归成"展的当代意义十分凸显:宾大归来的中国第一代建筑师首次以集体形式出现在公众面前,了解他们或许有助于唤醒中国20世纪建筑经典记忆,也会找到中国现代建筑大量呈现的理由,或许这也正是中国现当代建筑文化何以影响欧洲近现代开端的理由。

中国在现代化浪潮中应属于觉醒较晚的"后发"国家,因为承受了全球外来文化硬性进入,即称"西学东渐",但事实上,东西方文化的交流并非是单向的,作用与反作用同时存在。包括建筑文化在内的中国文化对欧美的影响远超人们的想象,这就是"中学西传"。"归成"展的建筑学家正是"中学西传"的贡献者。且不说到了21世纪的今天,就是在20世纪30年代,随着人类交往能力和深广度的增加,西方对中国的认识发生了质的变化,特别开始扭转传统"东方学"中过于强调东方自身的视角,中国作为独立的发展中大国在东方巍然屹立。应注意到,"二战"期间,在工业文明的裹挟下,各种文化思潮开始从"古典"向"现代"转型,以实用为原则,强调立足于事实的学术研究动向受到关注。无论是"中国学"的理论奠基人费正清(1907—1991年),还是中国科技史家李约瑟(1900—1995年)都在彰显"中国学"的跨学科研究与历史全域认知,如跨文化的"汉学"研究关注文化的多元共存,质疑并审视"西方中心主义"的偏见,重构文化自信。以此自信,对于20世纪40年代中国建筑学人梁思成、刘敦桢在中国营造学社的贡献,来自四川宜宾李庄的费正清无比感慨"这个曾经接受过高度训练的中国知识界,一面接受原始淳朴的农民生活,一面继续努力于他们的学术研究";英国剑桥大学教授、科技史学家、《中国科学技术史》著作者李约瑟对传承中国建筑文化,尤其是解读20世纪中国营造学社的工作贡献很大。1943年3月,他与几位同事通过"驼峰"航线艰难到达中国,开始了四年之久的考察。李约瑟前往李庄造访梁思成,目睹了中国建筑学人艰苦卓绝的工作。心灵受到很大震撼的他在笔记本上写下的预言式文字恰恰充满对中国学人的敬仰之情:"如果战后中国政府真正大规模地从财政上支持研究和开发建筑,20年后,中国会成为主要的科学国度。"

2019年是五四运动100周年，100年前的元月15日陈独秀在《新青年》撰文，提出了"德先生"与"赛先生"，他认定只有民主与科学的"德""赛"两先生才可在道德、思想诸方面救治黑暗的中国。巧合的是，2019年也正值世界上最有影响的现代建筑与设计潮流之一、在业界堪称不朽的包豪斯创立百年。无论是否承认，包豪斯作为世界上第一所完全为发展现代设计教育而建立的学院，至今影响着全球的城市与生活，大到地铁站，小到我们每天都居住的房屋、使用的家具和装饰品。正如纪念包豪斯100周年网站所述："真正的包豪斯风格是实验的、多样的、跨界的，而且是完全当代的。"与此同时，著名旅游指南《孤独星球》（Lonely Planet）评选的2019年世界最值得前往的国家中，德国位居第二。

包豪斯让我们联想到设计理念。1919年4月，魏玛包豪斯学校第一任校长格罗皮乌斯在《包豪斯宣言》中说："一切创造活动的终极目标就是建筑！为建筑进行装饰一度是美术最高尚的功能，而且美术也是伟大建筑不可或缺的伙伴……老式的艺术院校没有能力来创造这种统一。让我们共同期待、沟通并创造出未来的新建筑，用它把一切——建筑、雕塑与绘画——都组合在一个统一的形式里，有朝一日，它将会从百万手工艺者的手中冉冉升向天空，水晶般清澈地象征着未来的新信念。"2014年9月29日—11月8日，"作为启蒙的设计——中国国际设计博物馆包豪斯藏品展"在中国国家博物馆展出。在引进设计思想库的大背景下，它盘活了以包豪斯学校及设计风格为代表的世界级文物资源，正如两位策展人宋建明、杭间在展览序言中所说："包豪斯被誉为欧洲创造力的中心，它不仅仅是所学校，也是一个精神象征，一场艺术改革，一种关于现代生活启蒙的哲学……它对当代的启示辐射到艺术、设计、教育、生产和生活等各个方面。"感悟百年包豪斯的理念、模式与影响，令人感到包豪斯的遗产价值不仅是西方文明从工业期到发展期的遗存表征，对中国而言也是工业文明重要的参照系，折射出设计改变生活的城市命题，设计为一个国家指出了文化科技复兴的新途径。包豪斯是现代主义的代名词，无数风靡全球的建筑师、设计师以及艺术家与这个名词息息相关，如保罗·克利（Paul Klee）、瓦西里·康定斯基（Wassily Kandinsky）、拉兹洛·莫霍利-纳吉（László Moholy-Nagy）、约瑟夫·阿尔伯斯（Josef Albers）和安妮·阿尔伯斯（Anni Albers）等。包豪斯代表的不仅是一种特定的风格，更是一种普适方法。深受财政和政治的压力，包豪斯学校百年以来被迫两次搬迁。第一次是从魏玛搬到了新兴的工业城市德绍。在那里，包豪斯第二任校长瑞士建筑师汉斯·迈耶（Hans Meyer）负责管理最新设立的建筑系，而格罗皮乌斯完成了他最为著名的建筑。1933年，在包豪斯第三任校长路德维希·密斯·凡·德罗（Ludwig Mies van der Rohe）的带领下，包豪斯迁往柏林，然而，仅仅一个学期后就在纳粹的政治压力下被迫关闭。

一个城市的兴起与影响力需要文化坐标，用设计完全可以启

动一个国家、城市。1919年德国的包豪斯学校，同时在中国迎来了科技化身"赛先生"。作为设计与艺术服务社会的启蒙，包豪斯强调，设计首要目的是人而非作品，技术与艺术要统一，设计要遵循自然与客观的法则。由此让我想到已经风靡全球的苹果系列产品，它的推陈出新努力打造的一定是人文情怀，与人方便的品质，这从一定意义上也暗合了包豪斯设计的不少遗产思想。"少即是多""上帝就是在细节之中""今天看上去非常奢华，后天就成正常标准了"。这些当时有点惊世骇俗之语，现在早已被设计、建筑、艺术界所熟用。要看到，如今包豪斯仍在向设计、建筑等圈层注入生机，甚至为时装盛典、发型趋势等贡献着灵感之源。同样，今天在历史的包豪斯基址上仍可见到：在魏玛，再也无人投诉包豪斯了，相反，有专门的博物馆，在2013年对外开放；在德绍，投资巨大的包豪斯大楼复建工作已经完成，历史上著名的设计遗产再次被展示、书写；在柏林，1960年建立的包豪斯档案馆（副题为设计博物馆）已是世界上最大的包豪斯收藏馆，展示了包豪斯全部历史藏品中的珍品。特别要提及的是，1996年，包豪斯学校建筑及其遗址被列入《世界遗产名录》。2000年9月，李振宇在《世界建筑》撰文《解读包豪斯中国城市建筑展》，文章介绍了由西门子文化工程和德昭包豪斯基金会主办的"北京·上海·深圳——21世纪的城市"中国城市建筑展，在德国德昭包豪斯校舍举行，并出版了包豪斯学校出版社的厚600页的展览专集。

中国"赛先生"与德国包豪斯本该自然相融。"赛先生"强调，对于百年中国，科学不单单是一种知识与技能，包含当代设计在内的科学技术具有一种独立理性的批判精神，然而时至21世纪的今天，中国建筑尚缺乏评论，建筑大国尚缺少大国应有的建筑话语平台。"赛先生"强调科学作为人类探索自然与社会的利器，潜能无限、创造力无穷，但当下最大的不足是城市与建筑界社会责任彰显不够，面对城市发展自我审视不足，面对城市与公众需求建筑师对自我的创作欲望限制不够。"赛先生"还强调，无论是从公众的素养还是社会的建筑文化看，城市与建筑不再是孤立的，其深厚的科学与文化根基筑成了不同的视野，这是"赛先生"建筑科学理性的必然体现。从科学价值看，20世纪比历史上任何时期都更依赖科技。对于建筑而言，钢筋混凝土、玻璃、电梯乃至智能机电技术的广泛应用，确为建筑从构造到形体、从功能到品质带来革命性变化。因此，在环境设计中无论是体现地域特色、主题建筑还是工业遗产，均展现了20世纪建筑与科技应用是城市精神再造的推进器。保护并发现遗产，不仅为了继承传统，还要通过中国建筑的文艺复兴，真正在促进东西方建筑文化整合交融时，确立中国建筑作品与建筑师在世界建筑文化体系中的地位。20世纪建筑遗产的当代价值还在于要通过中国近现代建筑的全面发展，让世界建筑界认同"中国建筑不只木"，这是以欧美建筑为镜鉴，对当代中国建筑创作的一个科学诠释。李诫在《营造法式》的《序》中说："五材并用，百堵皆兴。"更远的奴隶社会末期的中

国建筑已综合用上了土、木、石、陶、铜等材料。西方有"建筑是石头的史书"，而中国找不到建筑是木头的艺术之类的说辞，但这不可简单推断中国当代建筑就一定沿袭传统，缺少现代性。中国建筑文化的当代发展本质是要融国际化为一体的，是要东西方建筑文化整合的且具备全球视野的，为此才有以堂堂之鼓、正正之旗下的中国建筑文化的自信与自豪之作品。

在全球 2019 年纪念百年包豪斯的文化活动中，"走向现代：包豪斯建筑艺术之旅"互动网站令人瞩目。这并非因为其旅游线路纵贯德国，包括《世界遗产名录》中的包豪斯建筑外，还展示了从 1900 至 2000 年的百年建筑史。例如，位于德国弗尔克林根钢铁厂（Volklingen Ironworks），是从钢铁工业黄金时期至今唯一保存完好的一座综合的铁工厂（1994 年，成为德国第一个被联合国教科文组织列入世界遗产的现代主义建筑）。在《北京宪章》中，吴良镛院士阐述了全社会的建筑学观，特别强调无视建筑学发展的前提条件，就不难以理解建筑学所有的社会重要性。这里既体现"赛先生"的科学理性，也解读了包豪斯的当代设计价值：①包豪斯体现了"大建筑艺术"的追求；②当代建筑的发展，越来越离不开乡土文化与乡土建筑，精湛的手工艺技术；③虽然新技术为建筑师、艺术家创作带来更多可能性，但新技术在创作上并不可取代设计与鉴赏力。由"赛先生"与包豪斯百年，令人想到一系列东西方建筑文化的事件与人物：中国的祠堂与西方的教堂，有建筑文化的差异与通性；600 年历史的故宫紫禁城与 800 多年的卢浮宫建筑与文物展览可比较，可对话；其实，与包豪斯百年最有比较价值的当属 2019 年是朱启钤(1872—1964)创办中国营造学社九十周年。朱启钤及其中国营造学社在中国传统建筑文化传承上，在"西体中用"上，在 20 世纪中国城市（特别是北京）乃至 20 世纪遗产的发现与创新上的贡献是划时代的。

二、国际现代建筑大师作品入选《世界遗产名录》的启示

在 2019 年的第 43 届世界遗产大会上，美国 20 世纪最著名建筑师弗兰克·劳埃德·赖特的 8 项 20 世纪上半叶建筑入选《世界遗产名录》，这是缔约国历经数年"申遗"耕耘的结果，是全球建筑界、遗产界的又一 20 世纪丰碑级文化盛事。入选的 8 个项目为：团结教堂（伊利诺伊州，1906—1909 年）、罗比之家（伊利诺伊州，1910 年）、塔立耶森设计工作室（威斯康星州，1911—1959 年）、霍利霍克别墅（加利福尼亚州，1918—1921 年）、赫伯特与凯瑟琳·雅各布第一住宅（1936—1937 年）、流水别墅（宾夕法尼亚州，1936—1939 年）、西塔立耶森住宅（亚利桑那州，1938 年）、纽约古根汉姆美术馆（纽约，1956—1959 年）。2019 年，也正逢这位建筑大师辞世 60 载，或许他的作品的入选也是源于世界遗产大会对他的敬意。美国建筑大师赖特（1867—1959 年）在其 92 载设计生涯中共设计了数百座建筑，尤其是他的纽约古根海姆博物馆和宾夕法尼亚州被称为"流水别墅"的河畔避暑住宅等都获得了全球建筑界的赞誉。已故清华大学建筑学院汪坦教授 1947 年求学于赖特门下，他在评介赖特大师的作品时曾说："建筑竟能把人领到如此感人的境界，风格鲜明，内涵深湛，而和生活息息相关，非其他艺术可拟。"赖特是 20 世纪举世公认的建筑师、艺术家和思想家，作为现代建筑的创始人，他与勒·柯布西耶、密斯·凡·德·罗、格罗皮乌斯被并称为世界四大著名建筑师。

可喜的是，这四位20世纪大师，均已有作品入选《世界遗产名录》：2016年勒·柯布西耶有跨越7国的17个项目入选；2001年密斯·凡德罗布尔诺图根德哈特别墅入选；2011年德国包豪斯学校创始者格罗皮乌斯的法古斯工厂项目设计入选。这足以说明《世界遗产名录》越来越瞩目当代建筑。

如上所述德国包豪斯学校，它曾于1996年及2017年两度入选《世界遗产名录》，更因为格罗皮乌斯、密斯·凡·德罗先后是包豪斯学校第一任及第三任校长，所以品味岁月长河中沉淀下的20世纪建筑遗产才格外有价值。德国现代建筑师、建筑教育家格罗皮乌斯于1911年设计的德国法古斯工厂项目于2011年"申遗"成功。该设计遵循工业建筑设计师维尔纳（Werner）的平面工艺图，着力于项目的外部与室内设计的提升。他一直认为工业建筑不该模仿过去数十年的同类建筑，要演变并适应变革的社会文化之发展，他坚信自己采用震撼的体量及大面积的玻璃立面，会成为未来现代主义建筑的走向。100年后格罗皮乌斯项目入选《世界遗产名录》，恰好说明在有文脉支撑的环境下，营造艺术性的设计之道是成功的。2001年入选《世界遗产名录》的德国布尔诺的图根德哈特别墅是建筑大师密斯·凡·德罗于1928—1938年间设计建造的。这位世界级功能主义建筑师表示该建筑"无法计算出理解的空间维度"，由于设计与建造无预算限制，设计上大量采用了进口石材与木材，空间的布局处理上富于革命性，最耀眼的是主起居空间内一面整墙大小的巨型可延伸窗与花园相连，三层建筑自主搭配，实现了人与自然在空间上的自由交流。值得特别记忆的是，在这座精致别墅中发生了一系列随时代变迁的"事件"：作为犹太人的图根哈特夫妇1930年搬入别墅，1938年为躲避纳粹只能逃亡瑞士，该别墅被盖世太保没收，1945年"二战"布尔诺解放，这里又成为苏联士兵马厩，瑰宝般的家具变成燃料。后来它又成为当地舞蹈学校及儿童护理院，直到20世纪80年代才恢复原有样貌。有文献称这栋反映时代审美的别墅，是世界上最具价值的重要别墅之一，更重要的是，它成为二战中诸多事件的背景。

2012年，在阿联酋艾恩举办的文化遗产完整性专家会议，讨论遗产可持续发展服务社会与完整性的关联，还针对与20世纪遗产关系密切的文化景观、历史城镇、纪念物、建筑与建筑群等解读了社会——功能完整性、历史——结构完整性、视觉——美学完整性等保护与发展要则。世界遗产保护制度40多年的历史说明，无论是20世纪建筑遗产，还是工业遗产，中国的保护与项目认定都起步较晚，引进先进的遗产保护与管理理念很迫切。本文对《世界遗产名录》中20世纪建筑遗产项目的捡拾虽不完备，但它无疑让我国看到了国际上20世纪建筑遗产这一重要类型。赢得国家话语权并非口号，它需要建立在一个国度的社会文化影响力是否具有引领国际社会发展的基础上。从严格意义上讲，迄今我们并没有20世纪建筑遗产项目入选，这表明中国在世界遗产的国际舞台上有"空白"和再发展的契机，要特别提醒注重在城市化进程中应加强对文化遗产的完整性保护，随意对建筑"拆除"不行，历史建筑未经严格论证与社会认证就"搬家"移动也是不允许的。

三、中国20世纪建筑遗产项目的"分类说"

本文既有针对第二批20世纪遗产项目的分析与评述，也有结合更广泛 20 世纪建筑经典的评说，旨在深入研讨并促进联想。

1. 20 世纪遗产的"城市说"

新中国成全 70 多年，虽是悠悠历史长河中的一瞬，却给历史留下一串串意味着变化的数字，每串数字之变，都仿佛一轴新的历史画卷，展开后可看到文化传承与改写时代的城市方案。70 年多来，城镇化经历了三个阶段：1949—1978 年的探索发展阶段；1979—2011 年的快速发展阶段；2012 年至今的提升发展阶段。2018 年末，中国常住人口城镇化率达到 59.58%，比 1949 年末提高了 45.94%。在这快速城镇化发展的背景下，城市文脉传承与保护都有一系列矛盾与新问题呈现，主要是文化遗产的发展困境，这里用几个典型城市解读，重点是提示城市更新别丢了文脉。2014 年 10 月 2 日《南方周末》有个写真专栏，它用图像表达了"城市"何以是我们的居所，有城的《呼吸》、城的《身体与身份》、城的《感官与感情》等。在中国城市化的迅疾进程中，我们在见证城的繁华与昌盛时，也亲历了城的阵痛与转折，更让城的变迁直抵内心。

北京：一座城市的价值不在于它悠久的历史，而在于它对这悠久历史的汲取；不在于它的物质有多么丰盈，而在于它拥有的文化底蕴；重在要遏制住原生文化消逝已成城市之殇的趋势。梁思成早就指出"北京城是一个具有计划性的整体"，其意是北京是先有计划再造的城，宏伟与庄严的布局，在空间处理上创造卓越的风格，使北京城有丰富历史意义及艺术的表现。梁思成进一步强调"北京城市必须（是）现代化的，同时北京原有的整体文化特征和多数个别的文物建筑又是必须加以保存的，我们必须'古今兼顾，新旧两利'。"西安规划局老局长韩骥先生特别以北京、西安的古城规划及 20 世纪建筑遗产，分析了梁思成的建筑与城市遗产观（见《中国文物报》2019 年 4 月 5 日三版）。值得一提的是，2007 年北京公布第一批《北京优秀近现代建筑保护名录》，2019 年 6 月和 10 月北京先后公布了两批超 700 处"历史建筑"。笔者认为如同"国庆十大工程"（已是中国 20 世纪建筑遗产）一样，它确应成为全国重点文物保护单位，进而 20 世纪建筑遗产的体系化"申遗"也应启动，同样，20 世纪 50 年代"北京八大学院"的经典性与多元建筑特质，也不失为特别有价值的建筑遗产。2018 年 11 月末公布的第三批中国 20 世纪建筑遗产中的北京科技大学（原北京钢铁学院）于 1952 年建校，1953 年学生从清华搬回时，学校只建有教学楼和理化楼，主楼刚刚打地基（1955 年完工），其复古的巴洛克花纹让每位经过者都感受到苏式风情。苏式建筑首先是左右呈中轴线对称，主楼高耸，中间高两边低，此外特有明显的檐部、墙身、勒脚三部分设计。这座老建筑群，20 世纪 90 年代虽被粉刷一新，但其沉稳的底座仍如当年风采。

上海：历史悠久，文化底蕴深厚，各类建筑遗产丰富，尤其还有 1058 处优秀历史建筑，郑时龄院士说它们涉及左翼文化运动史迹、抗日救亡史迹、近现代工业史迹、近现代名人故居和旧居、近现代金融业史迹、近现代商业史迹、近现代宗教史迹、租界史迹以及犹太人史迹等。无论在坚持历史建筑的目标导向、核心价值导向，挖掘内涵与阐释，突出"内

容为王"的传播等方面确有一系列经验可供借鉴,它是在总结过去经验教训基础上完成的。2017年初,在上海市委办公厅牵头下,开展了上海中心城区50年以上历史建筑的全面普查,初步查明上海外环内现存50年历史建筑约31520幢(建筑面积2559万平方米)。上海强调只有列为地方历史建筑保护名录即成为文物建筑,其日常管理与维护才获得相对有利的法律保障。2015年,拥有近百年历史的上海石库门里弄公益坊,就在被列入文物建筑前夕遭拆除,这是教训。保护历史建筑,摸清家底既是立法的数据基础,也是创造条件让公众走得进去的前提。让公众"走得进"历史,就是要"听得见""看得到",这是激活城市公共记忆的绝好方式。如在千余米长的上海武康路上,有许多记录百年沧桑的建筑,其中武康路与兴国路交会处的"武康大楼"是由邬达克设计的精美建筑。上海用了一年的时间,以架空线入地与合杆整治的方法,改造了武康大楼周身的"蜘蛛网",这是一个巧用"绣花针",精细化20世纪建筑管理的好示例。俄罗斯作家果戈里说:"建筑是世界的年鉴,当歌曲和传说都缄默时,只有它还在说话。"20世纪遗产是城市的生动面孔,建筑设计要面向未来,就要格外尊重现代遗产,这里有建筑师可学习的工艺水平、文化理念、审美取向。

厦门:2019年是陈嘉庚先生诞辰142周年,厦门是海上丝绸之路的支点城市及福建自贸区的组成部分。回溯集美往事,并不如烟,这里有80多年前与嘉庚精神密切相关的尚鲜为人知如大田第二集美学村的故事,它乃20世纪有抗战精神的文化遗产。厦门有万国建筑博览的鼓浪屿、最美的厦门大学,还有博物馆大观园集美的鳌园。总体说来,厦门有"四大"典型建筑风格:以厦门八卦楼为代表的欧陆建筑,以骑楼式街市为代表的骑楼建筑,以新垵红砖民居为代表的红砖民居,以厦大嘉庚楼、集美学村南薰楼为代表的嘉庚建筑等。在一座城市彰显的保护精神下,《厦门市鼓浪屿历史风貌建筑保护条例》自2000年4月1日施行,又于2018年出台《厦门历史建筑保护试点工作方案》,方案指出:"厦门不仅将加快全市域历史风貌建筑普查和认定工作,建立完善保护名录,还将重点推进鼓浪屿第三批约540栋历史风貌建筑的测绘工作,建立并不断完善历史建筑档案资料,实现'一栋一档'。"这些好经验是促使2017年鼓浪屿整体进入《世界遗产名录》的重要文化要素。

深圳:2019年7月24日《中共中央国务院关于支持深圳建设中国特色社会主义先行示范区的意见》,使得深圳再次肩负时代新使命。示范区关乎新时代全局,不可不察,因为其先行先试的"试验田"作用,已不是经济领域的制度创新,包括文化自信宏大框架下的创新。其作始为简,其将毕也巨,高质量的深圳发展也必须要在文化创新,乃至20世纪建筑遗产上有理念与行动有突破。深圳坚持20多年的城市阅读计划实施较好,2019年10月又提出要用"新十大文化设施"提升城市软实力,其良好愿望是高标准营造未来城市地标。无论"新十大文化设施"中的中国国家博物馆·深圳馆如何体现国家文化话语权、辐射港澳及东南亚,我们都尚未见到深圳在城市更新与微循环上的有遗产价值的文化举措,更缺少对有40年"市龄"历史建筑保护的创新性思路。尽管"不日新者必日退",但如果仅以拆旧建新为城市创意模式,认为只有老城才需保护历史建筑,新兴城市深圳可除外,那就

有负于文化城市建设的主旨。2018年11月1日,中国建筑学会在北京召开"深圳体育馆保护专题研讨会",希望用20世纪建筑遗产之思,展开深圳体育馆"拆与留"的城市与社会批评。笔者当时提交的书面文字是:1985年建成的深圳体育馆属深圳建市后"八大"文化设施,主持设计的建筑师们回忆到,当时为了体现体育建筑的力度与向上精神,设计摒弃了所有装饰附件,专门用建筑固有的构件表现它的形象美,高举的屋盖、自然坡起的看台体量与水平舒展的观众休息平台形成对比,从而表现出稳重有力的气势。该体育馆严谨、求新的设计风格无疑成为改革开放初期深圳建筑的标志,在荣获一系列建筑行业奖项外,1989年在国际建筑师协会举办的"体育与娱乐设施优秀设计"中该项目获银质奖,2009年中国建筑学会还授予其"建国60周年建筑创作大奖"。据此我的建言是如下。

其一,创造城市文化的深圳要特别珍视自身的文化。改革开放40载让深圳不断呈现新貌,其中坚持了23年的城市阅读是深圳文化建设的"亮点",反映了其渴望"文化立市"的信念,此外联合国教科文组织的"设计创意之都"也为深圳带来了国际文化创意之桂冠。问题是,深圳尚没有理解,其文化不仅要"创",还必须在回望中保护并珍爱,这是深圳文化自信的标志。所以,深圳相关部门应下决心留下深圳体育馆等"城市富矿"的建筑。其二,深圳的文化包括始于20世纪80年代的建筑文化,这是由深圳城市特性所决定的。20世纪80年代的深圳优秀建筑,就是象征深圳建筑史的"历史建筑",它是深圳乃至中国城市化改革的无价之宝。对于中国20世纪建筑遗产,住建部和国家文物局均有专门规定强调要予以保护:2017年9月住建部下发建规【2017】212号通知,从历史建筑普查、确定、建档、挂牌、不拆除、不乱建等方面提出明确要求;国家文物局早在2008年就发布《关于加强20世纪遗产保护工作的通知》,2018年7月8日又发布《不可移动文物认定导则(试行)》,其中第七条、第九条的内容规定了有关20世纪建筑遗产的保护要则。所以深圳要自识瑰宝,要从建筑当代遗产的载体中找到城市发展的脉络。但2019年7月,当第43届世界遗产大会公布《世界遗产名录》时,有35年历史的深圳体育馆终遭强拆且造成人员伤亡。很怪的是,广东省住建厅发文,强调拆楼要安全,基本未分析其为什么拆?该不该拆?因此,留住以深圳为代表的新兴城市及旧城新园区的城市记忆在全国也刻不容缓。

2. 20世纪建筑遗产的"工业说"

工业遗产作为文化遗产的重要组成部分,其构成与文化遗产相同。《世界遗产名录》中的工业遗产并非仅仅是传统意义上的厂房,其类型多样,包括如纪念物、厂房、设施(如都江堰水利设施),文化线路(如大运河),文化景观(如咖啡种植园、葡萄酒作坊与庄园)。20世纪遗产中的工业遗产,除了空间的价值外,对国家、历史、文明乃至城市建筑的推动作用巨大。事实上,没有工业化及工业文明的进步,何谈现代化国家与城市。跨越曼哈顿西区22个街区,共1.5英里(约2.41千米)长,分三期向公众开放的纽约高线公园,已是美国20世纪工业遗产经过精心设计且成功再生的示范精品。"高线之友"成立于1999年,是由约克亚·大卫和罗伯特·哈蒙德发起的公益组织,成功地将废旧的高架铁路进行保留和再利用,建成的高线公园使原本杂乱无章的工业废墟通过"生态架构"生出美感。如今，公众可漫步于没有过多修饰的有2.5米高架平台的蜿蜒小路上，俯瞰纽约街景。有人说，高线公园对纽约城市复兴作用很大，因为它带来的不仅是新与旧的转换，更有对20世纪建筑遗产有启迪意义的工业遗产之蜕变和重生，在这里行走，可读到纽约的历史，更感受到不同"节点"上的创新建筑。

我国工业遗产很丰厚，仅以北京为例就有太多的共和国足迹。2019年首钢百年，如今首钢园区内滑雪大跳台正在建设中，"二七厂"的1897科创城一期的建设已近尾声，它们对城市、对国家都意义非凡。回溯新中国初期，北京几乎没有工业，160多万城市人口竟有30万人失业，一座座工厂必然拔地而起。到20世纪70年代中后期，北京已是名副其实的工业城市了。如：1949年成立的北京新华印刷厂，专门印制传播到大江南北的领袖著作；1954年建成的国营751厂，由民主德国援建，系国家"一五"期间的156个大型项目之一，现已转型为时尚设计广场；1897年的卢保铁路卢沟桥机厂（新中国成立后更名为二七机车厂），是中共二七大罢工发源地，1958年试制出新中国第一台内燃机车，2017年已转型为以轨道交通、智慧交通为主题的"科创城"；1956年建成北京化工四厂，是新中国第一家研发金属钾、超氧化钾系列产品的化工厂，打破了国外垄断与封锁，现为东方1956文化创意园；1957年建北京第二棉纺织厂，是国内第一个采用全套国产设备的大型棉纺织厂，现为莱锦文化创意产业园；1964年建设的外贸三间房仓库，20世纪70年代美国前总统老布什在华任联络处主任时，曾三次携夫人骑自行车到懋隆位于三间房的样品间买中国工艺品。以上例子说明，在工业老厂房逐渐退出历史舞台后，它们又被改造为各具特色的文创园、科创城，这是20世纪工业遗产在时代的新起航。

2019年10月26日—28日，中国第十届工业遗产学术研讨会在郑州第二砂轮厂举行，主题为"新中国工业建设的发展历程、伟大成就、记忆及遗产"。刘伯英教授的文章《十年回顾》，强调工业遗产学术共同体始终坚守在工业遗产保护利用和城市更新的最前沿，已在推动工业遗产的社会认知及给城市复兴上产生作用。本人在会上做了"中国20世纪建筑遗产认定的几个联想——兼议20世纪建筑遗产的工业遗产问题"发言，其要点是：《世界遗产名录》中20世纪建筑遗产的比重与启示；全国第八期国保单位、历史建筑与20世纪建筑遗产面临的问题；工业文明对20世纪建筑遗产传承与创新作用巨大。演讲中特举例解读了第二批20世纪建筑遗产的池州祁红厂老厂房。

第一步，从发现到认定之旅（2017年8月—2017年12月）。第一次走进国润祁红厂，由对朴素之美的工业遗存的"活态"感悟，联想到中国20世纪建筑遗产的评定标准，体会着它传承至今的珍贵。2017年10月初在池州市人民政府支持下，考察组展开多次深度调研与访谈研讨。2017年12月1日，当我随一批院士、大师们走进祁红厂时，他们不约而同地认可着这个遗产的非凡价值，一致认同它能够成为中国20世纪建筑遗产。第二步，发现不忘科学评估（2017年10月—2018年元月）。委员会组织了专家团队，并没有简单以国润祁红旧厂房入选"第二批中国20世纪建筑遗产项目"为目标，而是希望用此"引爆"文化池州的"城市文化工程"：一方面，充分从学理上论证"国润祁红"的厂房建筑体现了新中国初创期的风采；另一方面，利用品质名扬天下的祁红茶，串联起中国正实施的"一带一路"发展之策，创造出有社会文化"大价值"的祁红格调及其生活方式。考察组提交了共计"12条"的《国润祁红贵池茶厂老厂区创意设计规划要点》，其中包括以祁红茶为切入点的"文化池州"长远发展之策。第三步，从创意设计考察到祁红品牌传播（2018年3月—2018年9月）。2018年元月发布的《国润祁红贵池茶厂老厂区创意设计规划要点》的第十一条"用故事思维传播'国润祁红'文化池州品牌"中分析了池州与祁红知名度不高的原因。而后于2019年4月3日在故宫博物院首发《悠远的祁红——文化池州的"茶"故事》一书。由池州的祁红茶厂房，联想到唐山的百年开滦矿区乃"洋务运动"时期兴办的企业，经过百年演化与蜕变，至今仍生生不息、充满生机。从1878年在"洋务运动"求新求变思潮下始建开平矿务局，有了近代中国内地第一眼采用西方开采技术的大型矿井，之后又创立了中国第一条标准轨铁路"唐胥铁路"，中国第一台蒸汽汽车，中国第一代"马牌水泥"……这些均堪称中国近代工业开先河之举，"仿西技、用其人"就是冲破中国封建板结的大地，与当时席卷世界的工业革命比肩看齐，在当时乃气魄辉煌的创举。尽管唐山经历了1976年大地震和特大透水灾害等数次困厄，但开滦闪耀着共和国骄子的光环及民族气节，体现了尊重历史的工业遗产发展的创意之路。

在第二批20世纪建筑遗产中，洛阳第一拖拉机厂早期建筑名列其中，它是洛阳新中国工业遗产的重要部分。"一五"期间156项工程中，有6项安排在洛阳，除了工业价值外，它们更有历史与文化价值。历史地看，当时的洛阳工业区规划科学合理，厂部大楼与厂前广场雄伟开阔，是在苏联列宁格勒国家城市设计院建筑师巴拉金的指导下完成的。该规划汲取长春、包头等工业区规划经验，得到国内外城市规划界高度评价，有"洛阳模式"之称。这6大项目是：第一拖拉机厂、轴承厂、矿山机器厂、铜加工厂、高速船用柴油机厂、

耐火材料厂均为国内最大型。第一拖拉机厂厂量长期占有中国拖拉机保有量的50%，承担了中国60%机耕地的作业量。以"一拖"为代表的工厂主楼，带有20世纪五六十年代特点，突出中轴线布局，体现宏大叙事风格，线条流畅，善于刻画细节，在洋溢崇高与自豪精神时，体现艺术性与观赏性。此外"一拖"的中国第一条拖拉机装配线、矿山厂的八千吨水压机，均在洛阳及中国工业史上占有特殊地位。洛阳乃中国著名古都，分布着夏、商、周、汉魏、隋唐五大都城遗址，有"五都贯洛"的气势，在世界上也罕见，而新中国洛阳工业区的建设，使洛阳进入城市发展与工业文明时代，揭开了洛阳3800年城市化的新篇。为此，保护新中国洛阳工业遗产除重要建筑、配套"教科研"体系建筑、居民街坊保护外，特别要关注工厂厂门及厂部办公楼及厂前广场的保护。在"一拖"厂、矿山厂、轴承厂及铜加工厂的四个厂区大门广场，一线相连，神脉相通，沿建设路延绵5.6公里，是一个完整的工业建筑风景带，构成洛阳工业区最重要、成系列的核心建筑群，保护它们并非仅仅为了载入史册，更为了满足20世纪中国工业遗产服务城市复兴的文化创意实践与当代发展。

3. 20世纪遗产的"民本说"

民生或称"民本"。设计的目的是满足广大人民群众的生活与舒适需要的设计。建筑师与设计机构作为设计服务的供给方，是当代社会创新的中坚力量。民本思想作为中华文明的重要资源由来已久，老子曰"治国有常，而利民为本"；《史记·赵世家》中说"衣服器械，各便其用"，明确提出衣食住行要方便民众使用。"民本"学说的本质是解决并优化人民生活中的问题，其设计从理性上应坚持：①设计创作应根植民众生活且回归设计本源；②设计创作应深入公众，聚焦社会底层与边缘；③设计创作还应着眼于节约发展，关注生态环境的未来。以工人新村为例，虽为过渡性质的居住模式，其遗产价值在于它介于20世纪20年代兴起的里弄模式与90年代兴起的小区模式之间，我颇同意它是即将"绝版"的历史记录，"工人新村"之所以鲜活与亲切，源于民本的平凡。上海的工人新村始建于20世纪50年代初，代表作是曹杨新村，其建筑风格与新式里弄、花园洋房经典住宅不同，最重要的是它服务是为工人阶层"量身定制"的。据统计，从1953年至1958年，上海共建工人新村201个，住宅面积有468万平方米，极大地改善了工人的居住状况。工人新村的选址很讲究，它表明新政权的执政基础及面向公众的价值取向，如普陀的曹杨新村临近华东师范大学，杨浦的控江新村、长白新村、鞍山新村周边有复旦大学、同济大学、上海理工大学等，这种选址充分考虑"工人子弟"的未来教育。作为全国最早的工人新村，上海市政府当时有《拟定普陀区曹杨村工人宿舍建设计划初步总结》，选择曹杨是因为"曹杨路两侧都是农地，虽不在普陀区之内，但接近普陀区边缘"，以距离计算，到国棉一厂仅3.6公里，到大自鸣钟也仅4公里。乘公共汽车仅需20分钟，步行不到一小时，不能算太远，空地多，环境也好。以工人阶层作为目标群体的曹杨新村规划极大地影响了上海这座城市，为这座城市增加了以前没有的新元素。六十多年来，仅曹杨一村就先后接待了超过150多个国家与地区的宾客。

以北京为例：1953年我们学习苏联引进了"街坊"为主体的工人生活区，如东郊棉纺厂、酒仙桥等生活区。此类生活区与旧街区相比，其突出特点在于住宅围合的内部庭院较宽敞，有较多绿色空间。1955年提出"成街成坊"建设原则，提高了建筑层数和密度，如永安路的住宅区。1956年兴建幸福村街坊，建筑布局顺应不等边地形及起伏地势，选用3种不同长度的3~4层坡屋顶外廊式住宅，自由围合成一系列不同形态、尺度的院子，每个院子又具有不同风格及特点。再看天津此时期仍以工人平房新村为主，如中山门新村是首个项目，它以"邻里单位"为依据，内部道路为八卦形，分成12个街坊，围绕中心公园布置了居民服务生活设施，"一五"期间（1953—1957年），天津在引入苏联居住区规划与住宅设计手法的前提下，建设了团结里、友好里、昆明路、德才里、佟楼、吴家窑等楼房居住街坊，"民族形式"住宅设计的典型代表有天津大学六村宿舍、尖山居住区等。

4. 20世纪遗产的"高楼说"

孙中山在《建国大纲》中有对上海的评价："上海现在虽已成为全中国最大之商港，而苟长此不变，则无以适合于将来为世界商港之需用与要求……故创造市宅中心于浦东，又沿新开河在左岸建一新黄浦滩，以增加其由此计划圈入上海之新地之价值……"20世纪30年代，上海是远东地区的一盏明灯，聚集了当时中国近40%的近代化工业、50%的贸易和60%的资本，是跨国公司开展贸易和商贸的枢纽。新中国以来上海进一步崛起，不但是中国城市化的传奇，也为上海重新加入全球分工发挥了关键作用，美轮美奂的高层建筑更是20世纪的特征，它承载着中国建筑的历程。1929年上海和平饭店（高77米）、1937年中国银行（高76米），曾经的远东第一高楼乃旧上海的地标即1934年建成的国际饭店，由建筑师邬达克设计，几乎是美国20世纪20年代摩天大楼的翻版，有美国艺术装饰主义的意蕴，其83.8米的总高度纪录，一直保持到1983年上海宾馆建成。

1949年以后，全国各行各业学习苏联的建设经验，中央决定于1955年3月在上海举办苏联经济及文化建设成就展，因此建设与之相适应的展览馆即中苏友好大厦一事就提上议程（1954年北京展览馆已建成，总高度87米）。经过反复调研，大厦选址在哈同花园旧址即上海静安区延安中路1000号，1954年5月动工兴建，1955年3月竣工，建筑高度110.4米，成为当时国内第一高楼。1972年美国前总统尼克松访华开启了中美之间的新关系，为扩大对外贸易，中央决定在广州兴建白云宾馆。白云宾馆是岭南派风格融入现代建筑的一个典范，总建筑师莫伯治院士表示："没有自然界配合的建筑，起码不是一个完美的建筑。"所以，莫伯治保留选址中的一座小山丘，特意将白云宾馆后退，山丘的土壤大树，也保留下来。如宾馆的中庭，利用原来的五棵古榕树，再通过设置瀑布、景石、水池，形成了一个在宾馆内的自然景观，是在此之前从未有过的设计思路。白云宾馆1973年动工，1976年初完工，共34层，楼高120米，又成为当时的中国第一高楼。深圳国贸中心大厦（建成于1985年，高160米）、深圳发展中心大厦（建成于1987年，高165米）、广州国际大厦（建成于1989年，高200米）、北京国贸一期（建成于1989年，高150米）、北京京城大厦（建成于1989年，高183米）、深圳地王大厦（建成于1996年，高384米）、北京京广中心（建成于1994年，高209米）、北京国贸二期（建成于1999年，高156米）……从1934年的国际饭店、上海中苏友好大厦到广州白云宾馆，建筑的不断升高蕴含着国内外建筑师的不凡贡献。作为20世纪建筑遗产，上海金茂大厦具有标志性意义，1994年5月兴建，1997年8月结构封顶，1999年竣工，地上88层，高420.5米，是中国高标准超高层建筑的"拓荒者"。金茂大厦设计采取国际征集方案的方式，最后美国SOM设计事务所的方案拔得头筹。在外观上，SOM借鉴中国宝塔的建筑风格，将金茂大厦设计成具有现代中国特色的超高层建筑；在细节上，88层、8根周边矩形柱、层面向内收缩每8层一单元等，符合中国人对吉祥数字的喜好；在结构设计采用钢材和混凝土的混合结构，其原理源于中国山西应县千年木塔的构造。尽管当时中国并没有超高层建筑设计规范，但反复的科学论证，为建造金茂大厦赢得了可能性。这虽是距今只有20年的新建筑，但它已进入20世纪中国建筑遗产的行列，是技术与设计风格乃至中西方文化交流创造的典范。

四、中国 20 世纪建筑遗产创新发展的智库之思

以智库的形式代表中国文物学会 20 世纪建筑遗产委员会向业界与社会发声是事业发展所需。2018 年 11 月 24 日第三批"中国 20 世纪建筑遗产"推介活动在东南大学举行，与会专家议到中国文物学会 20 世纪建筑遗产委员会的作为，大家对委员会要代表业界发声寄予期望。事实上，近年来委员会专家正从总结三批 20 世纪建筑遗产项目的特点所反映出的规律入手，研究这些项目对城市发展、对建筑创作乃至继承中国先辈建筑师思想的启示。尼克拉斯·佩夫斯纳（1902—1983 年）的著作《欧洲建筑纲要》一书（已出版七卷）是值得深读的 20 世纪经典建筑著作。作者系 20 世纪最著名的艺术家和建筑史学家，1967 年获 RIBA 皇家金质奖章，也是《企鹅建筑学词典》的合著者，还是"英格兰建筑"系列丛书及《建筑评论》的编辑。《欧洲建筑纲要》一书对编撰体现智库之思的《中国 20 世纪建筑遗产发展报告》有所启示："无论何人决定编写建筑简史、艺术简史、哲学简史……都必须最强烈地表达他至关重要的愿望及对事件的认知……建筑不是材料和意图的产物，也不依赖社会条件产生，而是变化的时代变幻精神的产物，它是时代精髓，渗透于整个社会生活、宗教、学术及艺术之中。"2011 年创办的《中国建筑文化遗产》及 2014 年诞生的中国文物学会 20 世纪建筑遗产委员会是相互衔接的平台，在过去的 9 年间专家们以求索精神，以改革开放的风格与勇气，依托中国文物学会、中国建筑学会及全国建筑设计研究及高效优质智力资源，通过每年"中国 20 世纪建筑遗产项目"评选认定的工作，有目标地面向全国和各级地方持续开展建筑遗产社会调查及理论分析，进行了多维度的测评分析，从而有效形成了调研分析与理论研究、实践探索与经验总结、内涵挖掘与传播分析、专题研讨与论坛峰会等内容。编研中的《中国 20 世纪建筑遗产发展报告》或称蓝皮书属性，不仅盘点中外 20 世纪建筑遗产年度资讯及成果，更将预测紧随时代变化的现当代建筑文化的前景，从而成为引导中国城市省思 20 世纪建筑遗产保护的"问题集"。

保护、传承、活态利用的理念，在 20 世纪建筑遗产与当代城市极其适用。不少旧城区、历史建筑、既有建筑都是几乎尚未挂上"文保"单位牌子的建筑。限于历史原因，这些蕴藏城乡文化与科技基因密码的建筑，并未得到认定，甚至被低估了文保价值。但近年来由于文化复兴的彰显，不少既有建筑（哪怕只有 30 年）都留下属于自身的文化标签和形象标志，简单粗暴的拆除成为城市建设的"败笔"，利用 20 世纪建筑遗产讲百年的特色故事，激发旧建筑的新活力，甚至嫁接新兴业态是潜力无穷的事。恰如英国皇家建筑师学会（RIBA）原主席乔治·弗格森所说："城市当然可建世界最高的楼，但最高纪录也只能保持一两年而已，很快就会被更高的楼代替而不存在独特性，但城市文化的独特性却不会消失，只会历久弥新。" 20 世纪建筑遗产于城市更新、于既有建筑活态利用，具有激活生根于城市的精神脉络之效，它能将改善民生与保护历史风貌并举，它能在强化优秀历史建筑严管制度的同时，有效避免人为毁坏。以全国为数不少的 20 世纪遗产中的工业遗址为例，若不认定只等待企业转型，旧厂遗址都会丢光，再过一些年，人们就将这些历史遗存忘记了。与放在博物馆、纪念馆中的器物、书法、绘画不同，20 世纪建筑很脆弱，一夜之间就能荡然无存。入选第三批中国 20 世纪建筑遗产的鞍山钢铁公司有"共和国钢铁工业长子"的称号，因钢而起、因钢而兴，鞍山与鞍钢命运相连。在这里，所有建筑都因时间的长久注视而赋有生命。在这里，烟囱是需要保护的"工友"，摩天轮是"钢城"当年的辉煌。城市更新是长期行为，微更新与有机更新是方向。挖掘并传承建筑文化，既要抓保留形态，更要保留基因，这是建筑师将传承与创新融为一体的设计新路径。据此，"蓝皮书"的智库研究，至少要围绕如下内容展开。

（1）主报告。主报告为中国20世纪建筑遗产项目的现状、分类体制与管理特征；重点围绕城市化建设与新型城镇化趋势，透彻解读20世纪建筑遗产概念的内涵与外延；要突出20世纪建筑遗产对城市社会经济发展的贡献，要分析20世纪建筑遗产的经济学属性等。主报告重在揭示管理体制机制，提出遗产保护视野下的政策、制度、立法等建言。

（2）分报告。分报告把控的是细节与微观，即要从各方面论证"遗产事业"怎样服务当下城市化建设：为什么说20世纪建筑遗产必须担当重任，20世纪建筑遗产对中国城市化的影响，城市管理如何提升对20世纪遗产的价值认知，高等教育与国民建筑文化普及如何从认知20世纪经典建筑入手等课题。其分报告至少有以下三个方向。

分报告一：国际经验借鉴与比较研究：从包豪斯学校到欧美诸国现代建筑教育的影响，从百年来中国建筑吸纳"洋风"到中国建筑的现代发展，也包括对以朱启钤为代表的中国营造学社对20世纪遗产的广义贡献与影响力等。

分报告二：在国家《文物保护法》等法律与政策语境下，要进行各城市适用于20世纪建筑遗产保护的相关法规分析，并在此基础上形成有针对性的《20世纪建筑遗产保护办法》（征求意见稿）文本，大胆探讨从立法层面上不允许"乱拆"的法规政策。

分报告三：本报告从建筑遗产技术管理的精细化入手，研究技术法规体系化，城市与建筑遗产评估方法，特别是拥有20世纪建筑遗产项目不同代表城市在修缮技术、建筑材料等方面的应用经验，以提升对当代建筑师有指导作用的技术规范、技术路线与设计修复策略的新思路等。

由《中国20世纪建筑遗产发展报告》的智库研究，让人想到如何在建筑创作与市场间找到文化保护的平衡点，如何建构一个城市文化与市场产业化正常的良性关联机制，如何为新中国城市建设累积更多堪称一流的建筑或艺术作品，如何在世界建筑文化舞台上推高中国20世纪建筑文化的国际地位等，都是亟待深化的作品、事件、人物、思想等问题研究的必要命题。

中国文物学会20世纪建筑遗产委员会副会长、秘书长
中国建筑学会建筑评论学术委员会副理事长
《中国建筑文化遗产》《建筑评论》主编
金磊
2019年11月

第一篇

第二批中国 20 世纪建筑遗产项目

虽然 20 世纪已过去，但 20 世纪的建筑纪念碑尚存。本篇展示的第二批中国 20 世纪建筑遗产项目有 100 项，它们在一定意义上代表了 20 世纪至今中国建筑的发展及主要成就。如果用中国和美国两位建筑前辈张钦楠先生和弗兰姆普敦教授的著述说"现代建筑：一部批判的历史"。他们的论断是在提醒读者，20 世纪技术与艺术驱动下的环境与结构工程及其创新形式，将建筑技艺带到了一个全新水准的文化高度，建筑在城市社会中扮演了越来越重要的角色。若从历史大文化的视角看，20 世纪经典建筑是人造的丰碑，创作者的思想、目标和行动，不但力求超越自己，也努力希望是传承的遗产，一切最有生命力的作品是最可表现"民本"感受与思想的作品，它们继承了现代精神与城市情感。薛加平的《建筑遗产保护概论》中归纳了国际遗址理事会与联合国教科文等国际机构的建筑遗产保护文件，从《关于历史性纪念物修复的雅典宪章》（1931 年 10 月，雅典）到《文化线路宪章》（2008 年 10 月，魁北克），至少表明早在 90 年前，20 世纪建筑遗产概念已在国际层面展开，尤其是"二战"之后宪章、宣言、公约等密集出台，使建筑经典的内容与外延不断拓展，特别是对建筑遗产的认定呈现当代态势，如巴西首都巴西利亚（1960 年建）于 1987 年进入《世界遗产名录》；更有甚者，波兰奥斯维辛集中营（1940—1945 年建）于 1979 年进入《世界遗产名录》。需要说明的是，在第二批"名录"中既有代表工业进步的百年首都钢铁公司（1919 年建）入选，也有反映新中国自力更生建设示范的中小型遗产项目如安徽池州润思祁红工厂（1959 年建），它们充分表现了在《中国 20 世纪建筑遗产认定标准（2014）》下，专家委员会委员顾问从实际出发，充分从历史、社会、城市、艺术、科技、经济诸方面对项目认定推介所作的扎实努力。历史环境是联合国教科文组织在 1976 年便给出的定义，即"过去存在的表面"，它映射出中外 20 世纪建筑遗产项目，哪个好作品不是利用历史参照物，挖掘其文化价值和文化魅力所为。这些作品中有 1919 年包豪斯建筑学派的设计影响力，更不乏勒·柯布西耶新建筑五点的传播力，更有梁思成、杨廷宝等建筑宗师们的开拓贡献之导引。中国 20 世纪建筑经典作品必然成为世界建筑舞台的佼佼者，至少从中可综合梳理出如下特征：①中国传统建筑的主流在接受西方思潮中进步；②图变创新的中国现代建筑特色明显；③民族式及本土化建筑风格积极呈现；④项目在新技术与新材料中融入艺术风格；⑤节制豪华、不事张扬的设计风气传承；⑥多元共存的开放创作理念形式等。

Unit One

The Second List of the 20th-Century Chinese Architectural Heritage

Despite the passage of the 20th century, the architectural monuments of this century still remain. The Second List of the 20th Century Chinese Architectural Heritage Projects to be presented in this chapter consists of 100 items, which, in a certain sense, represent the development and main achievements of Chinese architecture from the start of the 20th century to the present. "Modern Architecture: A Critical History", in the words of a masterpiece of Mr. Zhang Qinnan and Prof. Kenneth Frampton, two predecessors of architecture in China and the United States. Their conclusion is reminding the readers that the environmental and structural engineering and its innovative forms driven by technology and art in the 20th century have carried architectural techniques to a cultural height by a brand new standard. Architecture is playing a more and more important role in urban society. From the perspective of history and culture, the classical architecture in the 20th century has become artificial monuments, reflecting the ideas, goals and actions of the creators, who not only strove to transcend themselves but also worked hard to create heritages that could be passed on to future generations. All the works with the greatest vitality can best reflect the "people-oriented" feelings and thoughts, which inherit modern spirit and urban sentiments. *A Survey of the Protection of Architectural Heritage* by Xue Jiaping summarizes the documents on architectural heritage protection of international organizations such as the International Council on Monuments and Sites (ICOMOS) and the United Nations *Educational, Scientific and Cultural Organization (UNESCO), from the "Athens Charter for the Restoration of Historic Monuments"* (October 1931, Athens) to the *"ICOMOS Charter on Cultural Routes"* (October 2008, Quebec), demonstrating that at least 90 years ago the concept of the architectural heritage in the 20th century already spread at the international level. In particular, with the intensive launch of charters, declarations and conventions after the Second World War, the intension and extension of architectural classicals have been constantly developed. The accreditation of architectural heritages has taken on a contemporary trend, in particular. For example, Brasilia, the capital of Brazil (built in 1960), entered the *World Heritage List* in 1987; also, the Auschwitz Concentration Camp in Poland (built in 1940-1945) entered the *World Heritage List* in 1979. It should be noted that the Second List of Architectural Heritages in the 20th Century not only includes the century-old Shougang Iron and Steel Company (built in 1919) representing industrial progress but also medium and small heritage projects reflecting independent architectural models in the People's Republic of China such as Chizhou Runsi Keemun Factory in Anhui (built in 1959), which fully reflect the solid efforts made by the Expert Committee members and consultants under the *"Accreditation Standard for Architectural Heritages in the 20th Century in China"* (2014, Beijing), and from a comprehensive perspective including history, society, city, art, technology and economy. Historic environment is a definition given by the UNESCO as early as in 1976, which means "surface of past existence", reflecting that all masterpiece in the architectural heritage projects in the 20th century at home and abroad used historical references to explore their cultural value and charisma. These works reflect the design influence of the Bauhaus School of Architecture in 1919, the spreading power of Le Corbusier's Five Points of Architecture, and the guidance of pioneering contributions of architectural masters such as Liang Sicheng and Yang Tingbao. The 20th-century chinese classical architectural works will inevitably become outstanding masterpieces on the world architectural stage, from which at least the following characteristics can be summarized: (1) the mainstream of traditional Chinese architecture makes progress while accepting Western thoughts; (2) modern Chinese architecture aiming to reform and innovate stands out distinctively; (3) national and localized architectural styles keep emerging; (4) projects integrate artistic style in new technologies and materials; (5) moderately luxurious and low-keyed design style is passed on; and (6) creative ideas and forms are open to cultural diversity, among others.

国立中央博物院（旧址）
The Old Site of National Central Museum

工程地点｜南京市
建筑面积｜17000平方米
竣工时间｜1951年
设计机构｜兴业建筑师事务所
主要设计人｜徐敬直

国立中央博物院今为南京博物院主楼（习称南京博物院大殿，现为南京博物院接待大厅和历史馆），位于南京市中山门内。兴业建设师事务所徐敬直、李惠伯等设计，顾问为梁思成、刘敦桢，先后由江裕记营造厂、华东建筑工程公司、南京市建筑工程公司承建。初步设计于1935年提交，1936年修订设计方案并投入施工，因抗战爆发而工程中断，1946年复工，1948年初步完成第一期工程，1951年正式竣工。博物院总占地面积为13万平方米，大殿为砖混结构，为大型博物馆建筑。原设计方案中，中央博物院为多个单体建筑组合而成的大型庭院，大殿及背后的工艺馆是其中之一，也是唯一完成的一组。

The National Central Museum, now the main building of the Nanjing Museum (formerly known as the main hall of the Nanjing Museum, now the reception hall and history hall of the Nanjing Museum), is located within the Zhongshan Gate of Nanjing. With Xu Jingzhi, Li Huibo, and others as designers and Liang Sicheng and Liu Dunzhen as consultants, the museum was built successively by Jiangyuji Construction Factory, East China Construction Engineering Company and Nanjing Construction Engineering Company. The preliminary design was submitted in 1935, and the design scheme was revised and put into construction in 1936. The project was interrupted due to the outbreak of the War of Resistance against Japanese Aggression, and resumed in 1946. The first phase of the project was tentatively completed in 1948 and formally completed in 1951. The museum covered an area of 130,000 square meters, and the main hall was a large museum building with a brick-concrete structure. In the original design scheme, the Central Museum was a large courtyard composed of several single buildings, and the main hall and the craft museum behind it were one of the buildings and

建筑正立面

设计者力图体现中国盛唐建筑风格以别于同时期其他仿古类建筑。在最初的规划中，国立中央博物院的设计规模十分宏大，但采用明清官式建筑的外观并不理想，故以梁思成、刘敦桢为首的顾问建议采用唐辽建筑风格，并具体提供了技术指导，最终形成目前造型朴实雄厚的唐辽皇家大殿的建筑外观，而其内部装修也部分采纳了唐辽式藻井装饰，展厅、库房、办公用房等布局疏密有致，其库房采用了当时最先进的安全设备，展厅设计则尽量为参观者考量。此建筑在建造过程中几经周折，但得到各界的关注与支持，基本上达到了设计预想。

作为中国固有式建筑的典范之一，国立中央博物院大殿之独到之处在于：以唐辽风格取代同时期的明清宫殿建筑形式，此举堪称对中国古代文化精神的深化解读。

the only set completed.

The designers had done their best to embody the architectural style of the flourishing Tang Dynasty in China to differentiate it from other antique buildings in the same period. In the initial planning, the National Central Museum was designed to be very grand, but adopting the appearance of the official buildings in the Ming and Qing Dynasties could not perfectly present the design scheme. Therefore, the consultants headed by Liang Sicheng and Liu Dunzhen advised the adoption of the architectural style of the Tang and Liao Dynasties, while providing specific technical guidance. Thus, the architectural appearance resembling the royal halls of Tang and Liao was eventually developed, unadorned yet powerful. The interior decoration of the hall also partially followed the caisson ceiling of Tang and Liao. The exhibition hall, warehouse and office space were also aptly structured. The warehouse adopted the most advanced safety equipment then, and the design of the exhibition hall catered to the needs of visitors as much as possible. This museum experienced many twists and turns in the construction process, but the design expectations had been basically met with the support from all walks of life.

As one of the paragons of Chinese inherent architecture, the main hall of the National Central Museum is unique in that it replaces the palace architectural form of the Ming and Qing Dynasties with the Tang and Liao styles, which can be celebrated as a deep interpretation of the ancient Chinese cultural spirit.

徐敬直所绘设计图纸

屋脊吻兽

建筑与环境的结合

建筑室内局部

建筑外观局部

屋顶藻井装饰

鼓浪屿近现代建筑群
Gulangyu Modern Architectural Complexes

鼓浪屿位于福建省厦门市的九龙江出海口，19世纪中期至20世纪中期成为中西方文化交融的前沿。2017年7月8日，在第41届世界遗产大会上，"鼓浪屿：历史国际社区"被列入《世界遗产名录》。其遗存包括931栋当地和国际风格的历史建筑、自然风景及园林，其中核心要素53处。

其建筑特色与风格体现了中国、东南亚、欧洲建筑和文化价值观的交融，产生于岛上居住的外国人和归国华侨的多元性；是厦门装饰风格的起源地和最佳代表，体现了当地传统建筑灵感与西方早期建筑风格的融合，是现代主义和闽南移民文化影响之结晶。

由于1903年《厦门鼓浪屿公共地界章程》生效，鼓浪屿开始进入新的发展期。它由外来的多国侨民、本土居民和还乡华侨群体共同营建和管理，各种风格的建筑、宅院都呈现

Gulangyu Island, located at the sea exit of the Jiulong River in Xiamen City, Fujian Province, became the forefront of blending Chinese and Western cultures from the mid-19th century to the mid-20th century. On July 8, 2017, at the 41st World Heritage Conference, "Gulangyu: Historic Internation Community" was included in the World Heritage List. Its remains include 931 historic buildings, natural landscapes and gardens with local and international styles, of which 53 places are most essential.

The architectural features and styles of Gulangyu reflect the blending of architectural and cultural values in China, Southeast Asia and Europe, resulting from the diversity of foreigners and returned overseas Chinese living on the island. The island is the origin and best representative of Xiamen's decorative style, which embodies the fusion of local traditional architectural inspiration and the early Western architectural style, and reflects the product of the infutence of modernism and Minnan migrant culture.

As the Charter of Xiamen Gulangyu Public Land came into effect in 1903, Gulangyu ushered in a new period of development, when it was jointly built and managed by foreign expatriates, native residents and returned overseas Chinese. Buildings and houses with a great diversity of styles showcase the characteristics of multiculturalism, evolving into the unique urban historic

远眺鼓浪屿建筑群

了多元文化的特质，演化成当今独有的城市历史建筑景观。八卦楼（现鼓新路43号）是台湾商人林鹤寿1907年投资兴建的大型别墅。八卦楼在体量和高度上都是鼓浪屿别墅之最。1924年日本领事馆接收后在此开办"旭瀛书院"，1938年日军占领厦门后也曾在此开设避难所。再如番婆楼，系1927年菲律宾华侨许经权所建（现安海路36号），属华侨建造的大型殖民地外廊风格别墅建筑，在其建筑布局、立面装饰、园林庭院设计中，表现了本土建筑文化及工艺的强烈影响。进入大门，沿台阶拾级而上，可见院墙上贴墙而置的中国传统园林漏窗，顶上还有一座巴洛克风格的灰塑山花，中西合璧的动植物元素丰富地混搭其中。该建筑由主楼与东北角的牌楼构成，建筑面积约1600平方米。英商亚细亚火油公司旧址（现中华路21号），原为住宅建筑，建于20世纪初。1903年7月于伦敦成立的亚细亚火油公司曾垄断亚洲特别是20世纪上半叶中国的销售市场，1937年迁至该建筑的二层办公，建筑整体上采用英国维多利亚时期风格，采用有代表性的清水红砖墙及哥特式尖券拱窗等。

鼓浪屿历史建筑群丰富多样，其极具特色的近代园林艺术

architectural landscape today. Bagua Building (now at No.43 Guxin Road), a large-scale villa invested by Taiwan businessman Lin Heshou in 1907, is the largest villa on Gulangyu Island in both volume and height. In 1924, the Japanese Consulate took over the building opened the "Xuying Academy" here, and in 1938, after the Japanese army occupied Xiamen, it also set up a refuge here. Another example is Fanpo Building, which was built by a returned overseas Chinese Xu Jingquan from the Philippines in 1927 (now at No.36 Anhai Road). It is a large-scale villa of the colonial veranda style. In its architectural layout, facade decoration and garden courtyard design, a strong influence of local architectural culture and techniques can be discerned. Entering the gate and ascending the stairs, we can see latticed windows typical in traditional Chinese gardens on the courtyard walls, and a Baroque-style gray plastic molding pediment on the top, in which many Chinese and Western animal and plant elements are mingled. This architectural complex consists of the main building and the archway in the northeast corner, with a floor area of about 1,600 square meters. The former site of Asian Petroleum Company (now at No.21 Zhonghua Road), originally a residential building, was built in the early 20th century. Asia Petroleum Company, established in London in July 1903, once monopolized the sales market in Asia, especially in China in the first half of the 20th century. In 1937, the company's office moved to the second floor of this building, which as a whole assumes the Victorian style of England, with representative plain red brick walls and Gothic pointed arch windows.

Gulangyu Island boasts a great diversity of historic building complexes, with their unique modern garden art demanding our attention. In the garden layout, most adopt the axial layout of buildings or courtyard gates,

从鼓浪屿回望厦门城市建筑

俯瞰鼓浪屿建筑

（含私家园林）不可不提。在园林布局上，多以庭院中建筑或庭院大门的轴线式布局、几何形水池与花坛、修建整齐的灌木植栽体现西方园林设计的影响，也配以中国传统园林模式依自然地形布置；在造园手法上，中西园林元素拼贴与借鉴，装饰风格东西方混搭更加淋漓尽致，如番婆楼有南狮、寿桃等吉祥图案，又有黑人、白人男孩的小天使图案。菽庄花园位于鼓浪屿南面的港仔后海滨（现港仔后7号）。侨居鼓浪屿的台湾富绅林尔嘉，为怀念台北板桥花园，于1913年建造菽庄花园。其设计手法既运用中国传统文人园林的造园术，又融合近代建筑式样及西方造园风格，采用钢筋混凝土等材料，以明快简洁的形式构建拱桥、栈道等，成为我国20世纪罕有的文人园珍贵实例。该园择地于僻静的鼓浪屿南端的石墈顶，依崖临海，其中藏海园借海水与礁石造景，曲折蜿蜒的桥有44座，使菽庄花园成为近代园林中真正的"海上花园"。需要提出，其设计总体上注意与台北板桥花园景致相对应，如菽庄花园中建成的小板桥、渡月亭、千波亭、熙春亭等均与台北板桥花园景致相对应，虽板桥花园与菽庄花园分别出自林维源与林尔嘉父子之手，但造园意境在

decorated with geometric pools and flower beds, and neat shrubs reflecting the influence of western garden design, while the traditional Chinese garden model of following the natural terrain for scenic arrangements is observed. In the gardening techniques, Chinese and Western garden elements are collaged with mutual illumination, and the decorative style more vividly presents the mingling between East and West. For example, Fanpo Building features auspicious patterns such as South Lion and Longevity Peach, as well as little angel patterns of black and white boys. Shuzhuang Garden at Gangzihou Waterfront (now at No.7 Gangzihou) on the south side of Gulangyu Island was built by Lin Erjia, a wealthy gentry of Taiwan living on Gulangyu Island. To miss Banqiao Garden in Taipei, Liu built the Shuzhuang Garden in 1913. Its design not only drew on the gardening techniques of traditional Chinese literati gardens, but also integrated modern architectural styles and Western gardening styles. Reinforced concrete and other materials were adopted to build arch bridges and plank roads in a bright and concise form, making it a rare example of literati gardens in China constructed in the 20th century. Located at the top of a stone ridge at the south end of the secluded Gulangyu Island, the garden faces the sea and leans on cliffs, featuring the Canghai Garden landscaped with seawater and reefs as well as 44 winding bridges, which render the Shuzhuang Garden stand out as a real "Garden on the Sea" among modern gardens. It should be noted that the design of this garden on the whole corresponds to the Banqiao Garden in Taipei in terms of exquisiteness. For example, the Little Banqiao Bridge, the Duyue Pavilion, the Qianbo Pavilion and the Xichun Pavilion built in the Shuzhuang Garden all correspond to the sceneries of the Banqiao Garden in Taipei. Although Banqiao Garden and Shuzhuang Garden

共通点上有所不同，精华各异。此外，菽庄花园设计还与菽庄吟社的诗歌文学相呼应，成为菽庄吟社成员寄托家国情怀的雅致之所。建于1920年的黄荣远堂（现福建路32号），属鼓浪屿华侨花园洋房的代表作。1937年，华侨黄仲训将其买下作为地产公司黄荣远堂总部新址，这三层洋楼建筑受到曾在东南亚及殖民建筑中流行的帕拉迪奥风格的影响。建筑南端是宽敞的庭院，设计成中西合璧的园林：庭院中央为中轴对称的西方园林特有的双圆形嵌套形式水池，与主建筑相对，在水池正中放置了中国传统造园常用的仿太湖石假山等，庭院中还种植有岭南园林风格的植物等。

were designed by Lin Weiyuan and his son Lin Erjia respectively, the artistic conception of gardening differs essentially while sharing some commonality. In addition, the design of the Shuzhuang Garden echoes the poetry and literature created by the Shuzhuang Poetry Society, making it a refined place for the members of the Society to express their feelings of home and country. The Huangrongyuan Hall (now at No.32 Fujian Road), built in 1920, is a representative of returned overseas Chinese's garden houses in Gulangyu Island. In 1937, Huang Zhongxun, a returned overseas Chinese, bought it as the new headquarters of Huangrongyuan Hall, a real estate company. This three-story building of the Western style was influenced by the palladio style then popular in Southeast Asia and colonial buildings. At the south end of the building is a spacious courtyard designed by integrating Chinese and Western styles. In the center of the courtyard and opposite to the main building, there is a double-circular nested pool unique to Western gardens with axial symmetry. In the middle of the pool there is a rockery like Taihu stones commonly used in traditional Chinese garden landscaping. In the courtyard there are also plants with Lingnan garden style.

建筑局部 1

建筑局部 2

鼓浪屿建筑与厦门城市建筑的联系

开平碉楼
Kaiping Diaolou

该建筑是广东开平市境内的一种中西合璧式民居建筑，是中国乡土建筑中的一个集防卫、居住和中西建筑艺术于一体的特殊类型，是中国岭南农村大量移植西方建筑艺术之典范，是中国岭南华侨住宅中的一绝。经多次文物普查，现存单体建筑2500余座，以雁平楼、开平立园、方氏灯楼、马降龙碉楼群、锦江里瑞石楼、自力村碉楼群、日升楼、翼云楼、赤坎古镇景辉楼等为代表性遗存。作为罕见的建筑组群，开平碉楼于2001年列入第五批全国重点文物保护单位名单，2007年列入《世界遗产名录》，是首个属于华侨文化的世界遗产项目。

开平碉楼大规模兴建于20世纪初，原是海外华侨为国内眷属兴建的高层防御性住宅。受西方建筑影响，这批房屋汇集

Diaolou, a combination of Chinese and Western residential buildings in Kaiping City, Guangdong Province, is celebrated as a special type of Chinese vernacular architecture that integrates the functions of defense, residence and Chinese and Western architectural arts. After many cultural relics general surveys, it has been found out that there are more than 2,500 extant single buildings, and representative diaolous are Yanpinglou, Kaiping Liyuan, Fangshi Denglou, Majianglong Diaolou Cluster, Jinjiangli Ruishilou, Zilicun Diaolou Cluster, Rishenglou, Yiyunlou and Jinghuilou of Chikan Ancient Town. As rare building clusters, Kaiping Diaolou was included in the fifth batch of key cultural heritage units under state protection in 2001 and in the World Heritage List in 2007. It was the first World Heritage project from the culture of returned overseas Chinese.

Kaiping Diaolou were first built on a large scale in the early 20th century, originally as high-rise defensive residences built by returned overseas Chinese for their families. Influenced by Western architecture, these houses have drawn together the architectural styles such as ancient Greek colonnade, ancient Roman pillar, arch and vault, medieval Gothic

开平碉楼

建筑局部

建筑与周边环境

古希腊柱廊、古罗马柱式、拱券和穹隆、中世纪哥特式尖拱、伊斯兰风格拱券、欧洲城堡构件、葡式骑楼、文艺复兴式和巴洛克风格等建筑样式；局部装饰又时有中国传统建筑元素；在建筑材料和建筑技术方面，则大量使用进口水泥、木材、钢筋、玻璃等材料和钢筋混凝土的结构，可谓中国乡村民众主动接受西方建筑艺术并与本土建筑艺术融合的产物。

从目前所搜集的建造历史资料看，开平碉楼大多没有正式的建筑设计、施工图纸，本地工匠仅凭国外明信片等间接的西方建筑图像资料，即可按业主所提功能要求，即兴建造完成且质量上乘。中国工匠之聪明才智由此可见一斑。

pointed arch, Islamic arch, European castle component, Portuguese arcade, Renaissance and Baroque style. Traditional Chinese architectural elements are also frequently found in local decoration. In terms of building materials and technology, imported cement, wood, steel bars, glass and other materials and reinforced concrete structures have been used in large amounts. Therefore, Kaiping Diaolou can be described as the products of Chinese rural people's active acceptance of Western architectural art and integration with local architectural art.

In the light of the construction history data collected by far, most of Kaiping Diaolou buildings have no formal architectural design or construction drawings. Local craftsmen could improvise the construction with high quality according to the functional requirements put forward by the owners only by using Western architectural image such as foreign postcards, which illustrated the ingenuity of Chinese craftsmen.

若干年后仍不失建筑本色

黄埔军校入口处

工程地点	广州市
建筑面积	14180平方米
建成年代	20世纪30年代

黄埔军校旧址

The Former Site of the Whampoa Military Academy

黄埔军校位于广州市黄埔区长洲岛内，原为清朝陆军小学堂和海军学校校舍，1924年6月16日起，孙中山在此创办"中国国民党陆军军官学校"（后更名为"中华民国陆军军官学校"），习称"黄埔军校"。通常理解"黄埔军校旧址"为军校校园部分，但从建筑文化遗产角度看，则为包括黄埔军校校园旧址以及校园周边之八桂山孙中山纪念碑、教思亭、北伐纪念碑、万松岭东征阵亡将士墓等在内的整体性的历史建筑组群。1988年黄埔军校旧址被定为国家级文物保护单位，2016年12月入选《全国红色旅游景点景区名录》。

The Whampoa Military Academy, located on Changzhou Island, Huangpu District, Guangzhou, was originally the campus of the Qing Army Primary School and Naval School. Since June 16, 1924, Sun Yat-sen founded the "Chinese Kuomintang Army Military Academy" (later renamed as "Army Military Academy of the Republic of China"), commonly known as the "Whampoa Military Academy". It is generally understood that "the former site of Whampoa Military Academy" is a part of the military academy campus, but from the perspective of architectural cultural heritage, it is also an integral historic architectural complex including the former site of the Whampoa Military Academy campus, the Sun Yat-sen Monument in Bagui Mountain, the Jiaosi Pavilion, the Northern Expedition Monument, and the Tomb of Soldiers martyred in the Eastern Expedition of Wansongling. In 1988, the former site of the Whampoa Military Academy was designated as a cultural heritage site under state protection, and in December 2016, it was

黄埔军校校园区域含军校大门、走马楼（校本部，四排建筑包含总理、校长、党代表办公室，政治、教授、教练、管理、军需、军医等六部机构和学员宿舍、饭堂、展览室等）、教职员宿舍（后为总理纪念室）、平岗分教点、俱乐部、游泳室等，原多为两层砖木结构的岭南风格建筑，抗战期间遭日军轰炸，损毁严重，1996年由广州市政府斥巨资修复。孙中山纪念碑于1928年竖立，碑顶之孙中山纪念铜像为1930年补设，纪念碑身正面刻胡汉民书"孙总理纪念碑"，东面刻中山遗言"和平、奋斗、救中国"，此纪念碑坐南朝北，隐含着孙中山北定中原、统一中国的遗愿。东征阵亡将士墓于1926年落成，内含一座埋葬两次东征战役阵亡的516名将士公墓和一座仿巴黎凯旋门式建筑的纪念广场。校园南面之北伐纪念碑系为纪念1929年北伐中阵亡将士而建。

黄埔军校旧址承载着北伐前后的光辉革命历程，而作为建筑遗产，则展示出广州地区在20世纪20年代的独特历史风貌。

included in the National List of Revolutionary Tourist Destinations.

The campus of the Whampoa Military Academy included the military academy gate, the Zouma Building (academy headquarters, four rows of buildings including the offices of chancellor, president and party representative, six institutions including politics, professors, coaches, management, military supplies and military doctors, as well as student dormitories, dining hall, exhibition rooms, etc.), faculty dormitories (later the Chancellor's Memorial Room), Pinggang teaching position, clubs, swimming pool, etc., which were originally of the Lingnan style with a two-story brick and wood structure and badly damaged by Japanese bombing during Anti-Japanese War. The Sun Yat-sen Monument was erected in 1928, and the bronze statue of Sun Yat-sen at the top of the monument was supplemented in 1930. The front of the monument is engraved with Hu Hanmin's calligraphy, "Monument to Chancellor Sun", and the eastern side is engraved with Sun Yat-sen's last words "Peace, Struggle and Saving China". This monument faces north, implying Sun Yat-sen's last wish to recover the Central Plains in the north and unify China. The tomb of the soldiers killed in the Eastern Expedition was completed in 1926, which contains a cemetery burying 516 soldiers martyred in two Eastern Expedition campaigns and a memorial square built as an imitation of the Arc de Triomphe in Paris. The Northern Expedition Monument to the south of the campus was built to commemorate the soldiers martyred in the Northern Expedition in 1929.

The former site of the Whampoa Military Academy, which witnesses the glorious revolutionary course before and after the Northern Expedition, is also an architectural heritage presenting the unique historical features of Guangzhou in the 1920s.

奠基纪念石

长洲岛黄埔军校大礼堂

东江阵亡烈士墓墓碑、碑亭

黄埔军校北伐阵亡士官纪念碑

黄埔军校教室

东征烈士陵园纪念

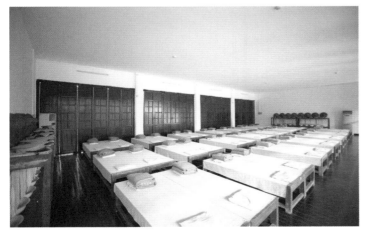

黄埔军校校舍 1

黄埔军校校舍 2

黄埔军校之长洲岛孙中山故居

校舍内院入口处

黄埔军校游泳池

校舍内院走廊

中国人民革命军事博物馆
Military Museum of the Chinese People's Revolution

工程地点：北京市

建筑面积：60557平方米

竣工时间：1959年

设计机构：北京市建筑设计研究院

主要设计人：欧阳骖

中国人民革命军事博物馆是是新中国成立十周年"国庆十大建筑"之一。主体建筑呈"山"字形，东西总长214.4米，南北总长144.4米，上部塔座2层，塔顶上为军徽，总高97.47米，共设有大小陈列室10个。中央大厅为30米×30米的方厅，是博物馆主入口部分的标志性空间，大厅两侧为交通厅，设有楼梯和电梯。中央大厅两侧及两翼陈列室基本布局为桶形，展览路线以一翼一层为一个段落，顺序循环路线。博物馆外立面采用渐次升高的体形，外墙勒脚为花岗石、蘑菇石，假石墙面。所有入口均为嵌有石刻旗帜花纹和带有勋章图案的花岗石柱头门罩，建筑造型雄伟稳重。博物馆全部为钢筋混凝土框架结构，中央顶部承托军徽的塔筒为钢骨架、钢筋混凝土筒壁。

The Military Museum of the Chinese People's Revolution is one of the Ten Great Buildings built in celebration of the 10th anniversary of the founding of the People's Republic of China. Its main building, laid out in the shape of an inverted "E", is 214.4 meters long from east to west, 144.4 meters wide from south to north, and 94.47 meters tall including a two-story steeple topped with the emblem of the Chinese People's Liberation Army. There are ten display halls inside altogether. The central hall with a 30-meter square forms the iconic space of the museum's main entrance, and either side of it is a conveyance hall equipped with staircases and elevators. The display halls on both sides of the central hall and inside the wings are basically in a layout of a barrel shape, each leading to another on the same floor. The museum has a facade of a gradually ascending style, with plinth walls built of granite and mushroom rock and curtain walls of artificial rock. All entrances have a granite frame carved with flag and medal designs, making the whole construction look magnificent and stately. The museum is built on the reinforced concrete framework, and the steeple in the middle that supports the army emblem has a steel frame with reinforced concrete walls.

建筑主立面及楼前广场

主入口西侧门廊

主入口内门厅局部

主入口立面

建筑顶部高耸的军徽

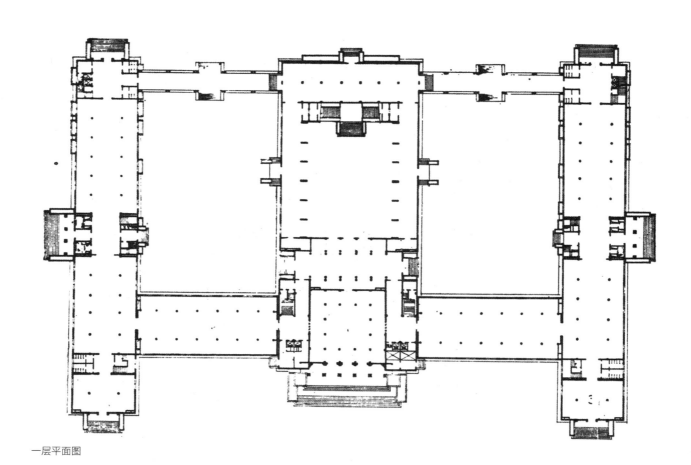

一层平面图

浮雕

从建筑室内看镂空花窗

主展厅一层室内

主展厅二层室内

故宫博物院宝蕴楼
Baoyun Building of the Palace Museum

工程地点｜北京市

建筑面积｜1646.9平方米

竣工时间｜1915年

主要设计人｜由朱启钤组织策划

宝蕴楼位于北京紫禁城西南隅、武英殿以西之原清廷咸安宫旧址，总建筑面积为1646.9平方米，建于1914年，1915年竣工，设计者不详。1911年清逊帝溥仪退位后，紫禁城外朝由北洋政府接管，于1914年成立的古物陈列所即设在外朝部分。宝蕴楼系古物陈列所为贮藏原盛京故宫、热河离宫20余万件珍贵文物而建的大型库房。因其收藏集历代古物之萃，定名曰"宝蕴楼"。古物陈列所于1947年并入故

The Baoyun Building is located at the former site of the Xian'an Palace of the Qing Dynasty in the southwest corner and west of Wuying Hall in the Beijing Forbidden City, with a total floor area of 1646.9 square meters. The construction was commenced in 1914 and completed in 1915, with the designer unknown. After Pu Yi, the last empenr of the Qing dynasty, abdicated in 1911, the imperial court outside the Forbidden City was taken over by the Northern Warlords government, and the Antiquities Exhibition Center established in 1914 is located in the Outer Palace. The Baoyun Building is a large-scale warehouse built by the Antiquities Exhibition Center to store more than 200,000 precious cultural relics from the Shengjing Palace and the Rehe Palace. Because of its rich collections of antiquities from past dynasties, it was named "Baoyun Building" (Treasure Storing

自东南侧看内庭院

东侧连廊

西侧入口阶梯及门上雕饰

宫博物院。

宝蕴楼采用封闭的周边式布局，由南面门楼与东、西、北三面楼房合围庭院。南面沿袭原咸安门为庭院正门，庭院之北、东、西三面各建一座二层带半地下室的砖木结构楼房，以北楼为主，东、西两楼对称相峙。除正面楼维持传统样式外，北、东、西楼之屋顶虽为四坡式屋顶，但舍弃传统式大屋顶反曲线挑檐，不铺琉璃瓦，而以绿灰两色片石形成西式瓦面。北楼的正中开门，门外有高大的四柱式雨棚，雨棚上有女儿墙，构成二层的一个室外平台。东西两楼，亦在正中开门，但门外雨棚较小，不设柱子。北楼与东西二楼在拐角处以二层的外廊相接。白色的栏杆、白色的廊柱、空透的走廊，使整组建筑增加了虚实对比和轻快的感觉。宝蕴楼的外窗窄长，所有窗户的线脚均饰白色，与红色墙身恰成对照，而内部装修如天花板、楼梯等处，均采用西洋样式。如此，因庭院式布局和建筑尺度的把控，北、东、西楼的西洋样式与南门的

Building). The Antiquities Exhibition Center was incorporated into the Palace Museum in 1947.

The Baoyun Building with a closed peripheral layout is an enclosed courtyard surrounded by the south gatehouse and the east, west and north buildings. In the south, the original Xian'an Gate is adopted as the main entrance to the courtyard, and a brick-and-wood structure building with a semi-basement has been built on the north, east and west sides of the courtyard respectively, with the north building as the main building and the east and west buildings symmetrically facing each other. Except that the front building maintains the traditional style, the four-slope roofs of the north, east and west buildings have replaced the traditional large roof with reverse curve overhangs and glazed tiles, but instead adopted a western-style tile surface with green and gray flake stones. In the middle of the north building, a door is opened, with a tall four-pillar canopy with a parapet on it, which constitutes an outdoor platform on the second floor. The east and west buildings also have doors in the middle, but the canopy outside the door is small without pillars. The north building is connected by a two-story veranda at the corners of the east and west buildings respectively. White railings, white colonnades, and empty corridors add contrast between reality and virtuality and an airy feeling to the whole buildings. The external windows of the Baoyun Building are narrow and long, with all the window mouldings

传统样式得以融合一体，形成极为独特的建筑景观。

在中国传统古建筑群里，宝蕴楼独树一帜，是按照西洋建筑的式样设计建造的。"20世纪是中西方文化激烈碰撞的时代。这些优秀建筑既是精彩的作品，也是重要的国家与城市记忆。"此言恰是宝蕴楼的文化内涵所在。

painted white, just in contrast to the red wall body. The interior decoration such as ceiling and stairs, on the other hand, is in the Western style. In this way, due to the control of the courtyard layout and architectural scale, the Western style of the north, east and west buildings and the traditional style of the south gate can be harmoniously integrated, presenting a very unique architectural landscape.

Among the traditional ancient Chinese architectural complexes, the Baoyun Building is unique because it has been built according to the Western architectural style. "The 20th century is an era of intense collision between Chinese and Western cultures. These excellent buildings are not only wonderful works, but also important national and urban memories." This statement precisely describes the cultural heritage of the Baoyun Building.

檐口细部构造

建筑连廊入口

会议空间

故宫文物南迁时所用储物箱

入口西侧檐口细部

主楼东侧窗饰及连廊

会客空间

室内中央阶梯

主楼二层展陈

金陵女子大学旧址
The Former Site of Ginling College

工程地点 | 南京市

竣工时间 | 1923—1934年

主要设计人 | 亨利·墨菲 吕彦直

南京市鼓楼区宁海路122号（原袁枚随园旧址），今南京师范大学随园校区，原为金陵女子大学之校区所在。这个中国第一所女子大学于1915年由美国基督教会、长老会、英国伦敦会等七个教会创办，校区规划及建筑设计系美国建筑师亨利·墨菲主持，陈明记营造厂承建。我国著名建筑师吕彦直在美留学期间曾参与此校园的部分规划设计工作，今校园设计总图上留有吕氏签字。

No.122 Ninghai Road, Gulou District, Nanjing (rly the former site of Yuan Mei's Suiyuan Garden), now the Suiyuan Campus of Nanjing Normal University, was originally the campus of Ginling College, the first women's college in China founded in 1915 by seven churches, including the American Christian Church, the Presbyterian Church and the London Church. Henry Killam Murphy, an American architect, presided over the campus planning and architectural design, while Chen Mingji Construction Factory built it. Lv Yanzhi, a famous Chinese architect, participated in part of the planning and design of this campus during his study in the United States, and Lv's signature can be found on the general layout of the campus design today.

主楼前广场

校区规划按东西向的轴线对称布局，入口采用林荫道加强空间的纵深感，主体建筑物以大草坪为中心，对称布置，100号楼后面设计一个以人工湖为中心的花园，中轴线的西端结束于丘陵（西山）制高点的中式楼阁。单体建筑造型采用中国传统宫殿式建筑风格，建筑材料和结构采用西方钢筋混凝土结构，建筑物之间以中国古典式外廊相连接，形成中西合璧的建筑景观，素有"东方最美校园"之誉。建筑工程于1922年开工建设，1923年落成第一批建筑，计有7幢宫殿式的建筑：100号（会议楼，建筑面积1431平方米）、200号（科学馆，建筑面积1541平方米）、300号（文学馆，建筑面积1492平方米）、400号~700号（学生宿舍，总建筑面积4603平方米）。1934年，按原规划续建完成校图书馆（建筑面积1397平方米）、大礼堂（建筑面积1444平方米）。

The campus planning is symmetrically arranged along the east-west axis, and the entrance uses a tree-lined avenue to enhance the sense of depth in space. The main buildings are symmetrically arranged with the big lawn as the center. A garden centered on an artificial lake is designed behind No.100 building, and the western end of the axis is a Chinese-style pavillion at the commanding height of a hill (Xishan). The architectural style of single buildings adopts the traditional Chinese palace style. The building materials and structures are of the western reinforced concrete structure, and the buildings are connected by classical Chinese verandas, forming an architectural landscape perfectly integrating the Chinese and Western styles. Thus it has been acclaimed as "the most beautiful campus in the East". The building construction commenced in 1922, and the first buildings were completed in 1923. There are seven palace-style buildings: No.100 (Conference Building, with a floor area of 1,431 square meters), No.200 (Science Museum, with a floor area of 1,541 square meters), No.300 (Literature Museum, with a floor area of 1,492 square meters), and No.400 ~700 (student dormitories, with a total floor area of 4,603 square meters). In 1934, the college library (with a floor area of 1,397 square

此建筑群为西方建筑师在华"适应性建筑设计"的尝试之作。囿于当时中外建筑师对传统建筑尚不谙熟,故某些装饰细节(如斗栱位置问题)有失当之处,但瑕不掩瑜,此建筑对日后诞生的"中国固有式建筑"流派,有着直接的影响。

meters) and the auditorium (with a floor area of 1,444 square meters) were completed according to the original plan.

This architectural complex is an attempt of "adaptive architectural design" by Western architects in China. As Chinese and foreign architects then were not familiar with traditional architecture, some decorative details (such as the location of the corbel brackets) were improper, but the defects could not obscure the splendor of the entire architecture. This building complex exerted a direct impact on the school of "Chinese Inherent Architecture" to be born in the future.

前广场校园建筑

校园建筑局部

原金陵女子大学设计效果图

人工湖及校园建筑

建筑局部（组图）

校园建筑及校园生活旧影

银杏树下的校园建筑

校园建筑旧影

全国农业展览馆
National Agricultural Exhibition Center

全国农业展览馆是新中国成立十周年"国庆十大建筑"之一,由综合馆、南北作物馆、水产馆、畜牧馆、特产馆、科学馆、气象馆及附属用房组成。以中国建筑传统风格的琉璃瓦屋顶、重檐、亭阁、栏杆等宫殿和庭园处理手法巧妙结合,形成融中国建筑传统形式与现代功能于一体的民族风格新建筑。综合馆位于东西主轴线上,面向东三环北路设主出入口。中央为八角大厅,上部为高33米三层重檐绿琉璃瓦顶的八角形亭阁及四角方亭。外墙粘贴黄色面砖。综合馆与南北作物馆之间形成中心广场。东北方由畜牧馆、特产馆等组成北部广场。东南方由科学馆、气象馆等组成南部广场。各馆均采用砖墙与钢筋混凝土混合结构。

The National Agricultural Exhibition Center, one of the Ten Great Buildings built in celebration of the 10th anniversary of the founding of the People's Republic of China, comprises the Integrated Hall, the South and North Crop Halls, the Aquatic Hall, the Livestock Hall, the Specialties Hall, the Science Hall, the Meteorological Hall, as well as annexes. A perfect combination of elements of traditional Chinese palace and garden architecture, this exhibition complex represents a new style of achitecture that integrates traditional Chinese architecture with modern functionality. The Integrated Hall is on the east-west axis, with its main exit facing the East 3rd Ring Road. In the center of it is a massive octagonal hall, and at the top is a 33-meter-tall octagonal pavilion with three tiers of eaves and a glazed tile roof, as well as quadrangular pavilions. The exterior walls of the Integrated Hall are decorated with yellowish bricks. The Integrated Hall and the South and North Crop Halls form the central plaza. The Livestock and Specialties Halls in the northeast form the north plaza, and the Science and Meteorological Halls in the southeast form the south plaza. All the halls are of a reinforced concrete structure.

工程地点	建筑面积	竣工时间	设计机构	主要设计人
北京市	30770平方米	1959年	中国建筑设计研究院	严星华

综合管入口处

建筑檐口细部

分馆入口处 1

建筑前广场上的雕塑(组图)

二号馆室内局部

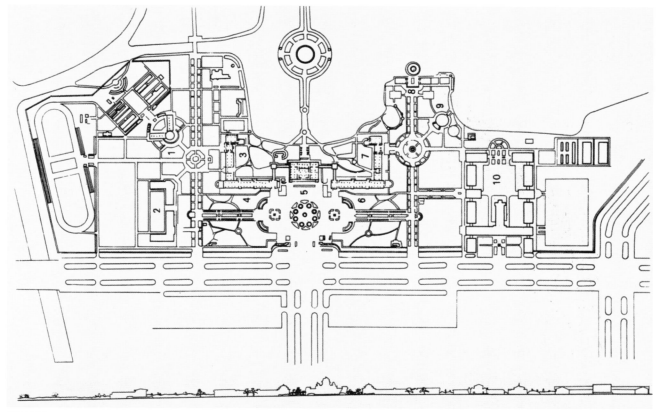

总平面图

分馆入口处 2

北平图书馆旧址
The Former Site of the Peiping Library

工程地点｜北京市
建筑面积｜13438 平方米
竣工时间｜1931 年
主要设计人｜莫律兰（V. Leth Moller）

北平图书馆，后称北京图书馆，即今国家图书馆之前身。其旧址即今国家图书馆分馆主楼——文津楼，位于北京市东城区文津街，由丹麦建筑师莫律兰（V. Leth Moller）设计，共耗费银元 240 余万元（系退还的庚子赔款），总占地面积 76 亩（1 亩 ≈ 666.67 平方米），于 1931 年落成。

北平图书馆建筑外观是华丽的中国传统宫殿式，建筑结构采用西式的砖混结构，内部设施则全部引进当时最先进的西式设备，较之美国国会图书馆亦不逊色。它不仅以总藏书量（落成之际即拥有 16 万册图书）和珍稀藏书量闻名遐迩（文津阁四库全书一部、唐人写经八千六百五十一卷、善本书二万二千余册，以及明清舆图、金石墨本等稀世珍宝），更采纳了科学的图书分类方式与借阅制度（改变我国传统的藏书楼的藏书法，将传统图书四库分类法改为近代分类法，按

The Peiping Library, later known as the Beijing Library, is the predecessor of the present National Library of China. The former site of the Peiping Library is the Wenjin Building, the main building of the present National Library Branch located in Wenjin Street, Dongcheng District, Beijing. Designed by Danish architect V. Leth Moller, the building cost a total of over 2.4 million silver dollars (funded by the returned Boxer Indemnity) and covered a total land area of 76 *mu* (1 *mu* eqnals to 666.67 square meters). It was completed in 1931.

The architectural appearance of the Peiping Library is of the gorgeous traditional Chinese palace style, while the architecture structure is of the Western brick-concrete structure. All its internal facilities were imported from the West and the most advanced then, not in the least inferior to the Library of Congress of the United States. The library is not only famous for its total book collections (160,000 books when it was completed) and rare book collections (one complete collection of four libraries in Wenjinge, 8,651 volumes of classics written by people of the Tang Dynasty, more than 22,000 rare books, as well as rare treasures such as maps of the Ming and Qing Dynasties, inscriptions on ancient bronzes and stone tablets and rubbings from stone inscriptions, etc.), but also for its adoption of a

建筑主楼及前广场

沿街主入口立面

建筑檐口

内庭院局部 1

书名第一个字的拼音及作者姓名第一字的笔画分目制成索引卡片），辅之以先进的库藏设备，无疑是当时亚洲最先进的图书馆之一。

设计者莫律兰曾谈及他的设计立意："现代图书馆之需要与中国宫殿式之建筑互相调和。"从竣工后的效果看，这个设计达到了设计初衷：北平图书馆西向与北海琼岛相望，南面与中海墙垣毗邻，其红墙绿瓦的建筑色调正与周围皇家园囿环境融为一体。

scientific book classification and borrowing system (changing the method of traditional Chinese book collections, replacing the traditional book four library classification by modern classification, making index cards according to the *pinyin* of the first word of a book title and the strokes of the first word of an author's name). Supplemented by advanced storage facilities, it was undoubtedly one of the most advanced libraries in Asia at that time.

The designer Moller once talked about his design conception, "The needs of modern libraries should harmonize with the Chinese palace-style architecture." From the effect after completion, this design idea has been achieved: The Peiping Library faces QiongDao in Beihai in the west, and adjoins the Zhonghai Wall in the south. With its architectural color of its red walls and green tiles, the building is harmoniously integrated with the surrounding royal garden environment.

建筑入口前丹陛

从北海公园远眺建筑组团

临琼楼局部

内庭院局部 2

青岛八大关近代建筑

Modern Architectural Complex of the Qingdao Eight Great Passes

山东省青岛市市南区汇泉东部街区的八条主要街道,以中国八大关隘命名,故统称为"八大关"。此地居太平山南麓,西临汇泉湾,南接太平湾,云集了各色西式建筑,汇聚了俄、英、法、德、美、丹麦、希腊、西班牙、瑞士、日本等20多个国家的各式建筑风格,举凡希腊式、罗马式、哥特式、文艺复兴式、拜占廷式、巴洛克式、洛可可式、田园风式、新艺术风格式、折中主义式、国际式等建筑风格,在青岛八大关近代建筑中皆有所见,遂有"万国建筑博物馆"之美誉。八大关近代建筑群作为整体性的建筑群,始建于20世纪30年代,面积70余公顷。2001年,青岛八大关近代建筑被公布为第五批全国重点文物保护单位;2005年,被评为中国最美五大城区之一;2009年,入选首届"中国历史文化

The eight main streets in the eastern block of Huiquan, Shinan District, Qingdao City, Shandong Province, are named after the eight major passes in China, so they are collectively referred to as "Eight Great Passes". Located at the southern foot of Taiping Mountain, bordering Huiquan Bay in the west and Taiping Bay in the south, the "Eight Great Passes" is home to various Western-style buildings, including a great diversity of architectural styles from more than 20 countries including Russia, the United Kingdom, France, Germany, the United States, Denmark, Greece, Spain, Switzerland and Japan. Architectural styles such as Greek, Roman, Gothic, Renaissance, Byzantine, Baroque, Rococo, pastoral, art nouveau, eclectic and international styles can all be found in the Modern Architectural Complex of the Qingdao Eight Great Passes, which is celebrated as "architecture expo". As a whole architectural complex, it started to be built in the 1930s with an area of more than 70 hectares. In 2001, the Modern Architectural Complex of the Qingdao Eight Great Passes was announced as being listed in the fifth batch of key cultural heritage units under state protection. In 2005, Eight Great Passes was rated one of the five most beautiful urban areas in China.

花石楼

公主楼

青岛基督教堂

蝴蝶楼

圣弥厄尔教堂

名街"。

其中别墅类建筑造形以坡屋顶为主,建筑面积以每栋 400 平方米左右为主,很少有超过 1000 平方米的别墅。代表性建筑物有花石楼,山海关路 1、3、5、9 号别墅,元帅楼,公主楼,蝴蝶楼,青扇寮,宋家花园和朱德别墅等。

花石楼,位于黄海路 18 号,建筑面积 753.7 平方米,主体共 5 层,顶层为观海台。1932 年由苏联建筑师格拉西莫夫设计,因用花岗岩和鹅卵石建成而得名"花石楼",为欧洲古城堡式风格,有古罗马哥特建筑的遗风。1949 年后,花石楼成为接待中外贵宾的馆舍,董必武、陈毅等党和国家领

In 2009, the block was selected into the first batch of the "Famous Historical and Cultural Streets of China".

Of this building complex, villa buildings mainly have sloping roofs, and the floor area is mainly about 400 square meters per villa. Only a very few villas have a floor area of more than 1,000 square meters. Representative buildings include Huashi Villa, Villas No.1, No.3, No.5 and No.9 on Shanhaiguan Road, Marshal Villa, Princess Villa, Butterfly Villa, Qingshanliao Villa, Song's Garden and Zhu De Villa, among others.

Huashi Villa at No.18 Huanghai Road has a floor area of 753.7 square meters, with 5 floors in the main building, and the top floor being a sea view platform. Designed by Soviet architect Gerasimov in 1932, it was named "Huashi (Patterned Stone) Villa" because it was built with granite and pebbles. With an ancient European castle style, it has stylistic traces of ancient Roman Gothic architecture. After 1949, Huashi Villa became a

建筑旧影

建筑建设旧影

青岛德国总督楼旧址

导人都曾在此下榻。

山海关路1号建筑为文艺复兴式别墅，建于1933年，刘耀宸、俄拉夫林且夫设计。此别墅院门底部为一米多高的粗石墙，上面为淡黄色的粉刷墙，漆黑的铁门中部有精美的盾形花纹装饰，铁门的两侧粉墙上，各有两个站立的黑色狮子。院墙铁栅栏内有木质围栏。山海关路1号建筑的四周均是宽大的花园，花园内除了各种植物外，修建有露天游泳池和喷泉。

公主楼，位于居庸关路10号，始建于20世纪30年代。相传1929年有一位丹麦王国的王子乘坐"菲欧尼亚"号豪华游轮来到青岛游览观光，委托首任丹麦领事在八大关海滨购置土地，按照安徒生童话中的意境设计这座丹麦古典式建筑，准备将其作为礼物送给丹麦公主。此事虽未成真，公主楼之名却因此不胫而走。

青岛八大关近代建筑群，为我们作了现代海滨建筑景观的有益尝试。

guesthouse to receive domestie and foreign distinguished guests. Dong Biwu, Chen Yi and other party and state leaders once stayed there.

Building No.1 Shanhaiguan Road is a Renaissance villa, built in 1933 and designed by Liu Yaochen and a Russian architect Labrinchev. At the bottom of the gate to this villa is a thick stone wall more than one meter high, above which is a pale yellow painted wall. The middle of the dark iron gate is decorated with exquisite shield patterns, and there are two black lions standing on the pink walls on both sides of the iron gate. There is a wooden fence inside the iron fence of the courtyard wall. Building No.1 Shanhaiguan Road is surrounded by a wide garden, in which there is an outdoor swimming pool and a fountain, in addition to various plants.

Located at No.10 Juyongguan Road, Princess Villa was built in the 1930s. Legend has it that in 1929 a prince of the Kingdom of Denmark took the luxury cruise ship "Flo Nia" to Qingdao for sightseeing, and he entrusted the first Danish consul to purchase land at the seaside of the Eight Great Passes, and design this classical Danish building according to the artistic conception in Andersen's fairy tales, preparing to give it as a gift to a Danish princess. Although the Danish prince's dream did not come true, the name of Princess Villa spread fast.

The Modern Architectural Complex of the Qingdao Eight Great Passes has made a valuable attempt for us to design modern seaside architectural landscape.

周志俊（周学熙后代）别墅

康有为故居

康有为故居入口

花石楼院落小品

公主楼室内

青岛基督教堂

青岛德国总督楼旧址室内

圣弥厄尔教堂内景（组图）

建筑主入口立面

工程地点｜昆明市
建筑面积｜7500平方米
竣工时间｜1910年
主要设计人｜李守光

云南陆军讲武堂旧址
The Former Site of the Yunnan Military Academy

云南省位于昆明市翠湖西岸之承华圃，总占地约1.44万平方米。云南陆军讲武堂自1909年开办，至1928年结束，共招收学生19批，分步、骑、炮、工四科，第15期增招归国华侨和朝鲜、越南留学生。历届毕业生中名人辈出，如第3期特别班的朱德，第15期的叶剑英等。

云南陆军讲武堂建筑的设计者、施工方、建造时间等史料不详，大致在1910年已成完整局面。其整体格局系由四面建筑物合围出一个规模宏大的庭院，呈正方形四合院，东、西、南、北楼外侧均约长150米，内侧各长120米，宽约10米，高均12米，对称衔接，对内形成周长480米的正方形广场，四角有拱形门洞可供出入，四面楼房浑然连为一体，层高通透，错落有致。其南北楼为学员宿舍，南楼中部突起为高约11.5米、宽13米的阅操楼，整体规模宏大。其建筑外观为

The Chenghua Garden is located on the west bank of Cuihu Lake in Kunming, Yunnan, covering a total area of about 14,400 square meters. The Yunnan Military Academy, founded in 1909 and closed in 1928, enrolled 19 terms of students for four arms of the services, namely, infantry, cavalry, artillery and sapper. In the 15th term, returned overseas Chinese and overseas students from Korea and Vietnam were recruited. The academy produced many famous graduates, such as Zhu De of the 3rd special class and Ye Jianying of the 15th term.

The historical data about the building of the Yunnan Military Academy, such as the designer, the construction party and the construction time, are unknown. The building was completed roughly in 1910. Its overall pattern is a large-scale courtyard surrounded by buildings on the four sides, taking the shape of a square quadrangle house. The outer sides of the east, west, south and north buildings are about 150 meters long, and the inner sides are about 120 meters long, about 10 meters wide and 12 meters high. These buildings are symmetrically connected, forming a square with a circumference of 480 meters inside, with arched door openings at the four corners for access. The buildings on the four sides are harmoniously integrated, structured in a distinct and orderly manner. The north and

中国西南乡镇传统的走马转角楼式样，以大木框架为受力结构，墙体混用砖石。东、西两楼外观大致相同，两楼各宽11米，但由于东楼距离翠湖和附近的沼泽地带较近，为了防止水浸受潮，墙体全用大青石砌成，而西楼距离水域较远，就用土基而建。南楼和北楼各宽9.4米，墙体则是青砖夹层；西楼是讲武堂教室，东楼为办公区域，对外一面朝向翠湖，居中门楼有西式带塔尖的装饰性墙体，形成整组建筑中西合璧风格的正立面形象。

云南陆军讲武堂旧址以建筑形式践行了军事教育的进步，也见证了中国在晚清民初由改良进而革命的历程。

south buildings are student dormitories, while the middle part of the south building is a reading and exercise building about 11.5 meters high and 13 meters wide. The Academy is magnificent on the whole. Its architectural appearance follows the style of stilted buildings typical in villages and towns of Southwest China, with a large wooden frame as the stress structure and mixed brick and stone mansonry walls. The east and west buildings look roughly the same, each building being 11 meters wide. However, because the east building is close to Cuihu Lake and the nearby swamp area, to prevent flooding and dampening, the walls are all made of bluestones. The west building is far away from the waters, so it is built with soil matrix. The south building and the north building are both 9.4 meters wide, with the walls having a blue brick interlayer. The west building houses classrooms of the military academy. The east building is the office area, with the external side facing Cuihu Lake, and the central gate building has western-style decorative walls with spires, forming the facade image of the Building Complex integrating both Chinese and Western styles.

The former site of the Yunnan Military Academy buttressed the progress of military education in the form of architecture, while witnessing the process of China's transition from reform to revolution in the late Qing Dynasty and the early Republic of China.

入口立面局部

内庭院出入口

内庭院局部及环境

光影下的建筑局部

民族饭店

Minzu Hotel

民族饭店是新中国成立十周年"国庆十大建筑"之一，占地0.99公顷，地下1层，地上多为10层，中央局部为12层。建筑平面呈"F"形。首层为公共用房，以门厅、交通厅为中心，设有办公室、中餐厅、西餐厅、厨房以及贵宾室等。2~10层共设客房597间。中间局部11层设有茶室、球房、理发室等公共用房。建筑外立面造型丰富，具有新颖的民族风格，外墙贴黄色面砖。首层为花岗岩饰面，上部有粗壮的束腰线，2层设有通长望柱栏杆式悬挑阳台，门扇两侧各有一扇镂空花饰的花隔扇窗。民族饭店是北京市第一座采用预制装配式钢筋混凝土框架结构的高层建筑。

Minzu Hotel, one of the "Ten Great Buildings" built in celebration of the 10th anniversary of the founding of the Republic of China, covers an area of 0.99 hectares, with 1 floor underground, 10 floors above ground for most parts and 12 floors for the central part. The building plane takes the shape of "F". The first floor is for public use, centering on the foyer and traffic hall, and having offices, Chinese and Western restaurants, kitchens and VIP rooms. There are 597 guest rooms on floors 2-10. In the middle part of the 11th floor there are public rooms such as tea room, billiard room and barber room. The facade of the building is diversified in shape, assuming a novel national style, and the exterior wall is covered with yellow tiles. The first floor is decorated with granite, with the upper part having a thick girdle line. The second floor is decorated with an overhanging balcony with a baluster railing, and there is a patterned partition window with hollow-out ornamental design on both sides of the door leaf. Beijing Minzu Hotel is known as the first high-rise building with prefabricated reinforced concrete frame structure in Beijing.

工程地点	北京市
建筑面积	34146平方米
竣工时间	1959年
设计机构	北京市建筑设计研究院
主要设计人	张镈、胡庆昌

建筑南立面

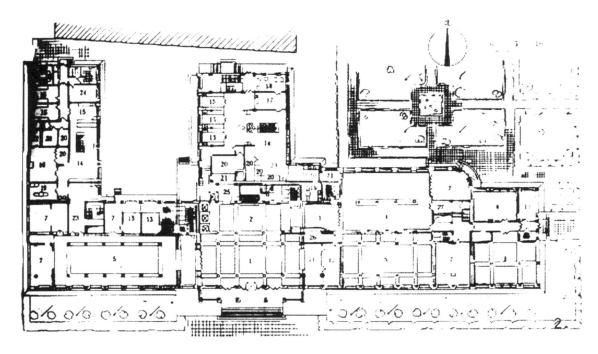

建筑一层平面图

建筑沿街立面旧影

建筑主入口旧影

建筑主入口现状

从西南侧拍摄建筑外立面

国立中央研究院旧址

The Former Site of the National Academia Sinica

国立中央研究院旧址，现为中国科学院南京分院、江苏省科学技术厅所在地。国立中央研究院于1928年6月成立，是中华民国最高学术研究机构。今南京市北京东路39号（原鸡鸣寺路1号）是中央研究院总办事处、地质研究所、历史语言研究所和社会科学研究所的所在地。

最初的建筑规划设计，始于1928年10月，设计者为我国著名建筑师杨廷宝。受抗战影响，此建筑群直至战后的1948年方大体告竣。各单体建筑均为仿明清官式建筑的"中国固有式建筑"形式，形成整体风格整齐划一的建筑景观。

总办事处大楼坐北朝南，建筑平面呈"T"字形，大门两侧及围墙东侧共建有三座方形攒尖顶警卫室，其风格与大楼一致。大楼建于1947年，由新金记康号营造厂建造。大楼高3层，建筑面积3000平方米，钢筋混凝土结构，单檐歇山顶，

The former site of the National Academia Sinica is currently the seat of the Nanjing Branch of the Chinese Academy of Sciences as well as the Science and Technology Department of Jiangsu Province. Established in June 1928, National Academia Sinica is the highest academic research institution in the Republic of China. Today, No.39 Beijing East Road (formerly No.1 Jimingsi Road) is the seat of the General Office of Academia Sinica, the Institute of Geology, the Institute of History and Language and the Institute of Social Sciences.

The original architectural planning and design began in October 1928, designed by Yang Tingbao, a famous architect in China. Due to the War of Resistance against Japanese Aggression, this Building Complex was not completed until 1948 after the war. Each single building follows the Inheremt Chinese Architectural Style, imitating the official buildings of the Ming and Qing Dynasties, and forming an architectural landscape with a uniformly neat style.

The general office building faces south, with a T-shaped building plane. There are three square pyramidal roof guard rooms on both sides of the gate and on the east side of the enclosing wall, which share the same style with the building. Built in 1947 by the New Jinjikang Construction Factory, the

原历史语言研究所建筑及庭院局部

屋面覆盖绿色琉璃筒瓦，梁枋和檐口部分均仿木结构，漆以彩绘，清水砖墙，花格门窗，具有浓郁的民族风格。

地质研究所大楼位于总办事处大楼西北方，建于1931年，是此建筑群中最早完成的单体建筑。它依山而建，高2层，钢筋混凝土结构，建筑面积约1000平方米，建筑平面呈"凸"字形。单檐歇山顶，屋面覆盖绿色琉璃瓦，梁枋及檐口部分为仿木结构，漆以彩绘。1932—1937年，李四光曾在此主持工作。

历史语言研究所大楼（习称史语所）位于总办事处大楼的正北面，建于1936年，高3层，钢筋混凝土结构，建筑面积1700平方米。建筑平面呈长方形，单檐歇山顶，屋面覆盖绿色琉璃瓦，外墙上部为清水青砖墙，下部采用水泥仿假石粉刷。大楼朝南，入口处建有一座仿木结构的单层门廊。在大楼的东西两侧，各辟有一个侧门。大楼两端为阅览室和小型书库，其余部分为办公、研究用房。

社会科学研究所大楼位于中央研究院总办事处大楼的东北面，建于1931年，高3层，"人"字顶，红砖墙，坐北朝南，西南角有蔡元培先生题写的奠基碑。

building has three stories, with a floor area of 3,000 square meters, featuring a reinforced concrete structure, with a single-eave hipped roof covered with green glazed tubular tiles. The beams and eaves are all imitation wood structures with color paintings, plain brick walls, and lattice doors and windows, showcasing a distinctive national style.

The building of the Institute of Geology is located to the northwest of the general office building and constructed in 1931. It is the earliest single building in this complex. Built along the mountain, the building has two stories with a reinforced concrete structure, having a floor area of about 1000 square meters and a convex building plane. The single-eave hipped roof is covered with green glazed tiles, and the beams and eaves are made of imitation wood structure and painted with colored drawings. Mr. Li Siguang presided over the work here from 1932 to 1937.

The building of the Institute of History and Language, located to the north of the general office building and built in 1936, has three stories with a reinforced concrete structure and a floor area of 1,700 square meters. The building has a rectangular plane, with a single-eave hipped roof covered with green glazed tiles. The upper part of the outer wall is made of plain blue bricks, and the lower part is painted with cement in the form of stones. The building faces south, with a single porch of an imitation wood structure at the entrance. There is a side door on each of the east and west sides of the building. The two ends of the building are respectively a reading room and a small-scale library, while the other parts are offices and research rooms.

The main building of the Institute of Social Sciences is located to the northeast of the general office building of Academia Sinica. Built in 1931, it has three stories, a herringbone roof and red brick walls. The building faces south and has a foundation stele inscribed by Mr. Cai Yuanpei in the southwest corner.

20世纪40年代拍摄的中央研究院总办事处

沿街入口现状

原总办事处建筑立面及环境

原社会科学研究所旧影

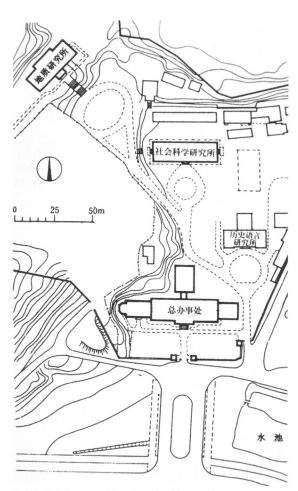
中央研究院总平面图（根据杨廷宝手绘图修改）

原历史语言研究所入口处

建筑主立面

工程地点	南京市
建筑面积	5100平方米
竣工时间	1936年
主要设计人	奚福泉

国民大会堂旧址

The Former Site of the Grand Hall of the Nationals

原国民大会堂（今南京人民大会堂）位于南京市长江路264号（原国府路），是民国时期规模最大、设施最为先进的会堂。1933年，国民政府为筹备原定于1935年3月召开的国民大会，提出在首都南京建造大会会场。陶记工程师事务所李宗侃建筑师设计方案中选，建筑于1936年5月建成。在施工过程中，设计者根据国民政府制定的《首都计划》中提出的首都建筑"要尽量采用中国固有之形式为最宜，而公署及公共建筑尤当尽量用之"的原则，对原设计方案作了局部修改。此建筑除具备议会性质外，根据相关决议，也同时是国立戏剧音乐院和国立美术陈列馆。因此，建成后的国民大会堂也称国立戏剧音乐院。这个功能要求加大了设计与施工难度。国民大会堂旧址是建设部、国家文物局评定的近代优秀建筑，现为江苏省重点文物保护单位。

The former Grand Hall of the Nationals (now the Nanjing Grand Hall of the People), located at No.264 Changjiang Road (formerly Guofu Road), was the largest hall with the most advanced facilities in the Republic of China. In 1933, to prepare for the National Assembly scheduled to be held in March 1935, the National Government proposed to build a conference venue in the then Chinese capital Nanjing. Li Zongkan, an architect of Taoji Engineering Office, was selected as its designer, and the building was completed in May 1936. During the construction process, the designer made partial modifications to the original design scheme according to the principle put forward in the Capital Plan made by the National Government that "it is best to follow the Chinese inherent form as much as possible, especially with regard to the office and public buildings". In addition to its parliamentary function, this building is also the National Academy of Theatre and Music and the National Art Exhibition Hall according to relevant resolutions. Therefore, the completed Grand Hall of the Nationals was also called the National Academy of Theatre and Music. This functional requirement stepped up the difficulty of design and construction. The former site of the Grand Hall of the Nationals is an excellent modern building evaluted by the Ministry of Construction and the National Cultural Heritage Admini stration,

国民大会堂主体建筑地上 4 层，地下 1 层，立面是西方近代建筑常用的勒脚、墙身、檐部三段划分，而檐口和门窗采用了传统的民族元素，建成时是当时中国最大的会堂之一。建筑坐北朝南，左右对称，分前厅、剧场、表演台三部分，建筑面积 5100 平方米，其中前厅为砖混结构，中部剧场为钢筋混凝土柱网结构，型钢屋架。建筑配有冷暖气、消防、通风、水电、卫生等设备，正面呈"凸"字形，内厅走廊宽畅，厅顶呈拱型。与同时期其他采用"中国固有式建筑"形式的作品相比，李宗侃建筑师更注重建筑的实用功能，并在美学趣味上，试图舍弃惯常的琉璃瓦大屋顶而能保持传统文化韵味。抗战胜利后，国民政府对国民大会堂进行了修复，增加了座位，添置了一些新设施。1949 年后，国民大会堂改名为南京人民大会堂。

and is now also a key cultural heritage site protected in Jiangsu Province. The main building of the Grand Hall of the Nationals has four floors above the ground and one floor below the ground. The facade is divided into three sections commonly used in modern western architecture: plinth, wall and eaves. Traditional national elements are blended into the eaves, doors and windows. It was one of the largest halls in the Republic of China when it was completed. The building faces south, symmetrical from left to right, consisting of the front hall, the theater and the performance platform, with a floor area of 5,100 square meters. The front hall is a brick-concrete structure, and the middle theater is of a reinforced concrete column grid structure with a structural steel roof truss. The building is equipped with heating, cooling, fire control, ventilation, water and electricity, sanitation and other facilities. The front is of a convex shape, the inner hall is connected with accessible corridors, and the top of the hall is arched. In comparison with other works in the form of "Chinese Inherent Architecture" in the same period, Architect Li Zongkan paid more attention to the practical function of architecture, attempting to abandon the usual glazed tile roof and maintain the traditional cultural taste.

Following the victory of the War of Resistance against Japanese Aggression, the National Government repaired the Grand Hall of the Nationals, adding more seats and some new facilities. After 1949, the Grand Hall of the Nationals was renamed the Nanjing Grand Hall of the People.

建筑东南立面旧影

建筑局部（组图）

建筑铭牌（组图）

建筑东立面现状

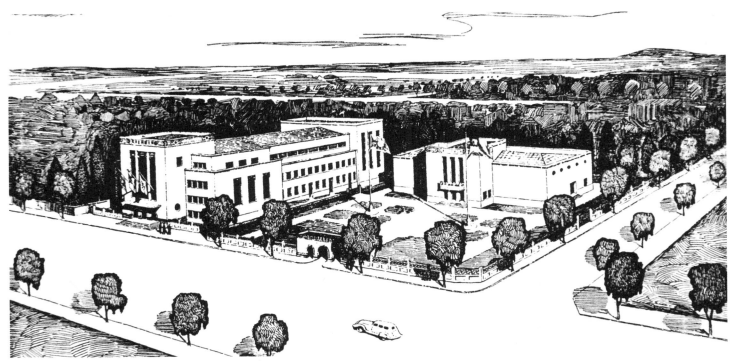

国立戏剧音乐院及美术陈列馆设计效果图（1935年绘制）

建筑入口处

三坊七巷和朱紫坊建筑群
Three Lanes and Seven Alleys and the Zhuzi Lane Architectural Complex

位于福州市老城区西南部的三坊七巷（衣锦坊、文儒坊、光禄坊，杨桥巷、郎官巷、塔巷、黄巷、安民巷、宫巷、吉庇巷）与朱紫坊以汇集明清民居建筑、保存古代里坊制街区闻名天下。但另一个值得重视的文化现象，即是这两处原本最具传统风貌的历史街区，同时也是中国自晚清（特别是1900年庚子事变之后）以来，率先反思传统，尝试现代化变革的发源地之一，如林则徐、严复、萨镇冰、沈葆桢、刘冠雄、陈继良等一批有识之士，都曾在此留下他们忧国忧民、力图革新的足迹。

这些思想者的思绪，也时时在建筑遗产中浮现印记。随着近现代的门户开放，三坊七巷许多人家开始在自家宅院的修建过程中，引进许多西洋建筑文化元素。无论首倡启蒙的严复，还是前清太傅陈宝琛，都不难在其庭院中看到这种时代痕迹：

Located in the southwest of the old city of Fuzhou, Three Lanes and Seven Alleys (Yijin Lane, Wenru Lane, Guanglu Lane, Yangqiao Alley, Langguan Alley, Ta Alley, Huang Alley, Anmin Alley, Gong Alley, Jibi Alley) and Zhuzi Lane are famous all over the world for the collection of residential buildings from the Ming and Qing Dynasties and preservation of the blocks in ancient Lifang system. However, another cultural phenomenon for noticed is that these two historical blocks characterized by the most traditional style are also known as one of the birthplaces of modernization in China since the late Qing Dynasty taking the lead in reflecting on tradition and attempting to reform. People of insight, such as Lin Zexu, Yan Fu, Sa Zhenbing, Shen Baozhen, Liu Guanxiong, and Chen Jiliang have left their footprints here, their concerns about the country and the people and their attempts to reform and innovate.

These thinkers' ideas have also left their traces in the architectural heritages. With the opening-up of China in modern times, many families in the Three Lanes and Seven Alleys began to introduce many Western architectural cultural elements in the process of building their own houses. Such traces of the times can be easily found in the residences of many people, whether Yan Fu, who initiated the Chinese Enlightenment, or Chen Baochen, Grand Preceptor of the Qing Dynasty: from decorative details in furniture and architecture to

建筑局部

建筑旧影

建筑现状卫星图

建筑室内

建筑局部

小到家具陈设、建筑细部装饰，大到索性在中式院落中增建西式小楼，均有细致的权衡，并不因舶来品的引进而妨碍原有意境。其经验在于：合适的建筑尺度控制与对引进物的理解与消化。时至今日，如衣锦坊41号之跳舞楼、文儒坊之陈季良宅小洋楼、朱紫坊之萨镇冰故居等，已成为见证"西风东渐"历史进程的重要建筑文化遗产。

尤其典型的例证，是宫巷11号刘冠雄故居。在这座传统民居的种种20世纪初之改新举措中，我们体会到的不是东西文化的冲撞，而是多元文化的相互适应。

simply adding western-style small buildings in Chinese courtyards. Meticulous balance can be found, in an attempt to prevent imported styles from shadowing the original traditional Chinese artistic conception. The lessons we can learn from such residences are: appropriate architectural scale control and the understanding and digestion of imported styles. By far, the dancing house at No.41 Yijin Lane, a small western-styled building of Chen Jiliang's residence at Wenru Lane, and a former residence of Sa Zhenbing at Zhuzi Lane have become important architectural cultural heritages, witnessing the historical process of "when east meets west".

A typical example is the former residence of Liu Guanxiong at No.11 Gong Alley. In various reform measures taken for this traditional civilian residence at the beginning of the 20th century, what we can perceive is not the collision between Eastern and Western cultures, but the mutual adaptation of multiple cultures.

中东铁路附属建筑群
Building Complex Attached to the China Eastern Railway

工程地点 | 遍布吉林省、黑龙江省、辽宁省

建造年代 | 20世纪10年代

设计机构 | 以俄罗斯设计机构为主，中国设计机构为辅

中东铁路是"中国东清铁路"的简称，1897年8月开始动工兴建，1903年2月全线竣工通车，全长近2500千米。全线由东西干线和南北支线构成"T"字形格局：干线西起满洲里，经哈尔滨，东至绥芬河，支线由哈尔滨起向南，经长春、沈阳直达旅顺口。这条铁路是在不平等条约背景下，沙俄为扩大在我国的侵略范围，控制远东而修建的。工程由俄国人设计、中方提供建设材料，并征用了大量中国劳工。通车后，该铁路线曾被俄方长期把控运营管理，后又被战略目标为东北全境的日本所把控，最终在抗战胜利后收归国有。这条铁路是列强侵华的产物，同时也是以中国劳工血汗筑就的伟大工程，更是中国人民从屈辱中赢得胜利的见证，其本

The China Eastern Railway, short for "China Dong Qing Railway", commenced construction in August 1897, and was completed and opened to traffic in February 1903, with a total length of nearly 2,500 kilometers. The whole railway consists of an east-west trunk line and a north-south branch line, forming a "T" shape: the trunk line starts from Manzhouli in the west, passes through Harbin and reaches Suifenhe in the east, while the branch line starts from Harbin, runs to the south and reaches Lvshunkou directly through Changchun and Shenyang. This railway was built by Russia against the backdrop of an unequal treaty to expand its scope of aggression in China and control the Far East. The project was designed by Russians. Its construction materials were provided by China and a large number of Chinese laborers were requisitioned to construct the railway. After being open to traffic, the railway line was controlled and operated by Russia for a long time, and then controlled by Japan whose strategic goal was to control the whole of northeast China, and finally recovered by China after the victory of the War of Resistance against Japanese Aggression. This railway, a

昂昂溪火车站

齐齐哈尔火车站

昂昂溪火车站局部

哈尔滨铁路博物馆（中东铁路俱乐部旧址）

富拉尔基江桥要塞（组图）

哈尔滨铁路博物馆

身拥有众多工业建筑遗产，被作为20世纪重要的线性文化遗产加以保护。

据近年文物普查显示，中东铁路沿线现存各类文化遗产近千处，建筑约1500余栋，代表性建筑遗产有：海林市横道河子建筑群（含铁路机车库、海林站旧址等）、绥芬河中东铁路建筑群（含火车站候车室、铁路大白楼等）、齐齐哈尔昂昂溪中东铁路建筑群（含火车站、铁路俱乐部、铁路医院等）、大庆喇嘛甸站中东铁路建筑（含俄式水塔、1号建筑等）、龙江县中东铁路建筑群（含龙江站、黑岗站等）、双城中东

product of the invasion by western imperial powers, is a great project built with the blood and sweat of Chinese laborers, and a witness to the triumph of the Chinese people over humiliation. Along the railway there are many industrial architectural heritages, which have been protected as important linear cultural heritages in the 20th century.

According to the cultural relics survey in recent years, there are nearly 1,000 cultural heritages and more than 1,500 buildings extant along the China Eastern Railway. Representative architectural heritages include: Hengdaohezi Building Complex in Hailin City (including the railway motor garage, former site of Hailin Station, etc.), Suifenhe China Eastern Railway Building Complex (including the waiting room of the railway station, railway big white building, etc.), Qiqihar Ang'angxi China Eastern Railway Building Complex (including the railway station, railway club, railway hospital, etc.),

哈尔滨铁路博物馆列车及站台陈列展示

富拉尔基火车站

铁路建筑群（含给水所旧址、双孔桥、双城堡站、同强村俄式站舍等）、哈尔滨中东铁路建筑（香坊火车站旧址、松花江铁路大桥、中东铁路管理局旧址、霁虹桥等）、尚志市一面坡中东铁路建筑群（车站办公楼、站内房舍等）、扎兰屯市中东铁路建筑群（吊桥、中东铁路俱乐部旧址等）、满洲里中东铁路建筑群（车站水塔、中东铁路技工学校、沙俄监狱等）、牙克石中东铁路建筑群（蒸汽机车水塔、兴安岭螺旋展线等）、长春市中东铁路建筑群（宽城子火车站、沙俄兵营旧址等）、中东铁路四平建筑群（第二松花江铁路大桥、中德站水塔、四平站机车司库等）、公主岭中东铁路建筑群（沙俄修建的机车修理库、机车厂等）、沈阳中东铁路建筑群（奉天驿旧址、南满铁道株式会社旧址等）、营口大石桥中东铁路建筑群（纯正寮房旧址等）、大连中东铁路建筑群（旅顺火车站旧址、东省铁路公司护路事务所旧址等）。

Daqing Lamadian Station China Eastern Railway Building (including Russian watertower, No.1 building, etc.), Longjiang County China Eastern Railway Building Complex (including Longjiang Station, Heigang Station, etc.), Shuangcheng China Eastern Railway Building Complex (including former site of Water Supply Institute, Double Orifice, Shuangchengbao Station, Russian-style station house in Tongqiang village, etc.), Harbin China Eastern Railway Building (former site of Xiangfang Railway Station, Songhuajiang Railway Bridge, former site of China Eastern Railway Administration, Jihong Bridge, etc.), Shangzhi City Yimianpo China Eastern Railway Building Complex (station office building, houses in the station, etc.), Zhalantun China Eastern Railway Building Complex (suspension bridge, former site of China Eastern Railway Club, etc.), Manzhouli China Eastern Railway Building Complex (station watertower, China Eastern Railway Technical School, Tsarist Prison, etc.), Yakeshi China Eastern Railway Building Complex (steam locomotive watertower, Xing'an Mountain spiral exhibition line, etc.), Changchun China Eastern Railway Building Complex (Kuanchengzi Railway Station, former site of Tsarist Barracks, etc.), China Eastern Railway Siping Building Complex (second Songhua River Railway Bridge, Sino-German Station Watertower, Siping Station locomotive garage, etc.), Gongzhuling China Eastern Railway Building Complex (locomotive repair depot and locomotive factory built by Tsarist Russia, etc.), Shenyang China Eastern Railway Building Complex (former site of Fengtian Post, former site of Nanman Railway Company etc.), Yingkou Dashiqiao China Eastern Railway Building Complex (former site of Chunzheng Liaofang, etc.), Dalian China Eastern Railway Building Complex (former site of Lvshun Railway Station, former site of East Railway Company Road Maintenance Office, etc.).

大庆站局部（组图）

哈尔滨铁路桥局部

昂昂溪苏军烈士陵园1

昂昂溪苏军烈士陵园2

昂昂溪铁路配套住宅

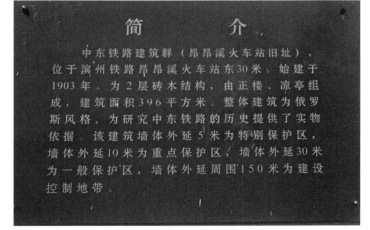
中东铁路建筑群铭牌

广州沙面建筑群
Guangzhou Shamian Building Complex

广州沙面建筑群位于广州珠江岔口白鹅潭畔，占地面积330亩（约22公顷）。1859年，第二次鸦片战争爆发后，英、法两国凭借不平等条约，以"恢复商馆洋行"为借口，强行租借沙面，雇工修护河堤，填土筑基，形成沙面岛。英、法两国在岛内分设领事馆，以今沙面一街为界，西为英租界，占全岛五分之四；东为法租界，占五分之一。从19世纪末到20世纪初，沙面租界内形成较完整的建筑街区，含领事馆、教堂、银行、邮电局、商行、医院、酒店、住宅、网球场和游泳场等，其住户多是各国领事馆、银行、洋行的人员以及外籍的税务官和传教士。1996年，国务院公布其为全国重点文物保护单位，并定为整体性文物保护区，其中53栋建筑被列为全国重点文物保护单位。其代表性建筑分布如下。

沙面大街：法国巡捕房旧址、美国领事馆旧址、露德天主教圣母堂、法军兵营旧址、英国医院旧址、法国传教士住宅旧址、汇丰银行旧址、粤海关馆舍旧址、广州俱乐部旧址等。

沙面一至五街：东方汇理银行旧址、三井洋行旧址、沙面西桥、亚细亚火油公司旧址、赫德爵士住宅旧址、泰和洋行旧址等。

The Guangzhou Shamian Building Complex is located at the bank of the White Swan Pond at a fork of the Pearl River in Guangzhou, covering an area of 330 *mu* (22 hectare). In 1859, after the outbreak of the Second Opium War, Britain and France, relying on unequal treaties and using the pretext of "restoring foreign trade and commerce", forcibly rented Shamian, hired employees to repair river banks, fill earth and built foundations to form Shamian Island. Britain and France set up consulates on the island. With Shamian Streetl One as the boundary, the island was divided into two concessions, with the British concession in the west, accounting for four fifths of the whole island; and the French concession in the east, accounting for one fifth. From the end of the 19th century to the beginning of the 20th century, relatively complete building blocks were developed in the Shamian Concessions, including consulates, churches, banks, post and telecommunications offices, firms, hospitals, hotels, residences, tennis courts and swimming pools, etc. Most of the residents were the staff of consulates, banks and foreign companies of various countries, as well as foreign tax officials and missionaries. In 1996, the State Council declared the build complex as an important heritage site under state protection; and designated it as an integral heritage site protection area, among which 53 buildings were designated as key units of cultural heritage under state protection. The representative buildings are as follows.

Shamian Street: the former sites of French patrol house, American Consulate, Notre Dame Cathedral, French barracks, British hospital, French missionaries' residences, HSBC, Guangdong Customs House and Guangzhou Club.

Shamian Street One to Shamina Street Five: the former sites of Credit

街景立面

台湾银行旧址

英太古洋行旧址

街景雕塑

露德天主教圣母堂

沙面街景

沙面南街：英国公园旧址、英国网球场旧址、法国公园旧址、法国海军办事处旧址、法国领事馆旧址、中法实业银行旧址、圣公会基督教堂旧址等。

沙面北街：捷克领事馆旧址、香港牛奶公司制冰厂旧址、沙面北街私人住宅旧址、沙面游泳池等。

岛内建筑有西方风行的新古典式、折中主义式、新巴洛克式、拱券廊式、仿哥特式等形式的建筑群体，至今尚基本保存。如：沙面大街14号露德天主教圣母堂，规模虽小，但以总体构图比例、尖顶、尖券状门窗等，成为仿哥特式建筑的佳品；沙面大街54号之汇丰银行，以粗大石材砌筑底层，以古典柱式控制立面构图，南面和西面分别有进出门，二楼的外墙砌有通柱到三层顶，在西南面楼顶建有穹窿顶的亭子，是西洋折中主义建筑的典范；沙面大街2号、4号、6号楼（俗称红楼），红砖砌筑，东西两端有高耸的圆锥形塔，有具有法国特色的弧形阳台，充满浓郁的法国塞纳石堡风格；沙面大街48号之旗昌洋行，高3层，立面四周为简洁的连续拱券的券廊，维多利亚风格，为券廊式建筑的代表。

Agricole Indosuez, Mitsui Foreign Firm, Shamian West Bridge, Asia Petroleum Company, Sir Hurd's Residence and Reiss Bradley & Co. etc.

Shamian South Street: the former sites of British Park, British Tennis Court, French Park, French Navy Office, French Consulate, Sino-French Industrial Bank and Anglican Christian Church.

Shamian North Street: the former sites of Czech Consulate, Ice Factory of Hong Kong Milk Company, Private Residences of Shamian North Street, and Shamian Swimming Pool, etc.

Buildings in the island include architectural complexes of various styles, such as Neoclassical style, Eclectrical style, Neobaroque style, Arcade style and imitated Gothic style, which are popular in the West and still basically preserved up to now. For example, the Catholic Notre Dame Cathedral at No.14 Shamian Street, small as it is, has become a paragon of Gothic architecture with its overall composition ratio, spire, pointed arch doors and windows, etc.; HSBC, located at No.54 Shamian Street, is a model of Western eclectical architecture, featuring thick stones for the ground floor, classical column-controlled facade composition, entrance and exit in the south and west respectively, the columns extends to the top of the third floor built on the outer wall of the second floor, and a pavilion with a dome roof on the southwest roof. Buildings No.2, No.4 and No.6 of Shamian Street (commonly known as Red Houses) were built with red bricks, with lofty conical towers at the east and west ends and curved balconies with a powerful style of French Sennastone castles; Russell & Company at No.48 Shamian Street is a three-story building, with its facade surrounded by a simple and continuous arcade in the Victorian style, which is a representative of arcade architecture.

沙面街景（组图）

自中坡炮台俯视马尾船政

马尾船政

Mawei Shipbuilding

马尾船政文化遗址群位于福州郊外闽江下游之马尾港一带，包括：马限山中坡炮台、昭忠祠及马江海战烈士公墓、英国领事分馆、马尾船政局（后称马尾造船厂）、船政绘事院、船政学堂等多个船政遗址，集中展现了自19世纪60年代至20世纪30年代以来，中国在科学技术、新式教育、工业制造、国防建设、西方经典文化翻译传播、东西方文化交流等方面的丰硕成果。

马尾船政局（又称福州船政局）是晚清政府经营的制造兵船、炮舰的新式造船企业，先后由晚清名臣左宗棠、沈葆桢主持，任用法国人日意格（1835—1886年）、德克碑（1831—1875年）为正、副监督，总揽一切船政事务。船政局主要由铁厂、船厂和船政学堂三部分组成。1869年6月10日，船局制造的第一艘轮船"万年清"号下水。船政学堂（前

The Mawei Shipbuilding Cultural Site Complex is located in Mawei Port in the lower reaches of the Minjiang River on the outskirts of Fuzhou, including: Maxian Mountain Middle Slope Fort, Zhaozhong Temple and Majiang Battle Soldiers' Cemetery, British Consulate Branch, Mawei Shipbuilding Bureau (later known as Mawei Shipbuilding Plant), Shipbuilding Painting Institute, and Shipbuilding School, etc., showcasing China's fruitful achievements in science and technology, new modern education, industrial manufacturing, national defense construction, spreading of western dassic culture and cultural exchange between East and West, and so on from the 1860s to the 1930s.

Mawei Shipbuilding Bureau (also known as Fuzhou Shipbuilding Bureau) was a modern shipbuilding enterprise operated by the late Qing government to making warships and gunboats. It was presided over by Zuo Zongtang and Shen Baozhen, famous ministers of the late Qing Dynasty. Frenchmen Prosper Giquel (1835-1886) and Paul Alexandre Neveue d'Aigwebelle (1831-1875) were appointed as the chief and deputy supervisors, taking charge of all Shipbuilding affairs. The Shipbuilding bureau is mainly composed of Iron Works, Shipyard and Shipbuilding School. On June 10th, 1869, the first ship "Wan Nianqing" manufactured by the Shipbuilding

马尾船政船厂旧影——法军企图占作其远东舰队修船基地

船政所造远东最大的木质巡洋舰——扬武号

马限山中坡炮台

马尾船政遗址

马尾船政、天后宫、衙门及十三厂旧影

造船厂车间外景局部

原船政绘事院

造船厂车间内景

身为求是堂艺局）设制造、航海两班，要求学员分别达到能按图造船和任船长的能力；并派员留学英、法，学习驾驶和造船技术。所制造海军舰船历经近代历次海战，直至最后一艘"通济号"钢胁钢壳练船于1937年抗战期间自沉。船政局在民国初年逐渐萧条，1949年后，才以福建马尾造船股份有限公司的形象重获新生（已具备设计、建造和修理3.5万吨级以下各类船舶的生产能力，是我国南方重要的船舶生产基地）。船政学堂已培养出严复、萨镇冰、詹天佑、邓世昌、刘冠雄等风云人物。目前，遗址尚存船政轮机车间、船政绘事院等20世纪20年代前的遗迹，与今日蒸蒸日上的福建马尾造船股份有限公司生产线交相辉映。

位于马限山麓的中法马江海战烈士墓和昭忠祠，安葬着福建水师1884年8月23日阵亡官兵约1200具遗体，系1920年由海军界和船政学校校友募捐兴建，也是重要的20世纪文化遗产。

Bureau was launched. The Shipbuilding School (formerly known as the Qiushi Tangyi Bureau) has two classes of manufacturing and navigation, requiring students to achieve the ability to build ships according to drawings and be captains respectively. Students were dispatched to study in Britain and France to learn navigating and shipbuilding techniques. The naval ships manufactured went through all previous naval battles in modern times, until the last ship, the Tongji steel-clad training ship, sank itself during the War of Resistance against Japanese Aggression in 1937. In the early years of the Republic of China, the Shipbuilding Bureau gradually fell, and it was not until in 1949 that it was reborn as Fujian Mawei Shipbuilding Co., Ltd. (with the production capacity to design, build and repair all kinds of ships below 35,000 tons and becoming an important ship production base in Southern China). Shipbuilding School has trained famous figures such as Yan Fu, Sa Zhenbing, Zhan Tianyou, Deng Shichang and Liu Guanxiong. By far relics before the 1920s are still extant, such as Shipbuilding and Marine Engineering Workshop and Shipbuilding and Mapping Institute, which set off the flourishing production line of Fujian Mawei Shipbuilding Co., Ltd. today.

In the Tomb of Martyrs of Majiang Battle between China and France and Zhaozhong Temple located at the foot of Maxian Mountain, about 1,200 officers and soldiers of Fujian Navy sacrificed on August 23, 1884 are buried. Built in 1920 by donations from the naval circles and alumni of Shipbuilding School, the sites are also important cultural heritages of the 20th century.

马江海战将士公墓（组图）

南京中山陵音乐台
Nanjing Sun Yat-sen Mausoleum Music Station

中山陵音乐台位于紫金山中山陵广场东南,是中山陵的配套工程,主要用于孙中山先生纪念仪式时的音乐表演及集会演讲。音乐台占地面积约为 4200 平方米,由关颂声、杨廷宝设计,1932 年秋动工兴建,1933 年 8 月建成。音乐台为钢筋混凝土结构,平面布局呈半圆形,圆心位置设一座弧形钢筋混凝土结构舞台及照壁,舞台长约 22 米,宽约 13.33 米,高出地面 3.33 米。台后建有大照壁,是音乐台的设计主体,仿中国传统五山屏风样式,表面以水泥斩假石镶面,既为舞台背景,又可起到反射声波作用。照壁坐南朝北,宽约 16.67 米,高约 11.33 米,水平截面为圆弧形。照壁上部及两侧雕刻有云纹图案。上部云纹图案之下,雕有三个龙头,突出墙体之外。在舞台边缘处有数道波浪形台阶,阶内填土以栽花草。乐坛正前方设一汪月牙形莲花池,池底有伏泉,池半径为 12.67 米,用以汇集露天场地的天然积水,同时可以增强乐坛的音响效果。乐坛两翼筑有平台,有台阶与花棚衔接,上砌钢筋混凝土花棚。舞台下辟有休息室、盥洗

Sun Yat-sen Mausoleum Music Station, located in the southeast of the Sun Yat-sen Mausoleum Square in Zijinshan Mountain, is a supporting project for the Sun Yat-sen Mausoleum, mainly used for music performances and assembly speeches during Dr. Sun Yat-sen's memorial ceremonies. The music station covering an area of about 4,200 square meters was designed by Guan Songsheng and Yang Tingbao. Its construction commenced in the autumn of 1932 and was completed in August 1933. The music platform is of a reinforced concrete structure, with a semicircular plane layout. There is an arc-shaped reinforced concrete structure stage and a screen wall at the center of the circle. The stage is about 22 meters long, 13.33 meters wide and 3.33 meters higher than the ground. There is a big screen wall behind the stage, which is the main part of the music stage design, imitating the traditional Chinese five-mountain-screen style, and its surface is inlaid with cement artificial stones. The big screen wall not only serves as the stage background, but also plays a role in reflecting sound waves. The screen wall faces north, with a width of about 16.67 meters and a height of about 11.33 meters. The horizontal section is a circular arc. The upper part and both sides of the screen wall are engraved with cloud patterns. Under the upper cloud patterns, there are three dragon heads protruding out of the wall. There are several wavy steps at the edge of the stage, filled with soil to plant flowers and plants. In front of the music altar there is a crescent-shaped lotus pond with a hidden spring at the bottom. Having a radius of 12.67 meters, the pond is used to collect the natural stagrant rainwater in the open

冬日的音乐台建筑与和平鸽

舞台局部

舞台建筑后的立面局部

从舞台看向观众席

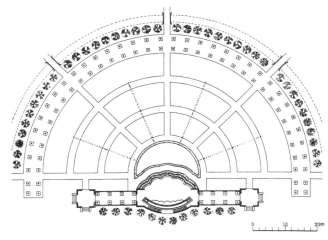
平面图

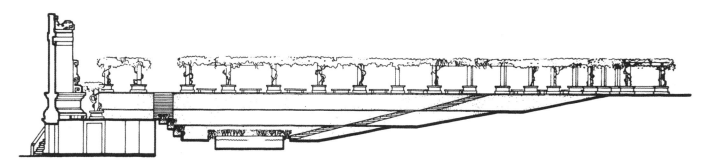
剖面图

俯瞰音乐台

室和贮藏室等。

设计者巧妙利用地势的自然起伏，将舞台前的草坪设为观众席。此半圆形坡面状草坪半径约 56.67 米，沿同心弧方向展开 3 条小径；沿扇面坡道圆心方向建有 5 条宽 2 米的放射形走道，每条走道有 45 级台阶。这个草坪席可容纳观众 3000 余人。环绕草坪观众席外缘，建一条宽 6 米、长 150 米的以 36 根立柱构成的紫藤花架走廊，廊道两侧有 36 个花盆、30 张石凳，可供观众品茶休憩。

音乐台建筑风格为中西合璧，平面布局、立面造型等借鉴古希腊露天剧场建筑特点，而在照壁、乐坛等装饰细部处理上，则采用中国江南古典园林的表现形式。

area and enhance the sound effect of the music altar. There are platforms on both wings of the music altar, which are connected by steps with reinforced concrete flower stands above. There is a lounge, a washroom and a storeroom under the stage.

The designers skillfully rely on the natural ups and downs of the terrain to make the lawn in front of the stage for audience seats. The radius of this semi-circular sloped lawn is about 56.67 meters, and three small paths spread in the concentric arc direction. There are five radial walkways with a width of 2 meters along the center of the fan ramp, each walkway having 45 steps. This lawn audience area can accommodate more than 3000 spectators. Around the outer edge of the lawn auditorium, a wisteria flower rack corridor with a width of 6 meters and a length of 150 meters and 36 columns has been built, along both sides of which there are 36 flower pots and 30 stone benches for the audience to enjoy tea and rest.

The music platform perfectly integrates the Chinese and Western architectural styles. The architectural feature of ancient Greek open-air theater have been used for reference in plane layout and elevation modeling, while decorative details such as screen walls and music altar are expressed in the form of classical gardens in south of the Yangtze River in China.

走廊局部（组图）

理学院

重庆大学近代建筑群
Modern Architectural Complex of Chongqing University

重庆大学近代建筑群位于今重庆大学 A 区（现重庆大学与重庆建筑大学、重庆建筑高等专科学校三校合并而成，A 区为本部老校区）。重庆大学始建于 1925 年，1933 年时，校址由菜园坝迁入沙坪坝，并建造了第一栋教学楼即理学院楼。抗日战争期间，中央大学借用校园和重庆大学联合办学，又陆续增建一批建筑，虽因日军空袭而致部分建筑遭到损毁，但是有相当一部分幸存至今。其中以理学院楼、工学院楼、文字斋最具代表性。

理学院楼为"中国固有式建筑"风格。平面呈"工"字形，中轴对称。建筑主体部分为双层带阁楼式，屋顶为重檐歇山形式，屋面开老虎窗且坡度较陡。此楼被评价为"中国建筑师探索中式建筑风格的典型例证之一，为研究民国时期建筑风格的演变及民族风格建筑的发展，提供了重要实物资料。"

The Modern Architectural Complex of Chongqing University is located in today's Zone A of Chongqing University (a merger of former Chongqing University, former Chongqing University of Architecture and Chongqing Architectural College; Zone A means the old campus of headquarters). Chongqing University was founded in 1925. In 1933, the campus moved from Caiyuanba to Shapingba, and the first teaching building, the Building for the School of Science, was built. During the War of Resistance against Japanese Aggression, the Central University borrowed the campus of Chongqing University and ran the university jointly. A number of buildings were added then. Although some buildings were damaged due to Japanese air-strikes, quite a few survived to this day. Among them, the Building for the School of Science, the Building for the School of Engineering and Wenzizhai Dormitory Building are the most representative ones.

The Building for the School of Science is in the "Chinese Inherent Architectural Style". The plane is I-shaped with a symmetrical central axis. The main part is a two-story building with an attic, with a double-eaves hipped roof provided with a dormer window with a steep slope. This building has been rated as "one of the typical examples of Chinese architects exploring Chinese architectural style and providing important material information for

校园细部

入口及裙房

工学院楼始建于1935年，由留法学者刁泰乾设计。工学院楼为中西合璧式样，尤其注重借鉴重庆的地方特色。建筑平面呈曲尺型布局，入口处设六边形塔楼，各层屋架均用杉木制作，墙体全部用条石砌筑。工学院楼曾在1939年9月、1940年5月和7月三次遭到日寇战机轰炸，伤痕累累，房屋多处被破坏，至今材质的契合线上仍有当年的累累弹痕——记录下侵略者的暴行，而其整体之昂然屹立，也实证着建筑的坚固异常。

文字斋于1933年建成，以青砖黛瓦砌筑，红色立柱支撑，古色古香，是纯正的中国传统建筑，也是重庆大学至今保存最完好的近代建筑之一。

此外，重庆大学A区遗存的老图书馆（博雅学院楼）、寅初亭、七七抗战大礼堂、大校门等，也极具历史文化价值。

studying the evolution of architectural style and the development of national architectural style in the Republic of China."

The Building for the School of Engineering was built in 1935 and designed by Diao Taiqian, a scholar studied in France. The Building for the School of Technology has integrated both Chinese and Western styles, with a special stress on the local characteristics of Chongqing. The layout of the building is in the shape of a curved ruler, with a hexagonal tower at the entrance. The roof truss of each floor is made of Chinese fir, and the walls are all built with strips of stone. In September 1939, May 1940 and July 1940, the Building for the School of Engineering was bombed by Japanese invaders, leaving many scars and damages. Up to now, many bullet marks on the fit lines of building materials can still be found, which records the atrocities of the aggressors. That the building still stands proudly as a whole demonstrates the exceeding solidarity of the building.

Built in 1933, the Wenzihai Dormitory Building is one of the best preserved modern buildings in Chongqing University, which is an authentic traditional Chinese building with an ancient style, featuring with blue bricks and tiles being supported by red columns.

In addition, the old library (the Building for the School of Liberal Arts), Yinchu Pavilion, Auditorium in Memory of July 7th War of Resistance against Japanese Aggression, and the Grand School Gate, which remain in Zone A of Chongqing University, are also of great historical and cultural value.

寅初亭

建筑外廊

理学院侧立面

文字斋

北京工人体育馆
Beijing Workers' Gymnasium

工程地点	北京市
建筑面积	42000平方米
竣工时间	1961年
设计机构	北京市建筑设计研究院
主要设计人	熊明 孙秉源

北京工人体育馆是我国为举办第26届世界乒乓球锦标赛而建设的工程,是当时国内规模最大的体育馆,于1961年建成。

北京工人体育馆位于北京市朝阳区,建筑面积42000平方米,地下1层,地上4层,是一座圆柱形大厦,其基座直径120.3米,屋檐高27米,圆屋顶中心高38米,比赛大厅能容纳15000名观众,比赛场地可供10场乒乓球比赛同时,还可举办篮球、排球、羽毛球等比赛或承接表演等。比赛大厅直径110米,看台分三层。比赛大厅的屋盖净跨度94米,首次采用双层悬索结构,其边缘结构为内径94米的钢筋混凝土外坯,搁置在48根钢筋混凝土结

The Beijing Workers' Gymnasium, inaugurated in 1961 for the 26th World Table Tennis Championships, was the largest stadium in China at the time.
Located in Chaoyang District, Beijing, the Beijing Workers' Gymnasium, covering a floor area of 42,000 square meters, with one floor underground and four floors above ground, is a cylindrical building with a base diameter of 120.3 meters, a roof height of 27 meters and a dome center height of 38 meters. The competition hall can accommodate 15,000 spectators, and the competition venues can accommodate 10 table tennis competitions. Meanwhile, it can also hold basketball, volleyball, badminton and other competitions or undertake performances. The competition hall, with 110 meters in diameter, has a three-story stand. The roof of the competition hall has a net span of 94 meters, for which a double-layer suspension cable structure has been adopted for the first time. Its edge structure is a reinforced concrete outer blank with an inner diameter of 94 meters, which rests on 48 veranda pillars of reinforced concrete structures. The inner ring is a central ring composed of cylindrical

入口处局部

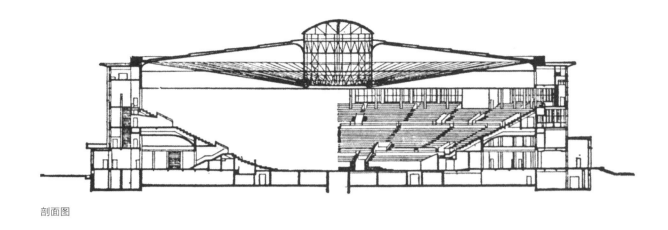

剖面图

室内空间

构的外廊支柱上,内环为直径 16 米、高 11 米圆筒形钢结构组成的中心环,连接内外环的上、下钢索沿径向辐射布置,上、下钢索各为 144 根。悬索上部为轻型屋面,下部为吊顶棚。

比赛场地人工照明分为两部分,篮球、排球比赛等照明用中心环上的镜面深罩灯。乒乓球比赛,每个台上在悬索下悬挂 5 个深罩灯,在中心环上操作升降。

馆内设有国内首次采用的霓虹灯记分牌 3 套及子母钟、广播电视系统。馆内设有冷、热、湿、净的空调送回风系统,大厅两侧采用高速喷口向下倾斜送风,回风口设于观众席前后部。

steel structures with a diameter of 16 meters and a height of 11 meters. The upper and lower steel cables connecting the inner and outer rings are radially arranged, with 144 upper and lower steel cables respectively. The upper part of the suspension cable is a light roof, while the lower part is a suspended ceiling. The artificial lighting of the competition field is divided into two parts. Games like basketball and volleyball games are illuminated by the lights with reflective chimney on the central ring. For table tennis competitions, five chimney lights are hung under the suspension cable over each table, which can be lifted and lowered on the central ring.

In addition, 3 sets of neon scoreboards, synchronized clocks, and the radio and television system have been installed in the stadium for the first time in China. Cold, hot, wet and clean air-conditioning, air supply and air return systems have also been installed as well. High-speed nozzles are used on both sides of the hall to supply air obliquely downward, and air return outlets are located at the front and back of the auditorium.

入口处局部雕饰

外立面局部

外立面远眺旧影

顶部悬索结构局部

梁启超像和饮冰室

梁启超故居和梁启超纪念馆（饮冰室）

Liang Qichao's Former Residence and Liang Qichao Memorial Hall (Yin-Bing Chamber)

工程地点 天津市

建筑面积 1100平方米

竣工时间 1914年、1924年

主要设计人 白罗尼欧（意）

梁启超故居及"饮冰室"书斋坐落在天津河北区民族路44、46号，是由两幢洋楼组成的院落，总占地面积为2500平方米。

其44号为住宅楼，建筑面积1100余平方米，建于1914年，是一幢带地下室的意大利式洋楼，一、二层各有9间居室。整体建筑分为两部分，东半部为梁氏专用，有小书房、客厅、起居室等；西半部是家属住房。后楼为附属建筑，有厨房、锅炉房、贮藏室、佣人住房等。主楼为砖木结构，水泥罩面，塑有花纹，采用异形红色瓦顶，石砌高台阶，双槽门窗，整体建筑相当讲究，有花园、汽车房、传达室。

46号即梁启超自题"饮冰室"的书斋。饮冰室由意大利建筑师白罗尼欧设计，为天井外廊式带封闭罩棚的浅灰色二层小洋楼，建于1924年，共有房屋34间，建筑面积949.5

Liang Qichao's Former Residence and the "Yin-Bing Chamber", located at No.44 and No.46 Minzu Road, Hebei District, Tianjin, is a courtyard composed of two Western-style buildings, covering a total area of 2,500 square meters.

No.44 is a residential building with a floor area of over 1,100 square meters. Built in 1914, it is an Italian-style building with a basement, and 9 rooms on the first and second floors. The whole building is divided into two parts. The eastern part was only used by Mr. Liang Qichao, with a small study, a sitting room, and a living room, etc.; the western half was for Liang's family members. The back building is ancillary, including a kitchen, a boiler room, a storage room, and servant's house, etc. The main building is of a brick-wood structure, covered with cement. It has decorative patterns, a special-shaped red tile roof, high stone steps, double-groove doors and windows. The whole building is quite exquisite, including a garden, a garage and a reception room.

No.46 is Mr. Liang Qichao's study which he named the "Yin-Bing Chamber". Designed by Italian architect Baronio and built in 1924, the Yin-Bing Chamber is a light-gray two-story small building with a patio, a veranda, and a closed canopy, having 34 rooms with a floor area of 949.5 square meters.

饮冰室入口局部

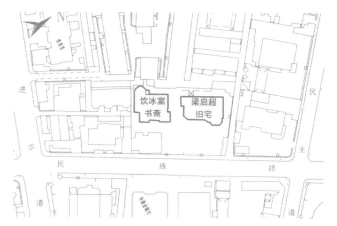

总平面图

饮冰室室内

平方米。此建筑门前两侧是石台阶，当中有蓄水池，池中雕一座石兽，口中常年喷水不断；楼内正中大厅实际是天井院的罩棚，罩棚高出屋顶，用花玻璃镶成；大厅周围5间房子除1间为杂房外，其余为书房和图书室；二楼西北角也是1间大厅，东南角有几间主要作卧室或图书资料室。整座建筑造型别致，具有文艺复兴时期意大利建筑风格。

此座建筑曾因年久失修而破损严重，近年按"修旧如旧"的原则予以修复。修复后的梁启超故居分为书房、起居室、家族纪念室等12个展室，再现了梁启超当年居住的环境。展室里陈列着梁启超的书信、书籍、历史文献以及活动照片等。梁启超纪念馆里一百多件家具，系按当年的陈设原汁原味复制，并根据梁启超后代反复回忆进行布置，力求贴近当年原貌。

There are stone steps on both sides of the front door of this building. Between the stone steps is a pool, in which a stone beast is carved with spraying water from its mouth all the year round. The hall in the middle of the building is actually formed by a canopy over the courtyard. The canopy is higher than the roof and inlaid with stained glass. The five rooms around the hall are all studies and reading rooms except for one miscellaneous room. There is also a hall in the northwest corner of the second floor, and several rooms in the southeast corner are mainly used as bedrooms or libraries. The whole building is uniquely structured, featuring with a Italian Renaissance architectural style.

This building was seriously damaged due to disrepair. In recent years, it has been repaired according to the principle of "restoring the old as the old". The restored former residence of Liang Qichao is divided into 12 exhibition rooms, including study rooms, living rooms and family memorial rooms, which represent the living environment of Liang Qichao. Liang Qichao's letters, books, historical documents and photos of activities are displayed in the exhibition rooms. More than 100 pieces of furniture in the Liang Qichao Memorial Hall were copied according to the original furnishings, and arranged according to the repeated memories of Liang Qichao's descendants, striving to be as close to the original appearance as possible.

饮冰室入口立面

饮冰室入口细部构造

梁启超故居外立面

143

石景山钢铁厂
Shijingshan Iron and Steel Plant

工程地点	北京市
占地面积	8.56平方千米
竣工时间	1919年
设计机构	龙烟铁矿公司策划筹建

1919年9月,官方(农商部和交通部)与商界(私人出资)集资500万元,创办了首钢集团的前身"龙烟钢铁厂"。在历史浪潮的跌宕中,首钢集团经历了4次更名(1937年更名为石景山制铁所,1945年更名为石景山钢铁厂,1966年更名为首都钢铁公司,1992年更名为首钢总公司),最后于2017年更名为现在的"首钢集团有限公司"。原有规模主要为有效容积389m³的一号高炉和516m³的二号高炉;而后扩大到963m³的五号高炉,1200m³的四号高炉,2536m³的三号高炉及焦炉、料仓等。为了响应环保政策,原首钢石景山厂区于2011年全部停产。随着2022年冬季奥运会奥组委办公园区落户首钢,首钢园区开始进行全面更新改造。

In September 1919, Longyan Iron Mine Co.Ltd., the predecessor of Shougang Group, was founded with an investment of five million yuan by government agencies (the Ministry of Agriculture and Commerce and the Ministry of Transport) and private investors. In the midst of historical tides, it was renamed four times: Shijingshan Ironworks in 1937; Shijingshan Iron and Steel Plant in 1945; Capital Iron and Steel Corporation in 1966; Shougang Corporation in 1992. And it was finally named Shougang Group Co., Ltd. in 2017. Original facilities were primarily #1 and #2 blast furnaces with 389 cubic meters and 516 cubic meters in effective volume respectively; the production capacity was later expanded to include #5, #4 and #3 blast furnaces (963 cubic meters, 1,200 cubic meters and 2,536 cubic meters respectively), coke ovens, bunkers, etc. In response to the country's environmental policy, the former Shijingshan plant area of Shougang stopped production entirely in 2011. With the Beijing Organizing Committee for the 2022 Olympic Winter Games moving to the Shougang High-end Industry Comprehensive Service Park, the former steel complex is now in a process of thorough transformation.

厂区远眺

厂区建设过程旧影

厂区风貌（组图）

中国银行南京分行旧址
The Former Site of the Bank of China, Nanjing Branch

中国银行南京分行旧址有两处,其一位于下关街区之大马路66号,可称为"中国银行南京分行旧址(一)";其二位于城南珠宝廊,即今白下路23号(又称珠宝廊中国银行新建行屋),可称为"中国银行南京分行旧址(二)"。

下关大马路66号的"中国银行南京分行旧址(一)",建造于1923年,平面呈"凸"字形,钢筋混凝土结构,高二层,带半地下室,顶部加一八角平面之望亭。外观以六根通高二层的爱奥尼式石柱形成柱廊,顶部石台的花纹简约精美;首层门厅以马赛克花砖铺地,内部装修精致而不奢华。该址现为长江水利委员会水文局。

白下路23号的"中国银行南京分行旧址(二)",于1916年兴建,1933年重建,由坐南朝北的临街营业厅用房和背后之分行礼堂组成院落。营业厅建筑面积2800平方

There are two former sites of the Bank of China, Nanjing Branch. One is located at No.66, Dama Road, Xiaguan Block, which can be called "The Former Site of the Bank of the China, Nanjing Branch (I)"; The other is located in the Chengnan Jewelry Gallery, at No.23 Baixia Road (also known as the new building of the Bank of China in the Jewelry Gallery), which can be called "The Former Site of the Bank of China, Nanjing Branch (II)".

"The Former Site of the Bank of China, Nanjing Branch (I)", located at No. 66, Dama Road, Xiaguan District, was built in 1923. With a convex-shaped plane, the building features a reinforced concrete structure, having two stories and a semi-basement, with an octagonal pavilion at the top. Six Ionian stone pillars with a height of two floors form a colonnade, topped by a stone platform decorated with elegantly simple patterns. The foyer on the first floor is paved with mosaic tiles. The interior decoration is exquisite but not luxurious. The site is now the domicile of the Bureau of Hydrology, Changjiang Water Resources Commission.

"The Former Site of the Bank of China, Nanjing Branch (II)" at No.23 Baixia Road was built in 1916 and rebuilt in 1933, consisting of the frontage business hall facing south and the branch auditorium behind it. The business hall covering an area of 2,800 square meters was designed by architects

沿街立面

建筑入口

建筑屋顶构造

建筑远眺旧影

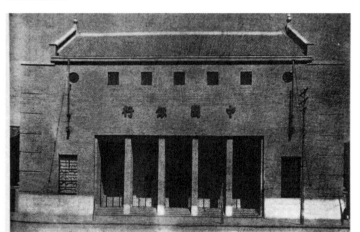
建筑旧影

入口阶梯一侧铭牌

檐口及柱头构造

米，由陆谦受、吴景奇建筑师设计，上海新亨营造厂承建。建筑坐北朝南，钢筋混凝土结构，高三层。外观墙面采用淡黄色泰山面砖，嵌缝用带颜色水泥以配砖色，勒脚用苏州石打光，屋顶用中国式青瓦，脊饰为漆金色人造石；内部营业厅的采光，部分来自天窗，分布匀和，同时还兼有通风功能。内部墙面多处采用壁画装饰。院内大礼堂为西方折中主义建筑，目前实测资料不详，有可能是1916年初建时的遗物。这两处建筑，前者表现了民国初年的西化时尚，而后者造型适用，布局合理，反映出20世纪30年代中国建筑师开始注重实用功能的思考过程。

Lu Qianshou and Wu Jingqi, and built by Shanghai Xinheng Construction Factory. The three-story building faces south, having a reinforced concrete structure. The exterior walls are made of light-yellow Mount Tai bricks, with the caulking made of colored cement to match the brick color, and the plinths polished with Suzhou stones. The roof is made of Chinese blue tiles, and the ridges are decorated with gold painted artificial stones. The evenly distributed lighting of the internal business hall comes partly from the skylight, which also has a ventilation function. Many interior walls are decorated with frescoes. The auditorium in the courtyard is a Western-style eclectic building, which may be a relic of the building first constructed in 1916 (field data unknown).

Of these two buildings, the former shows the westernization fashion in the early years of the Republic of China, while the latter, with practical styling and reasonable layout, reflects the process in which Chinese architects began to reflect on practical functions in the 1930s.

入口阶梯及柱础构造

门厅局部

田径场入口立面

中央体育场旧址
The Former Site of the Central Stadium

工程地点｜南京市
占地面积｜1000亩
竣工时间｜1931年
设计机构｜基泰工程司
主要设计人｜关颂声、杨廷宝

中央体育场旧址位于南京市玄武区南京体育学院内，是当时亚洲规模最大的运动场，包括田径场、国术场、篮球场、游泳池、棒球场及网球场、足球场、跑马场等，占地1000亩，一次可接纳观众6万人。

中央体育场于1931年8月基本建成，由基泰工程司关颂声、杨廷宝设计，利源建筑公司承造。全部工程采用钢筋混凝土结构，同时采用中国传统纹样装饰，进出口和主看台都采用中国式牌楼建筑风格。其代表性单体建筑简况如下。

田径场是中央体育场的主要建筑，可容纳观众35000余人，平面呈椭圆形，南北走向，占地面积约77亩。周围是看台，中间是田径场。田径场长300米，宽130米。外观正立面主要由三个大小相同的拱形门构成门楼，高5.5米；东西门楼均为中国传统牌楼式建筑，面阔九间，高3层，上部装饰

The former site of the Central Stadium is located in the Nanjing Institute of Physical Education, Xuanwu District, Nanjing. It used to be the largest sports ground in Asia, including a ground track field, a martial arts field, a basketball court, a swimming pool, a baseball field, a tennis court, a football field and a racetrack. It covers an area of 1,000 *mu* and can accommodate 60,000 spectators at one time.

The Central Stadium was basically completed in August 1931, designed by Guan Songsheng and Yang Tingbao of Jitai Engineering Company and built by Liyuan Construction Company. The whole project was supported by reinforced concrete structures and decorated with traditional Chinese patterns. The Chinese archway architectural style was adopted for the entrance and exit and the grandstand. Representative single buildings are as follows.

The ground track field, the main structure of the Central Stadium, can accommodate more than 35,000 spectators. With an oval plane, the field runs from north to south, covering an area of about 77 *mu*. Surrounded by stands, the ground track field is in the middle, 300 meters long and 130 meters wide. The facade mainly consists of three arched doors of the same size, with a height of 5.5 meters. Both the East and West Gate Towers is traditional three-story Chinese archway structures, nine-jian wide, with the upper part decorated with

从体育场内侧看主席台

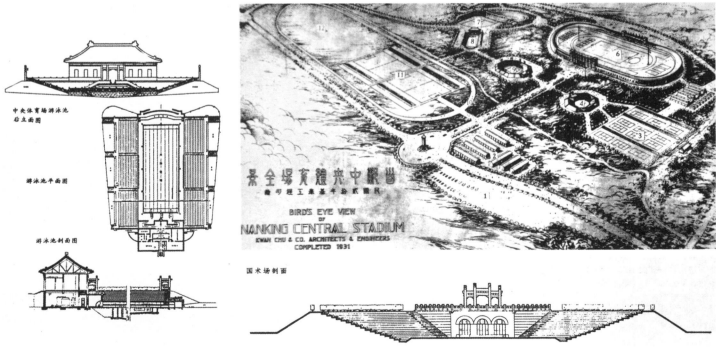

设计图纸（组图）

8个云纹望柱头和7个小牌坊屋顶。

游泳池包括露天比赛场及训练场。其入口处是一座中国古典宫殿式的建筑，地上地下各一层，面朝西南。该建筑平面为长方形，长26.28米，宽13.4米，钢筋混凝土结构，庑殿顶，屋面覆琉璃筒瓦，雕梁画栋。地上一层布置办公室、更衣室和淋浴室等；地下一层为滤水器房和锅炉房。建筑物的正前方布置游泳池，长50米，宽20米，设有9条泳道。游泳池的东南、西北、西南三面利用坡地构成看台，可容观众4000人。游泳池历经60余年而无渗漏现象。

国术场平面呈正八角形，四周视距相等，最远视距18.2米，满足国术比赛适宜近距离观看的要求。整个场地形似一个盆地，场上铺满黄土。正门朝北，迎着正门拾级而上，建有一座三开间牌坊。经牌坊达大平台，平台上陈列着各种武术器械，平台下建有办公室、运动员更衣室和厕所等。平台四周围以水泥假石雕纹栏杆。除正门一面外，其余七面均为看台，可容观众5400余人。

近年来，除田径场、国术馆外，其余场馆均被改造为室内场馆，颇存争议。

eight baluster capitals with cloud design and seven small archway roofs.

The swimming pool includes an outdoor competition field and a training pool. Its entrance is a classical Chinese palace-style building with one floor above and one floor below ground, facing southwest. The richly ornamented building is rectangular in plane, 26.28 meters long and 13.4 meters wide. Its structural frame is of reinforced concrete. Its hipped roof is covered with glazed tiles. Offices, dressing rooms and shower rooms are arranged on the ground floor. The floor underground is a water filter room and a boiler room. A swimming pool, 50 meters long and 20 meters wide, is arranged right in front of the building and has 9 swimming lanes. The sloping fields on the southeast, northwest and southwest sides of the swimming pool are used to build stands, and will be able to accommodate 4,000 spectators. The swimming pool has been free from any leakage for more than 60 years.

The plane of the martial arts field is octagonal, resulting in equal visual range around it, with the farthest visual range being 18.2 meters, meeting the close viewing requirement of martial arts competitions. The whole site looks like a basin, covered with loess. The main entrance faces north. Walking towards the main entrance and climbing up the stairs, we can see a three-jian archway. Through the archway, we can reach a large platform, on which various martial arts instruments are displayed, and under which offices, athletes' locker rooms and toilets are built. The platform is surrounded by cement artificial stone railings carved with patterns. Except for one side of the main entrance, the other seven sides are all stands, which can accommodate more than 5,400 spectators.

In recent years, with the exception of the ground track field and the martial arts field, all other venues have been controversially transformed into indoor venues.

看台局部及周边环境现状

奠基纪念铭文

主席台局部构造

田径场入口旧影

游泳池旧影

国术馆现状（组图）

篮球馆现状（组图）

游泳池现状（组图）

西安事变旧址

The Former Site of the Xi'an Incident

工程地点	西安市
建筑面积	3772平方米
竣工时间	1936年
设计机构	西北通济信托公司组织设计（张学良公馆）等

西安事变旧址包括张学良公馆、止园、新城黄楼、西京招待所、西安事变指挥部、高桂滋公馆、华清池五间厅、兵谏亭等8处。张学良将军公馆，始建于1935年秋，公馆以东、中、西三幢三层西式小楼为主体，附设传达室、卫士住房及西楼旁的中、西餐厅，总占地面积7703平方米，建筑面积3772平方米。公馆坐南朝北（现改为坐东朝西），四周有青砖花墙。三幢楼的外形、内部结构基本一致，小青瓦盖顶，仅在中楼的屋顶中部采用中式斗拱挑檐。

止园原为杨虎城将军的别墅，是一座坐北朝南的花园小楼，1934年动工，1936年初竣工。止园别墅现占地2331平方米，建筑面积约为1144平方米。初取名"紫园"，以

The former site of the Xi'an Incident include Zhang Xueliang's Mansion, Zhiyuan Villa, Xinheng Huanglou, Xijing Guesthouse, Xi'an Incident Command, Gao Guizi's Mansion, Huaqing Pool Five-Room Hall and Remonstration Pavilion. General Zhang Xueliang's Mansion, built in the autumn of 1935, mainly consists of three Western-style small buildings in the east, middle and west, in addition to a reception office, a guards' house and Chinese and Western restaurants beside the west building, covering a total land area of 7,703 square meters and having a floor area of 3,772 square meters. The residence faces north (now it is transformed to face west), surrounded by lattice walls of blue bricks. The three buildings share basically the same appearance and internal structure, with small blue roofing tiles covering the top, and only the middle part of the roof of the middle building is decorated with Chinese brackets and corniceo.

Zhiyuan Villa, formerly the villa of General Yang Hucheng, is a small garden building facing south. Commenced in 1934 and completed in early 1936, Zhiyuan Villa now covers a land area of 2,331 square meters with a floor area of about 1,144 square meters. At first, it was named "Ziyuan" (Purple Garden),

张学良公馆

张学良公馆局部（组图）

为紫气东来，后为了避蒋介石的猜忌，该园改名"止园"，其白底黑字，隽秀大方，醒目引人。进门后是花木繁盛、修筑茂林的小花园，石径从园中而过直达小楼前。小楼是二层一底的砖木混凝土结构建筑，风格是典型的中西合璧建筑。屋顶为琉璃瓦，顶部周围为复式彩绘斗拱挑檐。

新城黄楼，亦称"新城大楼"，因其通体黄色而得名，始建于1927年。全楼由一座中国宫殿式建筑和六座俄罗斯式亭子建筑联袂而成，浑然一体，风韵天成。

西京招待所始建于1935年10月，最初由中国旅行社负责经营管理，这在当时是西北地区唯一一所现代化酒店。整个招待所占地面积是6760.34平方米，其中主楼的建筑面积是2447平方米，由中央的三层八角形中楼、中楼后的一层八角大厅和二层的矩形东翼楼、南翼楼联袂而成。东翼楼和

meaning that purple air (auspicious omen) came from the east. Later, to prevent Chiang Kai-shek's suspicion, the garden was renamed "Zhiyuan" (Stop Garden). The name of the garden is written in black characters on white background, beautiful, elegant and conspicuous. Right behind the gate is a small garden where flowers, trees and bamboos flourish. A stone path passes through the garden, reaching the front of the small building. The small building is a brick-wood concrete structure with two floors and one basement, with a typical integration of Chinese and Western architecture styles. The roof is covered with glazed tiles, with the top surrounded by duplex painted cornices with brackets.

Xincheng Huanglou (Yellow Building), also known as "New City Building", was named after its yellow color. First built in 1927, the whole building is composed of a Chinese palace-style building and six Russian pavilion buildings, which are harmoniously and elegantly integrated.

Xijing Guesthouse, first built in October, 1935, was initially managed by China Travel Service, which was the only modern hotel in Northwest China then. The whole hotel covers a land area of 6,760.34 square meters. The main building, with a floor area of 2,447 square meters, is composed of three octagonal middle buildings in the center, a one-story octagonal hall behind the middle building, and two-story rectangular east and south wing buildings. There are 52 large

南翼楼的上下两层，大小房间总数为 52 间。为了减轻楼基承载力，整幢楼顶都用淡红色美国凹楞铁皮覆盖，代替了瓦片。在地下室还设有水卫暖管道及锅炉等设施，能为客房提供热水和暖气。当年客房的木门用的是南洋优质木材，虽历经 70 多年开启闭合，没有丝毫走样。客房内有会客室，顶上垂挂吊灯，摆放着茶几和圈椅。客房的套房是卧室，里面有绿色台灯以及宽大舒适的西洋床和床垫。卧室的墙壁上还有两扇小门，其后分别是衣橱、洗澡间和卫生间，卫生间的马桶和浴盆都是从意大利进口的。

西安事变指挥部，即杨虎城将军新城公馆，公馆面阔七间，进深五间，建筑面积 389 平方米。公馆周围为回廊，四周有院墙。

高桂滋公馆始建于 1933 年，占地面积 10 余亩，公馆除主楼外，还有位于主楼东侧的三座三进四合院，现已不复存在。主楼一共两层，占地面积 286.65 平方米。

西安事变旧址现为全国重点文物保护单位。1982 年，在张学良公馆建立西安事变纪念馆。

and small rooms in the upper and lower floors of the East Wing Building and the South Wing Building. To reduce the bearing weight of the building foundation, the whole roof of the building is covered with reddish American concave iron sheets instead of tiles. There are also water, sanitary and heating pipes and boilers in the basement, which can provide hot water and heating for guest rooms. The doors of the erstwhile guest rooms are made of high-quality wood from Southeast Asia. Opened and closed for more than 70 years, the doors are in the least deformed. There is a reception room in each guest room, with a chandelier hanging from the top and a tea table and round-backed armchairs. Inside is a bedroom with a green desk lamp, a spacious and comfortable Western bed and mattress. There are two small doors on a wall of the bedroom, inside which are respectively a wardrobe, a bathroom and toilet. The toilet and bathtub of the toilet were imported from Italy.

The Xi'an Incident Command, namely General Yang Hucheng's New Town Mansion, is seven-jian wide and five-jian deep, with a floor area of 389 square meters. The residence is surrounded by winding corridors and enclosed by walls.

Gao Guizi's Mansion was founded in 1933, covering a land area of more than 10 *mu*. Besides the main building, there used to be three ternary quadrangles on the east side of the main building, which is no longer extant. The main building has two floors and covers a land area of 286.65 square meters.

The Former Site of the Xi'an Incident is now key cultural heritage units under state protection. The Xi'an Incident Memorial Hall was established in Zhang Xueliang's Mansion in 1982.

止园

止园局部（组图）

保定陆军军官学校
Baoding Military Academy

保定陆军军官学校简称保定军校，创办于 1912 年河北省保定市，是中国近代史上第一所正规陆军军校，停办于 1923 年。校址前身为清朝北洋速成武备学堂、北洋陆军速成学堂、陆军军官学堂。中华民国成立后，北洋政府在原址上建立了"陆军军官学校"，习称保定军校，共办了 9 期，1923 年停办。学校位于保定旧城东北 2.5 千米，东西长 2 千米左右，南北长 1 千米左右，总面积约 1500 余亩。

此地原为关帝庙，后改为兵营，1900 年遭八国联军焚毁。

The Baoding Military Academy, founded in Baoding City, Hebei Province in 1912, is the first regular military academy in modern Chinese history, and was closed in 1923. The campus site was formerly known as the Beiyang Expedited Martial Studies Academy, the Beiyang Army Expedited Academy and the Army Officer School in the Qing Dynasty. After the founding of the Republic of China, the Beiyang government established the "Military Academy" on the original site, which was known as the Baoding Military Academy. It was run for nine terms. The school is located 2.5 kilometers northeast of Baoding Old City, about 2 kilometers long from east to west and about 1 kilometer long from south to north, covering a total land area of about 1,500 *mu*.

This place was originally a Guandi Temple, but was later turned into barracks, which were burned down by the Eight-Power Allied Forces in 1900. The Military

工程地点｜保定市
占地面积｜1500 余亩
竣工时间｜1912 年
设计机构｜北洋政府组织设计

主入口构造

内庭院建筑立面

建军校时，利用原庙产并征用邻近土地建起。全校分校本部、分校（包括小教场）、大操场和靶场四部分。校本部居中心，占地180余亩。其建筑格局系仿日本士官学校建成，中心操场正面及左右两侧，有序布置两组砖瓦结构的建筑群，四面有高大的围墙，墙外有河环护。军校大门在南侧居中，面阔三间，规制堪比直隶总督府大门。门前有石砌的高台阶，阶前道路直通河岸。路两旁有石狮一对，高丈许。河上架一平板桥，以通大操场。校本部分南北两院，北院是生活区。南院是军校的中枢和教学区，又分东、中、西三院。东、西院为教室与学生宿舍，各有十排带长廊的青砖瓦舍，布局对称，各排房舍之间有走廊相通，每两排组成一个独立的院落，院墙开月形门，每院住约一连学生，俗称一连道子。中院有校部办公室和尚武堂。尚武堂坐北朝南，四周环以石栏，雕梁画栋，气势宏伟。堂前有长廊直达校门。尚武堂北面是个大空院，因官长常在这里训话和发布命令，因此，这里被视为全校之中枢。校本部的东侧是分校，占地92亩。靶场在分校之北，占地330余亩。

2006年5月，保定陆军军官学校旧址被国务院批准成为第六批全国重点文物保护单位。

Academy was built by using the original temple property and requisitioning the adjacent land. The whole academy consisted of four parts: the main campus, the branch school (including the small training field), the big training ground and the shooting range. The main campus in the center covers a land area of more than 180 *mu*. Its architectural pattern is modeled after the Imperial Japanese Army Academy. On the front, left and right sides of the central training ground, two groups of buildings with brick and tile structures are arranged. The Academy is surrounded by tall walls and surrounded by a moat. The military academy gate is in the center of the south side, which is three-jian wide, comparable to that of Zhili Governor's Office. High stone steps are built in front of the gate, and the road before the steps leads directly to the river bank. There are a pair of stone lions about one-zhang tall on both sides of the road. A slab bridge is built on the river to connect with the big training ground. The main campus has two parts, the north court and the south court, with the former used as a living quarter.

The South Court, the center and teaching area of the military academy, is divided into the East, Middle and West Courtyards. The East and West Courtyards are classrooms and student dormitories, each with ten rows of blue brick houses with corridors, forming a symmetrical layout. Each row of houses is connected by corridors. Every two rows form an independent courtyard, and the walls of the courtyard have moon-shaped doors. Each quadrangle accommodates about a company of students, commonly known as a company of daozi. The Middle Courtyard has an academy office and the Shangwu Hall. The Shangwu Hall is a magnificent and richly ornamented building facing south and surrounded by stone railings. There is a promenade in front of the hall leading to the academy gate. There is a large empty courtyard to the north of the Shangwu Hall, which was regarded as the center of the whole school because the leaders often lectured and issued orders here. To the east of the main campus is a branch campus, covering a land area of 92 *mu*. The shooting range is located in the north of the branch campus, covering a land area of more than 330 *mu*.

In May 2006, the former site of the Baoding Military Academy was approved by the State Council to be listed in the sixth batch of national key cultural relics protection units.

纪念墙

环廊局部

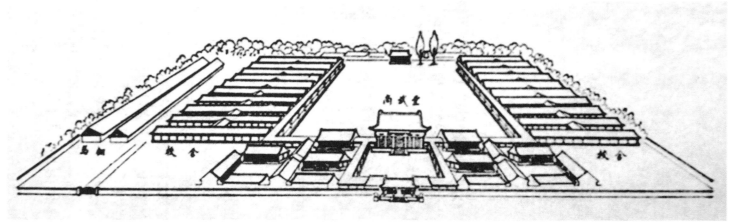
校区图纸

内庭院建筑侧影

内庭院环境

大邑刘氏庄园

Liu's Manor in Dayi

大邑县的安仁镇原是四川成都郊外的默默无闻的小镇，面积3.2平方公里。自清初从安徽徽州移民入川的刘氏家族迁居至此，安仁便产生了影响力。其中最有影响的是国民革命军24军军长刘文辉等。建筑群中的公馆是安仁镇的特色，发端于20世纪20年代后，公馆在民国时期专指达官显贵的府邸。经统计，安仁镇的公馆建筑数量有27座之多（同时期南京的公馆也只有26座）。名为老公馆建筑群即今天的刘氏庄园博物馆，包括刘氏祖居刘文成和刘文昭的刘氏兄弟公馆等。

刘文彩公馆是一个多重院落组成的四合院，1931年以来开始建设并扩建，形成了前院、逍遥宫、书房、花园、后罩院等。公馆前院是一个二进院落的四合院，后院南北为住房、贮藏室、佛堂等，其后为书房院，书房院之后是后罩院等。刘文彩公馆是最大的公馆，它与刘文成、刘文昭公馆都围绕祖屋兴建、当时面积最大的刘文彩公馆达6571平方米，有205间造型各异的房间，仅天井就有27个。公馆系一座不规则的

Anren Town in Dayi County was originally an unknown town outside Chengdu, Sichuan Province, with an area of 3.2 square kilometers until the Liu's family moved here from Huizhou, Anhui Province in the early Qing Dynasty, when Anren started to exert influence. The most influential figures were Liu Wenhui, commander of the 24th Army of the National Revolutionary Army. Mansions in the building complexes are features of Anren Town, which originated in the 1920s. Mansions in the Republic of China referred to the residences of high officials and dignitaries. According to statistics, there were as many as 27 mansions in Anren Town (there were only 26 mansions in Nanjing at the same time). The architectural complex named Lao Gongguan is today's Liu's Manor Museum, including Liu's ancestral home, and the mansions of Liu Wencheng and Liu Wenzhao.

Liu Wencai's Mansion is a quadrangle composed of multiple courtyards. Since 1931, it has been built and expanded to form a front yard, a liberty hall, a study, a garden, a back courtyard and so on. The front courtyard of the mansion is a quadrangle with two courtyards, bedrooms, storerooms and a Buddhist hall in the north and south of the backyard. Behind the quadrangle is a study courtyard, and behind the study courtyard is a rear courtyard. Liu Wencai's Mansion is the largest mansion, which was built around the ancestral home together with Liu Wencheng's Mansion and Liu Wenzhao's Mansion. Liu Wencai's Mansion, with the largest area of 6,571 square meters then, had 205 rooms of different shapes, and 27 patios. A mansion is an irregular closed modern architectural complex composed of

入口处局部

建筑局部（组图）

泥塑"收租院"（组图）

封闭式和多院落组成的近代建筑群，以刘文彩公馆建筑为代表。

整个安仁的公馆建筑都在细节上蕴含着中国传统的吉祥文化，如象征富贵的牡丹、象征长生的仙鹤、象征福禄的蝙蝠和鹿及象征文武双全的兵书宝剑等。再以1942年建成的刘文辉公馆为例，其面积8406平方米，由南北两个面积相等的独立三进院落组成。公馆大门设计成欧洲哥特式建筑风格，门

many courtyards, represented by Liu Wencai's Mansion.

All the mansions in Anren express traditional Chinese auspicious culture in their details, such as peonies symbolizing wealth, cranes symbolizing immortality, bats and deer symbolizing good fortune, and military works and swords symbolizing both civil and military affairs. Liu Wenhui's Mansion, for example, was built in 1942 with a land area of 8,406 square meters, consisting of two independent triple courtyards in the north and south with the same area. The gate of the mansion was designed in the Gothic architectural style of Europe, with the plaques inscribed in Chinese and Tibetan respectively. A large garden is used as a transition between the gate

建筑局部（组图）

匾分别用汉文和藏文题写，大门与前厅之间以一个大型花园作为过渡，花园中还各设置了一个网球场。刘元瑄公馆（刘文辉侄子）的隔壁，是安仁唯一的完全按西洋风格设计建造的建筑，在整个川西民居风格中标新立异。

大邑刘氏公馆建筑群灰塑的运用是其细部的表现。灰塑是以石灰为主要材料，在建筑物的墙壁、屋脊、门楼上创作形如圆雕和浮雕的工艺，它是对岭南传统建筑装饰工艺的借鉴，民国时期在四川民居中这种手法被广泛采用。灰塑工艺精细、立体感强、色彩丰富、题材广泛。迄今，公馆建筑中还有1965年四川美术学院集体创作的一组组泥塑群像《收租院》，体现了中国传统民间塑像手法与现代雕塑相结合的艺术典范。

and the front hall, and a tennis court is set in each garden. Next door to Liu Yuanxuan's Mansion (nephew of Liu Wenhui) is the only building in Anren designed and built completely in the Western style, unconventional among western Sichuan residential buildings.

The application of gray sculpture in the building complexes of Liu's Mansion in Dayi is a typical feature of architectural details. Gray sculpturing is a process of creating round carvings and reliefs on the walls, roofs and gatehouses of buildings with lime as the main material. This was learned from the traditional architectural decoration process in the Lingnan region, widely used in Sichuan dwellings during the Republic of China. The gray sculpturing process is exquisite, with a strong 3-D effect, rich colors and diversified subjects. Up to now, the mansions still have groups of clay sculpture figures "Rent Collection House" created by the Sichuan Academy of Fine Arts in 1965, which are artistic paragons of combining traditional Chinese folk sculpture with modern sculpture.

二院教学楼外观

湖南大学早期建筑群
The Early Building Complex of Hunan University

湖南大学位于长沙市湘江西岸之岳麓山东麓，是我国历史上最悠久的大学之一，迄今已有120余年的历史。与此同步，其校园建筑也在此百年历程中逐渐形成规模，并集中保留下一批20世纪初至50年代的建筑精品，其中多为我国近现代建筑名家之作。

湖南大学二院（今物理学院实验楼）为刘敦桢设计，1929年竣工。外立面是典型的折中主义建筑风格，正立面入口上部呈现古典工艺，而门廊柱子显然是现代风格；屋顶是西洋式的坡屋顶，却又适时采纳了一些中式曲线。

科学馆（今湖南大学办公楼），属折中主义建筑风格，由蔡泽奉设计，1937年竣工。建筑的东面、北面两座大门以及檐口、屋顶栏杆等部位都是典型的古典主义手法。此楼建成之后即遇抗战爆发，曾遭日军轰炸，但主体建筑未受大损失。1945年9月15日，长沙战区日军投降仪式在此楼中举行。

图书馆（已毁，仅存遗迹）曾是湖南大学历史上最早的一座西洋古典主义风格建筑，于1938年被日军飞机炸毁。今天湖南大学入口处矗立着的两根爱奥尼克石柱，是这座图书馆残存的遗物。

Hunan University, located at the eastern foot of Mount Yuelu on the west bank of the Xiangjiang River in Changsha City, is one of the oldest universities in China, boasting a history of more than 120 years. At the same time, campus buildings had been gradually developed over the course of one hundred years, leaving fine architectural works from the early 20th century to the 1950s, most of which were designed by famous modern architects in China. The Second Court of Hunan University (now the Laboratory Building of the School of Physics) was designed by Liu Dunzhen and completed in 1929. The facade is of a typical eclectic architectural style. The upper part of the facade entrance has been built with classical techniques, while the porch pillars are obviously modern. The roof is a Western-style sloping roof, but some Chinese curves have been aptly used.

The Science Building (now the office building of Hunan University) is a building of the eclectic architectural style, designed by Cai Zefeng and completed in 1937. The east and north gates, eaves, roof railings and other parts of the building have been built with typical classical techniques. Upon the completion of the building, the Anti-Japanese War broke out and the building was bombed by the Japanese army, but the main part did not suffer heavy losses. On September 15, 1945, the surrender ceremony of Japanese troops in Changsha was held in this building.

The Library (destroyed, only some remains are extant) was once the earliest building of the Western classical style in the history of Hunan University, which was blown up by a Japanese aircraft in 1938. The two Ionic pillars standing at the entrance of Hunan University today are the remnants of this library.

二院教学楼旧影

二院教学楼现状

科学馆立面

湖南大学工程馆（今湖南大学教学北楼），为现代主义建筑风格。由柳士英于1947年设计，1953年竣工。工程馆墙面上通长的水平线条以及楼梯间的圆弧形墙体、圆弧形窗檐等，具有典型的德国表现主义流动线条的造型特征。

学生一舍、二舍、四舍、七舍、九舍以及胜利宅教工宿舍属现代主义建筑风格。建于1945-1948年间，均为柳士英设计。每一座建筑都表达了设计者对现代主义建筑理念的理解。

老图书馆，为中国固有式建筑风格，1948年始建，1951年扩建，由柳士英设计。这是一座中西合璧的建筑，采用绿色琉璃瓦宫殿式大屋顶，正面墙上却是西方现代主义建筑手法，通贯数层的竖向长窗是典型的维也纳分离派的表现手法。

大礼堂，中国固有式建筑风格，1953年建，由柳士英设计。这座建筑是20世纪50年代"民族形式加社会主义内容"思潮的产物。

湖南大学建筑群的建筑风格正好与中国近现代建筑风格的发展演变同步，是中国近现代建筑史的缩影。

The Engineering Building of Hunan University (now the North Teaching Building of Hunan University) is a building of the modernist architectural style, designed by Liu Shiying in 1947 and completed in 1953. The long horizontal lines on the wall of the Engineering Building, the arc-shaped walls and window eaves of the stairwell are characterized by typical German expressionist flowing lines.

Students' first, second, fourth, seventh and ninth dormitories, as well as the Shenglizhai Teachers' Dormitory are buildings of the modernist architectural style, built between 1945 and 1948, and all designed by Liu Shiying. Each building expresses the designer's understanding of modernist architectural concepts.

The Old Library, a building of the Chinese inherent architectural style, was built in 1948 and expanded in 1951, also designed by Liu Shiying. This building is a perfect integration of Chinese and Western styles, covered with a palace-styled roof made of large green glazed tiles, but the front wall features a Western modernist architectural technique, and the vertical long window connecting several floors is a typical expressive technique of Vienna Separatists.

The Auditorium is a building with a Chinese inherent architectural style, built in 1953 and designed by Liu Shiying, and is a product of the ideological trend of "national form and socialist content" prevalent in the 1950s.

The architectural style of the Hunan University building complex, synchronized with the development and evolution of the modern Chinese architectural style, epitomizes the history of modern Chinese architecture.

中心位置的一组传统风格建筑

20世纪30年代的图书馆旧影

20世纪30年代的图书馆被炸毁后遗存石柱

一舍

七舍

现图书馆

现图书馆竖向长窗

现图书馆内阶梯

大礼堂立面

大礼堂瓦当

大礼堂侧面局部

工程馆西侧圆弧楼梯间

工程馆

北戴河近现代建筑群
Beidaihe Modern Architectural Complex

北戴河位于河北省秦皇岛市南部,晚清以来被开放建设为滨海避暑休闲胜地,建有数量繁多的近现代建筑群(含别墅类住宅、教会建筑、公共建筑、商务建筑等门类),尤以别墅区最为知名,与江西庐山、山东青岛和福建厦门并称为中国四大别墅区。

1900年以前的北戴河别墅建筑的设计与建造者已难考证。1901年以来,丁家立、胡佛等的先农公司,德国人魏迪锡、盖林的建筑事务所,美国人爱温斯与我国周志俊合办的平安公司等,都在北戴河建造了具有较高建筑艺术价值的别墅群。

北戴河别墅多为单层建筑,部分为二、三层建筑,大都以红顶、素墙、大阳台和整体的精致见长。门、窗、墙、顶、台阶等通过艺术造型和点缀,既显露出鲜明的异国格调,又凸现出迥异的建筑流派。鼎盛时期,北戴河有别墅建筑近千座,目

Located in the south of Qinhuangdao City, Hebei Province, Beidaihe has been developed as a coastal summer resort since the late Qing Dynasty. Beidaihe, with a large number of modern architectural complex (including villas, church buildings, public buildings, and business buildings, etc.), Beidaihe Villas in particular, are known as one of the four major villa groups in China together with Mount Lushan in Jiangxi Province, Qingdao in Shandong Province and Xiamen in Fujian Province.

It has been difficult to verify the designers and builders of Beidaihe villas before 1900. Since 1901, Tientsin Land Investment Co., Ltd of Tenney Charles Daniel and Herbert Hoover, the architectural firm of two Germans Weidixi and Gelin, Ping'an Company co-operated by American Evans and Chinese Zhijun Zhou have all built villas with high architectural and artistic value in Beidaihe.

Beidaihe villas are mostly single-story buildings, and only some are two, or three-story buildings. Most of them are celebrated for their red roofs, plain walls, large balconies and overall exquisiteness. Doors, windows, walls, roofs, steps, etc., with artistic modeling and embellishment, not only feature distinct exotic styles, but also highlight the styles of different architectural schools. During the peak period, nearly 1,000 villas were found in Beidaihe,

观音寺山门

蠖公亭内的朱启钤雕像

北戴河海滨的朱启钤雕像

蠖公桥

陶通伯别墅

陶通伯别墅曲折的外环廊

前保存完好的有130余幢近代建筑。章家楼（建于1925年，由奥地利工程师盖苓设计，毛泽东在此写下《浪淘沙·北戴河》）、何香凝别墅（建于1942年，建筑面积440.94平方米，原房主为日本人东金草燕）、东岭会教堂（建于1898年，建筑面积250平方米，纯美式的建筑，以花岗岩为墙，屋顶为灰绿色石片瓦）、瑞士小姐楼（建于1897年，二层欧式建筑）、马海德别墅（奥地利人白兰士建造，典型的四面廊式别墅）、五凤楼（并列五座前廊式欧风别墅，系周叔韬任启新洋灰公司总经理时为五个女儿建造）等为代表性遗存。

此外，1903年建造的标志着中国一段屈辱历史的"海关楼"，解读北方工业巨子周学熙心声的"趣园"，北戴河建设功臣朱启钤的"蠢天小筑"，少帅张学良作为行辕的张家大楼等，多层次记载着近百年来的晚清、民国的历史事件，也极具文化遗产价值。

more than 130 of which are currently in good condition. Representative remaining villas including Zhangjia Villa (built in 1925, designed by Austrian engineer Geyling, where Mao Zedong wrote a poem titled "Langtaosha Beidaihe"), He Xiangning Villa (built in 1942, with a floor area of 440.94 square meters, originally owned by Japanese Dongjin Caoyan), Dongling Church (a purely American building with granite walls and roof, built in 1898, with a floor area of 250 square meters), Swiss Miss Villa (a two-story European-style building built in 1897), Ma Haide Villa (a typical villa surrounded by corridors on four sides built by Austrian Brandeis), and Wufeng Villa (five adjoined front-porch European-style villas, built by Zhou Shutao for his five daughters when he was the general manager of Qixin Cement Company).

Additionally, some villas record the historical events of the late Qing Dynasty and the Republic of China in the past century from multiple levels, and are thus very valuable as cultural heritages, such as the "Customs Building" built in 1903, which marks a humiliating history in China, "Qu Garden" expressing the voice of Zhou Xuexi, an industrial magnate in North China, "Litian Xiaozhu" of Zhu Qiling, an official having rendered meritorious services for the development of Beidaihe, and the Zhang's Mansion, which was used as field headquarters by Marshal Zhang Xueliang.

陶通伯别墅外环廊及周边环境

何香凝别墅前廊

卡其别墅走廊

查克松别墅局部

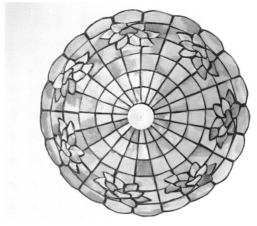
卡其别墅灯饰局部

瑞士小姐楼

卡其别墅

阿温太太别墅

国民党"一大"旧址(包括革命广场)
The Former Site of the First National Congress of the Kuomintang (including the Revolution Square)

国民党"一大"旧址位于广州市文明路215号,即原广东高等师范学堂钟楼(含楼前广场)。此地曾是清代广州贡院。1905年,两广总督岑春煊在贡院旧址上创建两广速成师范馆,1906年改为两广优级师范学堂,1912年改名为广东高等师范学校。

此建筑坐北朝南,面阔37米,进深51.1米,占地面积1890平方米,是一座"山"字形的中西合璧、仿罗马古典式建筑。楼高二层,中间部分加砌二层钟楼,长宽均为4.5米,加上穹隆顶共5层。总高24米的钟楼四面原设有时钟。正门是拱券形柱廊,廊宽4.9米,长4.9米,廊上有平台,廊

The former site of the First National Congress of the Kuomintang is located at No. 215 Wenming Road, Guangzhou, which used to be the bell tower of Guangdong Higher Normal School (including the square in front of the building). This place used to be Guangzhou Gongyuan (Examination Compound) in the Qing Dynasty. In 1905, Cen Chunxuan, Governor of Guangdong and Guangxi, founded the Expediated Normal School of Guangdong and Guangxi on the former site of the examination compound. In 1906, the school was renamed the Top Normal School of Guangdong and Guangxi, and in 1912, it was renamed Guangdong Higher Normal School.

This building faces south, 37 meters wide and 51.1 meters deep, covering an area of 1,890 square meters, and is a "E"-shaped classical building imitating the Roman style while integrating the Chinese and Western styles. The building is two stories high, above which is a two-story bell tower in the middle, which is 4.5 meters in both length and width. Counting in the dome top, it has a total

建筑外立面

室内展陈

建筑与环境

入口门廊构造细部

下是门厅，楼下四周是柱廊走道。底层是一个长方形礼堂（即国民党"一大"旧址），面积300多平方米。其外墙呈淡黄色，是当时岭南新派建筑的常用色调。作为岭南文化重镇的标志性建筑，钟楼后被用作中山大学校徽标志的主体设计图案。因孙中山先生于1924年1月在钟楼底层礼堂主持召开有共产党人参加的中国国民党第一次全国代表大会，因此被定为国民党"一大"旧址；又因鲁迅先生于1927年受聘为中山大学教务主任兼中文系主任时，也在此楼工作、生活，故同时也辟为广州鲁迅纪念馆。国民党"一大"旧址前面的广场，即原广东大学操场，是当时广州人民集会活动的重要场所之一，今被命名为革命广场。这里是第一次国共合作的诞生地，也是鲁迅1927年在中山大学任教时的工作生活场所，具有重大历史意义。

of five stories. The bell tower with a total height of 24 meters was originally equipped with clocks on four sides. The main entrance is an arched colonnade with a width of 4.9 meters and a length of 4.9 meters. There is a platform on the corridor, and a foyer under the corridor. Under the bell tower are colonnade walkways around. On the ground floor is a rectangular auditorium (the former site of the First National Congress of the Kuomintang) with an area of over 300 square meters. Its exterior wall is light yellow, a common color used by the Lingnan New Architectural Style back then. As a landmark building in the Lingnan region, the Bell Tower was later used as a main design pattern of the emblem of Sun Yat-sen University.

In January 1924, Dr. Sun Yat-sen presided over the first National Congress of the Chinese Kuomintang attended by Communists in the auditorium on the ground floor of the Bell Tower, so it has been designated as the former site of the First National Congress of the Kuomintang; Mr. Lu Xun also worked and lived in this building when he was appointed Dean of Academic Affairs and Head of Chinese Department of Sun Yat-sen University in 1927, so it has also been designated as the Lu Xun Memorial Hall in Guangzhou. The square in front of the former site of the First National Congress of the Kuomintang, namely the playground of Guangdong University, was one of the important venues for people's assembly activities in Guangzhou at that time, and is now named the Revolution Square. This site holds great historical significance because it was the birthplace of the first cooperation between the Kuomintang and the Communist Party of China, and also the working and living place of Lu Xun when he taught at Sun Yat-sen University in 1927.

门廊及室内局部（组图）

国民党"一大"旧址旧影

建筑立面外观

北京国会旧址

The Former Site of the Beijing Congress

北京国会旧址是中华民国成立后第一届国会的旧址，由德国建筑师罗克格（Curt Rothkegel）主持设计，现用作新华社礼堂。1912年4月，中华民国北京临时政府开始筹建国会，并选定原财政学堂为众议院建筑基址。众议院由东、西两部分组成，东部为财政学堂原有建筑与连廊合围而成的院落；西部为新建的众议院议场，通称"国会议场"。"国会议场"为方形建筑，坐北朝南，高三层，建筑外表朴素，灰砖砌成，至今保存完好。"国会议场"北侧为"圆楼"，同样为灰色砖墙，是当时国会办公楼，第二层为北洋政府总统和议长开会的地方。"圆楼"的外形则是方的，之所以叫"圆楼"，是因为楼内的北部为半圆门厅围合成的圆形会议厅。

The former site of Beijing Congress is the former site of the First Congress after the founding of the Republic of China. Its design was led by German architect Curt Rothkegel. It is now used as the auditorium of Xinhua News Agency. In April 1912, the Beijing Provisional Government of the Republic of China began to prepare for the Congress, and designated the former Finance School as the base of the House of Representatives, which consisted of two parts: the east and the west. The east part was a courtyard surrounded by the original buildings and corridors of the Finance School. The west part was the newly-built House of Representatives forum, commonly called the "Congress Forum", a square three-story building facing south. The building made of gray bricks with a plain appearance has been well preserved up to now. On the north side of the "Congress Forum" is the "Round Building", also with gray brick wall. It was the office building of the Congress at that time, and its second floor was the place where the President and Chairman of the Beiyang Government met. The "Round Building" is square, and it is so called because the northern part of the building is a round conference hall surrounded by a semicircular lobby.

建筑侧入口局部

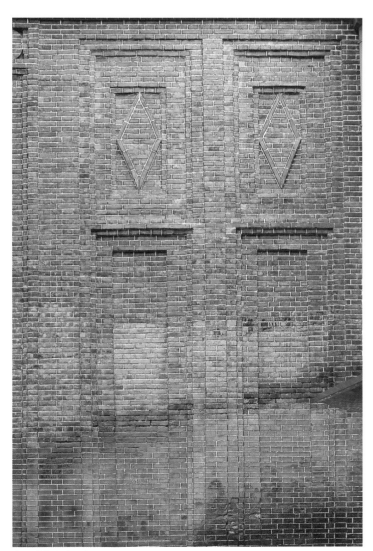
立面砖饰图案局部

正立面局部

建筑主入口

京张铁路南口段至八达岭段

The Beijing-Zhangjiakou Railway from the Nankou Section to the Badaling Section

工程地点 北京市　　全线长度 约20千米　　竣工时间 1909年　　主要设计人 詹天佑

京张铁路为詹天佑主持修建并负责的中国第一条铁路，1905年9月开工修建，于1909年建成，是中国首条不使用外国资金及人员，由中国人自行设计、投入运营的铁路。在技术方面，铁路关沟段采取了一系列创造性的措施，如利用青龙桥东沟的天然地形，采用"之"字形展线；又如关沟段共有4个隧道，其中八达岭隧道最长，为1091米，开凿该隧道时采用中间竖井法，加速成洞的速度等。在选线、设计方面，詹天佑不是单纯采取提高线路标准，增大工程量的办法，而是着眼于顺从自然，工、机配合的先进的选线设计基本原则，一方面顺从展线定坡，另一方面借重机车，以补不足，使京张铁路成为在当时情况下经济合理的铁路线。

The Beijing-Zhangjiakou Railway was the first railway designed, invested and operated by Chinese, not using any foreign fund or personnel. Presided over by Zhan Tianyou, the construction of the railway was commenced in September 1905 and completed in 1909. In terms of technology, a series of creative measures have been taken in the Guan'gou section of the railway, such as using the natural terrain of the East Valley of the Qinglong Bridge and adopting the zigzag line. Another example is that there are 4 tunnels in the Guan'gou section, among which the Badaling Tunnel is the longest with 1,091 meters in length. The middle shaft method was adopted when digging this tunnel to accelerate the speed of tunnel formation. In terms of route selection and design, Zhan Tianyou did not simply adopt the method of improving the railway standard and increasing the engineering quantity, but focused on the advanced basic principles of route selection design, which followed the natural terrain and coordinated engineering and machinery. On the one hand, the slope was set by following the extension line; on the other hand, locomotives were relied on to make up for the deficiency. In this way, the Beijing-Zhangjiakou Railway became an economical and reasonable railway line under the circumstances then.

建筑立面

詹天佑塑像

詹天佑之墓

远望建筑及铁路

线路规划图

"之"字形说明展示

齐鲁大学近现代建筑群
Modern Architectural Complex in Cheeloo University

原齐鲁大学校园即今山东大学趵突泉校区,始建于1911年。齐鲁大学时期建筑由卜道成筹划,美国工程师佩利姆(G. H. Perriam)设计,芝加哥三家公司负责建设。所有建筑均以欧美风格为主,并采用了大量中国传统民居建筑手法和符号,为折中主义建筑的代表,形成了特色鲜明的建筑文化。

主校园教学区南北轴线长达200多米,轴线北端为办公楼,南端为康穆礼拜堂,南区八条卵石铺成的道路呈放射形布置,为西方园林式布局。现存较完整的历史建筑可择要例举三项。

齐鲁大学校门,是1924年为庆祝建校60周年由校友捐资修建,又名"校友门"。门的造型采用中国传统的三间三叠牌楼的形式,大门牌楼主门顶较高,两侧门顶稍矮,形成一个"山"字形,而结构支撑则采用西方建筑技术。

柏根楼(今教学三楼),为纪念齐鲁大学校长柏尔根而命名,1917年建成,是校园中现存最老的建筑。该楼地上三层,

The former Cheeloo University campus, now the Baotuquan Campus of Shandong University, was first constructed in 1911. The buildings from the period of Cheeloo University were planned by Joseph Percy Bruce, designed by American engineer G. H. Perriam, and built by three Chicago companies. All buildings mainly followed European and American styles, while a large number of traditional Chinese residential architectural techniques and symbols were adopted. Thus, they have become representatives of eclectic architecture, developing a distinctive architectural culture.

The north-south axis of the teaching area on the main campus is more than 200 meters long, with the office building at the north end and Kumler Chapel at the south end. The eight pebble-paved paths in the south area are arranged in a radial way, a typical Western style of garden layout. The following are three examples of relatively complete historical buildings now extant.

The Gate of Cheeloo University, also known as "Alumni Gate", was built by alumni in 1924 to celebrate the 60th anniversary of the founding of the university. The shape of the gate is in a traditional Chinese form with three traditional triple arched buildings. The top of the main gate of the main entrance building is higher, and the tops of the two sides are slightly lower, forming a "E" shape, while Western architectural techniques are adopted for structural support.

建筑立面

旧影（组图）

地下一层。平面布局为西方近代建筑形式，中间为走廊，两面为房间。该楼为石墙基，灰砖清水墙体，歇山式、硬山式屋面灰瓦覆顶。入口处设单坡屋顶柱廊式门斗，垂花门罩。屋脊吻兽、山墙砖石雕刻为中西混合式样，门由石料仿木制成，并将古代彩绘雕刻成立体图案，是西方建筑师运用中国传统民居建筑手法、建筑符号来建造完全近代化的西方教学建筑的尝试。

考文楼（今教学五楼）为纪念登州文会馆创始人狄考文而命名，1919年建成，是一座中西合璧、庄重典雅的大体量教学楼。建筑风格、结构类似于教学三楼。大门由楼体凸出一个斜坡式前厦而成，两侧是高大宽厚的墙垛，门外两侧是中国传统的大型抱鼓石，屋顶为中国传统硬山顶，主楼一层和二层之间的砖雕上都雕着万字纹、寿字、中国结、菊花等图案；主楼两翼都有单层建筑作为附属向外延伸，使整体建筑错落有致、独具美感。

另有遗存如教学一、二、四、六楼，健康楼（圣保罗教堂），教授别墅区等，也各具特色，反映了自民国初期至共和国时代的历史变迁轨迹。

The Bergen Hall (currently No. 3 Teaching Building), named in memory of Bergen, former President of Cheeloo University, was built in 1917 and is the oldest building extant on campus. The building has three floors above ground and one floor below ground. The plane layout is a modern Western architectural form, with a corridor in the middle and rooms on both sides. The building features a stone wall base, walls of plain gray bricks, and saddle roofs covered with gray tiles. At the entrance is a colonnade foyer with a single slope roof and a festoon gate cover. The animal ornaments on the roof ridges and brick and stone carvings on the gables follow a mixed style of Chinese and Western. The gate is made of stones imitating the texture of wood, and ancient paintings are carved into 3D patterns. These are attempts made by Western architects to build a completely modern Western teaching building by using traditional Chinese residential building techniques and architectural symbols.

The Calvin Mateer Hall (now No. 5 Teaching Building) built in 1919 and named in memory of Calvin Wilson Mateer, founder of Dengzhou Wenhui College, is a large-scale, solemn and elegant teaching building perfectly integrating the Chinese and Western styles. The architectural style and structure are similar to No. 3 Teaching Building. The gate is formed by a sloping front room protruding from the building, with tall and wide buttress walls on both sides, large traditional Chinese drum stones on both sides outside the gate, and a traditional Chinese saddle roof without the overhang at the ends. The brick carvings between the first and second floors of the main building are engraved with swastika patterns and patterns of stylized character "shou" (longevity), Chinese knots, chrysanthemum patterns, among others. Both wings of the main building have single-story structures extending outward as annexes, rendering the whole architecture a uniquely well-proportioned appeal.

Other remains, such as No. 1, 2, 4 and 6 Buildings, the Health Building (St. Paul's Church), and the professor's villa area, also assume their own characteristics, reflecting the vicissitudes from the early Republic of China to the People's Republic of China.

齐鲁大学现状（组图）

东交民巷使馆建筑群
Embassy Architectural Complex in Dongjiaomin Lane

东交民巷使馆建筑群形成于1901—1912年，是一个集使馆、教堂、银行、俱乐部为一体的欧式风格街区。现存建筑有法国使馆、奥匈使馆、比利时使馆、日本公使馆和使馆、意大利使馆、英国使馆、花旗银行、东方汇理银行、俄华银行、正金银行、国际俱乐部和法国兵营等。现存建筑均保留原状，保持20世纪初欧美流行的折中主义风格，用清水砖砌出线脚和壁柱，采用砖拱券加外廊，木结构角檩架，铁皮坡顶。东交民巷使馆建筑群是北京仅存的20世纪初的西洋风格建筑群。

The Embassy Architectural Complex in Dongjiaomin Lane, developed from 1901 to 1912, is a European-style block integrating embassies, churches, banks and clubs. Extant buildings include the French embassy, the Austro-Hungarian Embassy, the Belgian Embassy, the Japanese Legation and Embassy, the Italian Embassy, the British Embassy, Citibank, Crédit Agricole Bank, Russia-China Bank, Specie Bank, International Club and French barracks, etc. Extant buildings are preserved as they were, retaining the eclectic style popular in Europe and America at the beginning of the 20th century, featuring plain bricks used to build architraves and pilasters, brick arches plus verandas, wooden angled purlin, and tin roofs. The embassy architectural complex in Dongjiaomin Lane is the only Western-style architectural complex in the early 20th century extant in Beijing.

日本横滨正金银行旧址

圣弥厄尔教堂

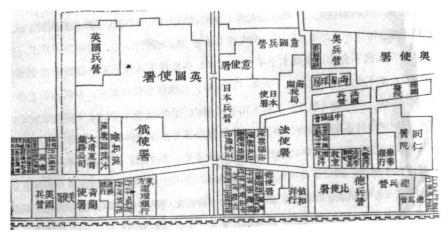

总平面图示意

花旗银行旧址

街景细部（组图）

法国邮政局旧址

庐山会议旧址

庐山会议旧址及庐山别墅建筑群
The Former Site of Lushan Conference and Lushan Villa Complex

庐山会议旧址位于江西省九江市庐山牯岭东谷长冲河畔、掷笔峰麓的火莲院。原名庐山大礼堂，为传习学舍会议礼堂。庐山会议旧址为国民党中央党部所建，耗资20万元大洋。当年国民党中央党部委托"华中公司"建筑工程师高观四对其进行规划与设计。1937年6月24日，"中央社"电称庐山大礼堂"为宫殿式，覆琉璃瓦，内分二层，上作膳厅，上为礼堂，可容数千百人，并可放映电影"。建筑门面带有中国宫殿风格的意蕴，门首的中国民族风格醒目地表现在蓝色的琉璃瓦的披檐上，同时檐下一楼中间也有三个圆拱形的大门，两旁均有八角形的窗子。而建筑的主体结构、装饰则采用西方建筑风格。大礼堂平面是由前部的横向矩形与礼堂部分的纵向矩形相结合而成，其占地面积约830平方米。新中国成立后这里改名为"庐山人民剧院"。

The former site of the Lushan Conference is located in the Huolian Courtyard at the foot of the Zhibi Peak, along the Changchong River in the Guling East Valley of Mount Lushan, Jiujiang City, Jiangxi Province. Formerly known as the Lushan Grand Hall, the site is the hall where the cadets study and attend meetings. The former site of the Lushan Conference was built by the Kuomintang Central Party Department at the cost of 200,000 silver dollars. The Central Party Department of the Kuomintang Party entrusted Gao Guansi, a construction engineer of Huazhong Company, to plan and design the academy. On June 24, 1937, the "Central News Agency" announced that the Lushan Auditorium had been built, which "is a building of the palace style, covered with glazed tiles and divided into two floors, with a dining hall on the ground and an auditorium on the top, able to accommodate thousands of people and play movies". The facade of the building has the implication of the Chinese palace style. The Chinese national style of the front gate is prominently manifested on the eaves made of blue glazed tiles. Meanwhile, there are three arched gates in the middle of the first gate under the eaves, with octagonal windows on both sides. On the other hand, the Western architectural style has been adopted for the main structure and decoration of the building. The plane of the auditorium consists of the

会场室内

建筑屋顶局部

庐山 359 号别墅

庐山 180 号别墅

庐山 175 号别墅

美庐别墅

雪后的牯岭别墅群

庐山别墅建筑群的建筑各自处于自然的随意状态，但整体上形成浑然一体的建筑环境。其建筑常设置庭院，庭院经过精心的绿化和美化，营造出深邃、宁谧的氛围。别墅室外常营建券廊，由券廊引导至主入口。别墅墙体大都由未打磨的不规则的粗石块砌筑，呈现出厚重朴实、质感强烈、色调沉着的美感。与别墅墙体那深褐色、灰色形成鲜明对比的别墅屋顶的色彩，称得上是浓墨重彩，或褐红，或青绿，或深蓝，形成独特而又动人的景色。庐山别墅大都有石头烟囱。石烟囱造型分为整块构件拼砌、不规则石头垒砌、卵石垒砌三大类。庐山别墅大部分有敞开式外廊，是连接别墅与大自然的一个纽带。外廊分石制与木制两大类，但外型上都采用了"拱券"，庐山别墅因此凸显出浓厚的欧洲建筑风格与情调。庐山别墅群的出现，是西方建筑文化的具象体现，又因是在庐山独特的地理环境中所出现的产物，因此创造出特有的错落有致的景观群落。

horizontal rectangle at the front and the vertical rectangle at the auditorium, covering a land area of about 830 square meters.

After the founding of the People's Republic of China, the site was renamed "Lushan People's Theatre".

The individual buildings of the Lushan Villa Complex, natural and random as they look, form an integrated architectural environment as a whole. The villas are often equipped with courtyards, carefully landscaped and beautified to create a deep and tranquil atmosphere. Outside a villa is often an arched corridor, leading to the main entrance. Most of the villa walls are built by rough and irregular stones, presenting the aesthetics of profound simplicity, striking texture and tonal equanimity. In sharp contrast with the dark brown and gray of the villa walls, the villa roofs stand out for their great diversity of striking colors, brownish red, turquoise, and deep blue, forming a uniquely touching scenery. Most villas in Mount Lushan have stone chimneys, which have three types of modeling: built with monolithic components, irregular stones or pebbles respectively. Most villas in Mount Lushan have an open veranda, serving as a link between the villa and the nature. Verandas fall into two categories: made of stones or wood. But "arches" are adopted for both types, which highlight the strong European architectural style and sentiment. The emergence of the Lushan Villa Complex, while embodying the Western architectural culture, has also created a unique landscape community because these villas have been produced in the unique geographical environment of Mount Lushan.

庐山 124 号别墅

马迭尔宾馆
Madie'er Hotel

工程地点 | 哈尔滨市
建筑面积 | 10803.77 平方米
竣工时间 | 1906 年
主要设计人 | 阿·勒·尤金洛夫（俄）

马迭尔宾馆位于中央大街中段的繁华街区，自 1906 年建成以来一直是哈尔滨最具代表性的建筑和文化符号，见证了这座城市百年兴衰变迁的风风雨雨。宾馆的平面功能十分复杂，自入口进入即为前厅和会客厅，北侧与过厅和餐厅相连，再向北走则是一座不大的舞厅。建筑北端的冷饮厅沿用至今，已成为了享誉全国的"马迭尔冰棍"的官方售卖窗口。会客厅东侧为主楼梯及剧场，南侧为南餐厅及各种配套辅助用房。沿阶梯拾级而上，走道两侧多有宾馆所保留下来的珍贵陈列，钟表、电话、铭牌等不一而足，均是在宾馆中使用过的物件。

Madie'er Hotel, located in a bustling block in the middle of Central Street, has been the most representative architectural and cultural symbol in Harbin since its completion in 1906, witnessing the ups and downs of the city in the past century. The hotel features very complex plane functions. Its entrance is the front hall and a reception room; the north side is connected with the lobby and the dining room; and in the further north, there is a small ballroom. The cold drink hall at the northern end of the building is still in use now, and has become the official sales window of the nation-wide famous "Madie'er Popsicle". On the east side of the reception room are the main stairs and the theater, and on the south side are the southern dining room and various auxiliary rooms. Up the stairs, there are many precious displays preserved by the hotel on both sides of the walkway, such as clocks, telephones and

从中央大街看马迭尔宾馆

斑驳的黄铜中倒映着岁月的沧桑，时光的流逝划下了真实的痕迹。建筑的立面颇具特色，浓重的色彩、拱形的窗棂、端庄的女墙和出挑严整的布局均给人留下深刻的印象。建筑转角处设计了阳台悬挑，由栱石托住，侧45度角的雕饰成为了建筑最具魅力的一面，结合顶部的穹窿造型更添建筑的富丽堂皇。

马迭尔宾馆于1987年恢复了历史原名，为俄语音译，意指现代的、时髦的，与modern同义。而在百年历程中，这座宾馆曾六易其名，每个名字的背后，都蕴含着深刻的历史，如中共中央东北局招待处、哈尔滨市革命委员会第二招待所等。现如今，宾馆保存有历史名人房间18间，记录着曾在这里下榻过的名人们所经历的点点滴滴。

nameplates, all of which have been used in the hotel. The mottled brass reflects the vicissitudes of time, and the passage of time leaves its marks. The facade of the building is uniquely decorated with rich colors, arched window frames and solemn parapet walls in a neat layout. At the corner of the building, an overhanging balcony has been designed, supported by an arched stone. The carving with a side angle of 45 degrees has become the most striking part of the building, enhancing the magnificence of the building together with the dome shape at the top.

In 1987, the original name of Madie'er Hotel was restored, which is a transliteration from Russian, synonymous with "modern" in English. In the course of one century, this hotel has changed its name six times, and each name implies a period of profound history, such as the Reception Office of Northeast Bureau of the CPC Central Committee and the Second Guesthouse of Harbin Revolutionary Committee. Nowadays, the hotel maintains 18 rooms that have received historical figures, recording the events of the figures who have stayed here.

马迭尔宾馆旧影

马迭尔宾馆旧影 2

沿街侧立面建筑及商铺

会议室室内

外立面局部

室内展陈（组图）

外立面局部（组图）

室内局部（组图）

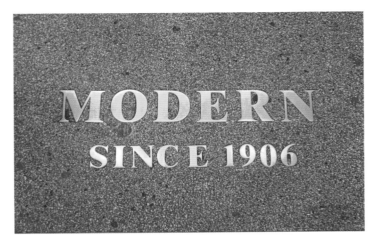

镶嵌于地面的铭牌（组图）

上海邮政总局
Shanghai General Post Office

工程地点	建筑面积	竣工时间	设计机构	主要设计人
上海市	25291平方米	1925年	英商思九生洋行	思九生

上海邮政总局坐落于中国上海市四川路桥北堍，门牌为虹口区北苏州路276号。大楼始建于1924年，造价为320万银元，英商思九生洋行设计，余洪记营造厂营建。邮政总局大楼建筑总面积为25291平方米，建筑高度为51.16米，整体建筑为"U"字形，共有地面建筑四层，地下室为一层，大小房间187间。大楼底层和二层为营业厅，负责收发国内和国际信件包裹，而三层为邮政总局的相关办公部门，顶层则为高级职员宿舍，地下室是包裹分拣封发部门，中庭天井为装卸往来邮件包裹的场地。

大楼拥有号称"远东第一大厅"的邮政营业厅，气势雄伟，曾经享有"远东第一大厅"的美名，现仍为上海市邮政局和四川路桥邮政支局所在地。

大楼整体风格为19世纪上半叶到20世纪初流行于欧美国

The Shanghai General Post Office is located at Qiaobeitu, Sichuan Road, Shanghai, China and its address is No.276, North Suzhou Road, Hongkou District. The building was built in 1924, with a cost of 3.2 million silver dollars. It was designed by a British company, Stewardson Company, and built by Yu Hongji Construction Factory. The building of the General Post Office has a total floor area of 25,291 square meters and is 51.16 meters high. The whole building is U-shaped with four floors above ground and one floor underground, having 187 large and small rooms. The first and second floors of the building are business halls, responsible for sending and receiving domestic and international letters and parcels. On the third floor are the relevant office departments of the General Post Office. On the top floor is the senior staff dormitory. In the basement is the parcel sorting and sealing department. The atrium patio is the place for loading and unloading incoming and outgoing mail parcels.

The building has a magnificent postal business hall known as "No. 1 Hall in the Far East", which is now still the location of the Shanghai Post Office and the Sichuan Road Bridge Post Sub-office.

The overall style of the building is the eclectic architectural style popular in European and American countries from the first half of the 19th century to the beginning of the 20th century. The main body has been designed with reference

建筑全景

外立面局部

入口上方构造局部

柱式细部

建筑外立面

家的折中主义建筑风格，主体参照英国古典建筑风格，融合了罗马式的大型科林斯立柱和巴洛克式钟楼。其中大楼主立面即南面和东面以及东北转角处共设有19根高数十米的仿罗马柱；此罗马柱系柯林斯柱式与爱奥尼柱式的组合，又将爱奥尼柱式之卷涡纹改造为本土化的回纹；主立面墙面使用细粒水刷石粉面，而北墙则是经典的机制红砖墙。邮政总局最具标志性之处，是东南角正门处顶层加建的钟楼。其中钟楼分二层，下层高13米，正面镶嵌有直径达3米的大钟，基座两边各有一座水刷石粉面的火炬台雕塑，此层又为二层塔式方亭的基座；方亭高17米，顶端另设置高8.2米的旗杆，基座两旁则各有一对希腊人物青铜雕塑群像。

to the British classical architectural style, while integrating the large Corinthian columns of the Roman style and the style of the Baroque bell tower. In the main facade of the building, that is, in the south and east and at the northeast corner, there are 19 Roman-style columns with a height of tens of meters. These Roman columns are a combination of Corinthian order and Ionic order, with the vortex pattern of the Ionic order transformed into a localized fretwork. The wall of the main facade is painted with fine-grained cement powder, while the north wall is a classic machine-made red brick wall. The most symbolic feature of the General Post Office is the bell tower built on the top floor at the main entrance of the southeast corner. The bell tower has two floors. The lower floor is 13 meters high, and the front is inlaid with a big clock with a diameter of 3 meters. There is a torch platform sculpture powdered with cement on both sides of the base and this floor is the base of the tower-styled square pavilion on the second floor. The square pavilion is 17 meters high, with a flagpole 8.2 meters high at the top and a pair of bronze sculptures of Greek figures on both sides of the pedestal.

室内局部（组图）

檐口及屋顶雕塑

四行仓库
Four-Bank Warehouse

工程地点	上海市
建筑面积	20000平方米
竣工时间	1935年
主要设计人	唐玉恩（复原改造设计）

这是一座钢筋混凝土结构的六层仓库建筑，位于上海原闸北区南部、苏州河北岸、西藏路桥西北角。它建成于1935年，占地0.3公顷，建筑面积20000平方米，屋宽64米，深54米，高25米。此地原是大陆银行和北四行（金城银行、中南银行、大陆银行及盐业银行）联合仓库，一般统称为"四行仓库"。1937年10月26日至11月1日，在这里发生的谢晋元团长率部英勇御敌的四行仓库保卫战（参战将士被称为"八百壮士"），重新振奋了因淞沪会战受挫而下降的中国军民的士气。此战还催生出著名抗战歌曲《中国不会亡》。

此建筑在战争中被重击受损，后曾简单修复、粉饰一新，2014年，上海对其重新进行整体性保护修缮，并将其约5000平方米的室内改建为"上海四行仓库抗战纪念馆"，与建筑西侧的纪念广场相互呼应，成为上海市民纪念抗战的

This is a six-story warehouse building of a reinforced concrete structure, located in the south of the former Zhabei District, the north bank of the Suzhou River and the northwest corner of Xizang Road Bridge. Built in 1935, the warehouse covers a land area of 0.3 hectares, with a floor area of 20,000 square meters and a house of 64 meters wide, 54 meters deep and 25 meters high. This place used to be the joint warehouse of Mainland Bank and four banks in the north (Jincheng Bank, Zhongnan Bank, Mainland Bank and Salt Bank), generally called the "Four-Bank Warehouse".

From Oct. 26 to Nov. 1, 1937, Xie Jinyuan, the regiment commander, led his troops to bravely defend the Four-Bank Warehouses against the enemy (the soldiers in the battle were called "eight hundred heroes"), which revived the morale of the Chinese soldiers and civilians defeated in the Battle of Songhu. This battle also inspired the famous song against Japanese aggression, "China Will Never Fall."

This building was severely damaged during the battle, and then it was simply repaired and whitewashed. In 2014, Shanghai carried out the overall protective renovation, converting its interior of about 5,000 square meters into the "Shanghai Four-Bank Warehouse Anti-Japanese Battle Memorial Hall", which echoes the memorial square on the west side of the building,

外立面及广场

四行仓库旧影（组图）

沿河立面

重要场所。

值得称道的是,这项修缮工程严格遵循"真实性""最小干预""整体性"和"可识别性"原则,对西立面、南立面、中庭、无梁楼盖结构体系等重点部位作复原性修缮和整治,最大限度还原了1937年淞沪会战期间的历史风貌。以西墙战斗遗迹保留区为例,修缮过程中采用红外热成像、摄影测量、定位剥除墙内面粉刷等方法探查原炮弹洞口遗迹,查明洞口位置实物印证,令人信服地展示了当年的战斗痕迹。可以说,这项修缮工程为今后同类文化遗产的保护与研究、展示等,提供了一份可资借鉴的先例。

forming an important place for Shanghai citizens to commemorate the War of Resistance against Japanese Aggression.

It is commendable that this restoration project has strictly followed the principles of "authenticity", "minimum intervention", "integrity" and "identifiability" to repair and rectify the key parts such as the west facade, the south facade, the atrium, and the beamless floor structure system, thus restoring the historical features during the Songhu Battle in 1937 to the greatest extent. With the West Wall Battle Remains Reserve, for example, infrared thermal imaging, photogrammetry, positioning and peeling off the old powder on the wall were used to explore the remains of the original bullet holes, identify the physical proofs of the hole positions, and convincingly present the battle traces in the past. This renovation project has arguably provided an excellent example to follow for future protection, research and exhibition of cultural heritages of the same type.

入口处

室内展陈（组图）

外立面局部

望海楼教堂
Wanghailou Church

工程地点	建筑面积	竣工时间	设计机构
天津市	812平方米	1904年（现存建筑）	法国教会组织设计

天津望海楼教堂旧称圣母得胜堂，位于天津市河北区狮子林大街西端北侧。此堂于1869年底建成，具有欧洲哥特式建筑风格，是天主教传入天津后建造的第一座教堂。此建筑于清同治八年（1869年）由法国天主教会建造，次年6月在"天津教案"中被烧毁；清光绪二十三年（1897年）在废墟原址重建，增建了角楼；1900年在义和团运动中再次被焚毁；现存望海楼为光绪三十年（1904年）用"庚子赔款"按原形制重建。

其建筑平面呈长方形，长55米，宽16米，建筑面积812平方米，占地3000平方米。堂身坐北面南，高10米，为石基砖木结构建筑，青砖墙面，面向西南，建筑正面有平顶塔楼三座，呈"山"字形，中间的塔楼最高，高12米，呈笔架式结构。除中间塔楼外大部分建筑为二层。在两侧塔楼的顶部还各镶有8个兽头，专门为下雨时排水所用，雨水可以从兽头口中涌出。教堂内部并列庭柱两排，无隔间与隔层，入口两侧设有扶壁，内部有三道通廊，中廊稍高，侧廊次之，属巴西利卡型。中塔楼正厅东西侧各有八根圆柱，支撑拱形大顶，北端为圣母玛利亚的主祭台，左右分别是耶稣和鞠养

The Wanghailou Church in Tianjin, formerly known as Shengmu Desheng Tang, is located on the north side of the west end of Shizilin Street, Hebei District, Tianjin. Built at the end of 1869, this church is characterized by a European Gothic architectural style and is the first church built after Catholicism was introduced into Tianjin. This church was built by the French Catholic Church in the eighth year of the Tongzhi Reign in the Qing Dynasty (1869), and was burned in the "Tianjin Missionary Case" in June of the following year. In the twenty-third year of Guangxu in the Qing Dynasty (1897), the church was rebuilt at the original site of the ruins, with a turret added. The church was burned again in the Boxer Rebellion in 1900. The Wanghailou Church extant now was rebuilt according to its original form with the Boxer Indemnity in the 30th year of the Guangxu Reign (1904).

Its building plane is in a rectangular shape, 55 meters long and 16 meters wide, with a floor area of 812 square meters and covering a land area of 3,000 square meters. Facing south and rising 10 meters high, the building is a stone-based brick-wood structure with blue brick walls, facing southwest. There are three flat-topped towers on the front of the building, forming a "E" shape. The tower in the middle, rising 12 meters high, is the highest, which is of a pen rack structure. Except for the middle tower, most structures have two floors. At the top of the towers on both sides, there are also 8 animal heads, specially used for drainage when it rains, as rainwater can gush out from the animal mouths. Inside the church, there are two rows of parallel court columns, without compartments or barriers. There are buttresses on both sides of the entrance, and three corridors inside, with the middle corridor slightly higher than the side corridors as a Basilica style. There are eight columns on the east and west sides of the main hall of the middle

建筑旧影（组图）

建筑正立面

像。另外，教堂四周的墙壁上悬挂着耶稣受难图，整个建筑的顶和壁都设有彩绘。教堂内窗券为尖顶拱形，窗面由五彩玻璃组成几何图案，上面嵌有耶稣遇难的故事。内部地面砌有瓷质黑白相间花砖。

望海楼教堂既是建筑史上值得研究的建筑实例，又是一处爱国主义教育基地，于1988年被公布为第三批全国重点文物保护单位。

tower, which support the arch top. The northern end is the main altar of Virgin Mary, and the left and right sides are the statues of Jesus and his mother respectively. In addition, there are pictures of the crucifixion on the walls around the church, and the top and walls of the whole building are painted with colorful pictures. The window arches in the church are spire arches, and the window surface is made of colorful glass of geometric patterns, embedded with the story of Jesus' crucifixion. The interior floor is paved with black and white porcelain tiles.

The Wanghailou Church is not only an example worth studying in the history of architecture, but also a base for patriotic education. It was announced to be listed in the third group of key cultural heritage units under state protection in 1988.

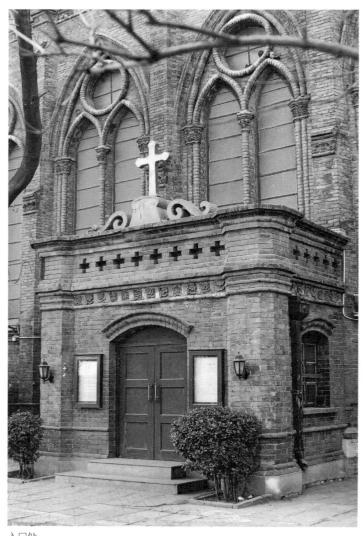

入口处

建筑立面局部

建筑旧影

窗饰构造及屋顶局部（组图）

建筑与环境

长春电影制片厂早期建筑
Early Buildings in Changchun Film Studio

长春电影制片厂是新中国第一家电影制片厂，堪称新中国电影的摇篮，先后拍摄故事影片900多部，译制各国影片1000多部。这个中外闻名的制片厂是在1937年伪满时期日本"株式会社满洲映画协会"原址基础上建立的。

早在1923年，日本满铁就设立了"映画班"（后来成为"满铁弘报课映画制造厂"），拍摄了大量满铁在东北的活动和当时中国东北的风土人情。伪满时期，日伪于1933年8月成立"满洲映画国策研究会"；1936年7月，起草《满洲国映画对策树立方案》，正式开始了"满映"的组建；1937年8月，由"满洲国政府"和"满铁"各占一半股分，正式成立。抗战胜利后，中共东北局于1945年9月成立了"东北电影工作者联盟"，率先从日本人手中复制了"株式会社满洲映画协会"，成立"东北电影公司"。后几经周折，"东

The Changchun Film Studio, the first film studio in the People's Republic of China, is arguably the cradle of the film industry of the PRC, having produced more than 900 feature films and translated more than 1,000 films from various countries. This studio widely acclaimed at home and abroad was established on the basis of the former site of the Manchukuo Film Association of Japan in the Puppet Manchuria Period in 1937.

As early as in 1923, Japan's Manchurian Railway set up a "Movie Program" (later turned into the "Movie Studio of the Manchurian Railway Propaganda Department"), filming a large number of activities of South Manchuria Railway Co. (SMR) in Northeast China and the local customs and practices of Northeast China back then. During the Period of the Manchuria Puppet Regime, the Japanese Puppet Regime established the "Manchuria National Policy Research Association for Movies" in August 1933. In July 1936, the "Plan for Establishing Manchukuo Movie Countermeasures" was drafted. Then the establishment of "Manchu Movie Studio" was officially launched. In August 1937, with the Manchukuo Government and SMR holding 50% shares each, the studio was formally established. After the victory of the War of Resistance against Japanese Aggression, the Northeast Bureau of the Communist Party of China established the "Northeast Film Workers' Union" in September 1945, which copied the

北电影公司"最终更名为"长春电影制片厂（简称长影）"。"株式会社满洲映画协会（简称"满映"）"位于伪满新京洪熙街（今红旗街），为院落式建筑群，于1937年10月兴建，1939年11月落成。在满映建筑群占地16万平方米的大院内，建有一栋办公楼、六座面积为600平方米的摄影棚，以及录音室、洗印间、道具加工厂等，按照德国乌发电影制片厂的建筑形式和布局设计。建成后的满映，因为场地宽阔、设备精良，成为当时亚洲最大的电影制作企业，其制片技术达到同期世界领先水平，号称远东最大最先进的电影制片厂。1949年后，长影在苏联专家的协助下，又建造了第七摄影棚，面积达1200平方米，是当时亚洲最大的摄影棚之一。

此外，建于1938年的长影招待所，时称"小白楼"，原系伪满治安部（后改为伪军事部）大臣于琛澂的别墅，建筑风格具有有明显的日本风格，亦是有一定艺术价值的历史建筑。小白楼分为上下两层，一层面积为616平方米，有半圆形花厅，共大小7间居室。楼上面积为320平方米，有大小房间9个。楼内房间结构各异，自成格局。楼后有一排平房，设有厨房、餐厅、卫生间及三大间居室。

长影老厂区正面主楼现已改建为长影旧址博物馆，已于2014年8月19日向大众开放参观，主要介绍新中国电影的发展、创业史。

"Manchukuo Film Association" of the Japanese to initiate the "Northeast Film Company". Going through many twists and turns, "Northeast Film Company" was finally renamed as "Changchun Film Studio".

The "Manchukuo Film Association", located in Hongxi Street (now Hongqi Street), of the Puppet Manchuria Regime, is a courtyard-style architectural complex, of which the construction was commenced in October 1937 and completed in November 1939. There is an office building, six studios with an area of 600 square meters, as well as a recording studio, a film processing room, and a props processing factory, etc. in the compound with an area of 160,000 square meters, designed according to the architectural form and layout of the Universum Film A.G in Germany. After its completion, the "Manchukuo Film Association" became the largest film production enterprise in Asia then because of its extensive space and excellent facilities, with its production technology reaching the world leading level of the same period, claiming to be the largest and most advanced film studio in the Far East. After 1949, with the help of Soviet experts, the Changchun Film Studio (Changying) built the seventh studio, covering an area of 1,200 square meters, which was one of the largest studios in Asia at that time.

In addition, the Changying Guest House, built in 1938, known as "Little White Villa" then, was originally a villa of Yu Chencheng, minister of the Puppet Manchurian Security Department (later changed to the Puppet Manchurian Military Department). With an obvious Japanese architectural style, the building is also a historical building with a certain artistic value. Little White Villa is divided into upper and lower floors. The first floor covers an area of 616 square meters, with a semi-circular parlor and 7 rooms. The second floor covers an area of 320 square meters, with 9 large and small rooms. Rooms in the building are of different structures with unique patterns. There is a row of bungalows behind the building, including a kitchen, a dining room, toilets and three large rooms.

The main building in the front of the former area of the Changchun Film Studio (Changying) has now been rebuilt into the Changying Former Site Museum, opened to the public from August 19, 2014, mainly introducing the history of development and pioneering of the film industry of the People's Republic of China.

建筑入口处

奠基石

入口及外立面

室内局部（组图）

室内局部

建筑与广场

798 近现代建筑群
798 Modern Architectural Complex

798近现代建筑群原为国家"一五"时期建设的北京华北无线电联合器材厂（718厂），是156项重点建设工程之一。当时同属社会主义国家的民主德国帮助中国完成了从基本建设到设备安装的全部设计、建造工作。1964年，718厂撤销，成立706、707、797、798等厂。2000年后，闲置的厂房被陆续租用。此处建筑群占地60多万平方米，设计时采用了典型的包豪斯风格，实用、简洁、灵活多样，讲究功能、技术和经济效益，适应现代大工业生产和生活的需要。随着城市发展和人们生活方式的改变，原有的工业厂房已变为更适合当代城市生活的场所，使用者在对原有的历史文化遗存进行保护的前提下，将原有的工业厂房进行了重新定义、设计和改造，带来了对建筑和生活方式的创造性的理解。

The 798 Modern Architectural Complex, originally the Beijing North China Wireless Joint Equipment Factory (Joint Factory 718), was built as one of the 156 key construction projects in the First Five-Year Plan of China. The German Democratic Republic, which was also a socialist country then, helped China complete all the design and construction work of the Complex from basic construction to equipment installation. In 1964, the Joint Factory 718 was abolished and the Factories 706, 707, 797 and 798 were established instead. Since 2000, idle factory buildings have been rented one after another. The building complex here covering a land area of more than 600,000 square meters has been designed in a typical Bauhaus style, practical, concise, flexible and diverse, with attention focused on functions, technology and economic benefits, to meet the needs of modern industrial production and life. With the development of the city and the change of people's life style, the original industrial buildings have been renovated to suit the contemporary urban life. On the premise of protecting the original historical and cultural relics, the former industrial buildings have been redefined, designed and transformed with a creative understanding of architecture and life style.

工程地点	占地面积	竣工时间	设计机构	主要设计人
北京市	约60万平方米	20世纪50年代	德国魏玛包豪斯设计机构	施耐德（德）等

锯齿状厂房侧立面

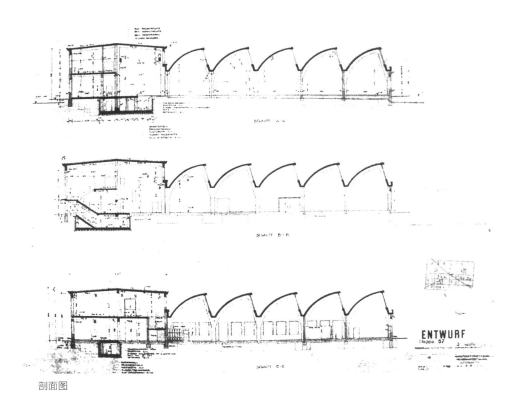

剖面图

建筑局部（组图）

建筑与环境

建筑局部（组图）

225

铁人第一口井说明

松基三井

大庆油田工业建筑群
Daqing Oilfield Industrial Architectural Complex

工程地点：大庆市
竣工时间：1959年
设计机构：松辽石油勘探局组织设计

大庆油田是我国目前最大的油田，也是世界上为数不多的特大型砂岩油田之一，因 1959 年 9 月在大同镇松基三井喷出工业油流，取为新中国十周年庆贺之意而得名"大庆"。油田现存主要建筑包括松基三井、"铁人第一口井"及铁人学院体验式培训基地、大庆油田北二注水站、西水源、铁人王进喜纪念馆等。

松基三井位于大庆市大同区高台子镇永跃村旁，于 1958 年勘探。它是大庆长垣构造带上的第一口油井，也是大庆油田的发现井。此井作为新中国现代重要的工业遗产得以永久保存，并在井附近增建展示性大型浮雕以及陈列室。

铁人第一口井位于大庆市红岗区解放二街八号，为铁人王进喜与队友们以人拉肩扛运钻机、破冰取水保开钻的方式，在物资极度缺乏的情况下用 5 天零 4 个小时打完的井，创造了当时世界石油钻井速度的最快纪录，因而作为重要的

The Daqing Oilfield, the largest oilfield in China at present, is also one of the few super-large sandstone oilfields in the world. It was named "Daqing" (meaning Grand Anniversary) because the industrial oil flew in Songji No.3 well in Datong Town in September 1959 when Chinese people were about to celebrate the 10th anniversary of the People's Republic of China.

The extant main buildings in the oilfield include No.3 Songji Well, "Iron Man No. 1 Well", experiential training base of the Iron Man College, No. 2 Northern Water Injection Station of The Daqing Oilfield, Western Water Sources, and Iron Man Wang Jinxi Memorial Hall, etc.

No.3 Songji Well is located near Yongyue Village, Gaotaizi Town, Datong District, Daqing City, for which exploration started in 1958. As the first oil well in the Daqing placanticline structural belt, it was also the discovery well in the Daqing Oilfield. As an important industrial heritage in modern China, this well has been permanently preserved, and a large-scale relief for display and an exhibition room have been built near the well.

Iron Man No. 1 Well is located at No.8 Jiefang Second Street, Honggang District, Daqing City. It was drilled by Iron Man Wang Jinxi and his teammates, who carried drilling rigs on their own and broke ice to keep the drilling hole open in 5 days and 4 hours despite the extreme shortage of materials, making the fastest record of oil drilling speed in the world at that time. Therefore, it has been

工业遗产得以保留。该井场仍然保留着当年的卸车坑、泥浆池、土油池、值班房和工人当年住过的地窖子等遗址与实物。井场西侧立有贝乌-40型钻机实物，与王进喜曾经使用过的钻机型号相同。

北二注水站隶属于大庆油田第一采油厂第二油矿北八采油队，1962年4月1日建成投产，是大庆油田岗位责任制的发源地。这里现存管理泵房、排涝站、锅炉房、采油一厂传统教育展览室。截至考察之日（2019年8月20日），已累计安全生产20918天。北二注水站现存完整的厂房及各种沿用时间很久的相关设备，其厂房为砖混结构，人字坡瓦顶，平面布置为曲尺形，立面造型简洁，表现出时代特色，是典型的20世纪工业建筑风格。

西水源位于大庆市让胡路区喇嘛甸镇，现存大庆第一口水井井址、原厂区办公室、文化长廊、职工技能培训场、通行值班室、水源投产纪念石、铭言牌等。这里保留的一栋建于1963年的干打垒，是大庆油田进行传统教育的生动教材，是大庆油田现存不多的历史遗迹之一，参观者络绎不绝。

铁人王进喜纪念馆是为纪念中国工人阶级的先锋战士王进

retained as an important industrial heritage. The well field still retains former sites and objects such as the unloading pit, the mud pit, the soil oil pool, the duty room and the cellar where workers lived then. On the west side of the well field, there is a Beiwu-40 drilling rig, the same type as the drilling rig used by Wang Jinxi.

No. 2 Northern Water Injection Station belongs to No. 8 Northern Oil Production Team of No.2 Oil Mine of No.1 Oil Production Plant of Daqing Oilfield, which was completed and put into operation on April 1, 1962, and also considered the birthplace of the post responsibility system of Daqing Oilfield. Sites still extant now include the management pump room, the drainage station, the boiler room and the traditional education exhibition room of No.1 Oil Production Plant. Up to the date of inspection (August 20, 2019), 20,918 days of safe production have been recorded. No. 2 Northern Water Injection Station still retains an intact factory building and all kinds of related equipment used for a long time. Its factory building is of a brick-concrete structure, with a gable slope and a tile roof, and a curved plane layout. Its facade is simple and concise, displaying the characteristics of its time as a typical industrial architectural style in the 20th century.

Western Water Source (Xishuiyuan) located in Lamadian Town, Ranghulu District, Daqing City retains the first well site in Daqing, the original factory office, the cultural corridor, the staff skills training field, the traffic duty room, the memorial stone for water production, and the inscription board. An adobe house built in 1963 is a vivid teaching material for traditional education in the Daqing Oilfield and one of the few extant historical sites in the Daqing Oilfield, attracting numerous visitors.

The Iron Man Wang Jinxi's Memorial Hall was built in 1971 in memory of Wang

浮雕及广场

浮雕组群

喜于1971年建成的。纪念馆原址在大庆市解放二街8号，2003年2月，由大庆地区石油石化企业、中共大庆市委、市政府共同协商，中共黑龙江省委、省政府支持，中国石油天然气集团公司批准，决定迁建铁人王进喜纪念馆。新馆由沈阳建筑大学建筑设计研究院设计，2006年9月26日大庆油田发现47周年纪念日开馆。时任中共中央政治局常委、国务院总理温家宝题写馆名。

铁人王进喜纪念馆占地11.6公顷，主体建筑面积2.15万平方米，展厅总面积4790平方米，展线总长度917延米。主体建筑外形为"工人"二字组合，鸟瞰呈"工"字形，侧看为"人"字形，象征这是一座工人纪念馆。主体建筑高47米，正门台阶47级，寓意铁人王进喜47年不平凡的人生历程。建筑顶部为钻头造型，象征大庆人奋发向上、积极进取的精神。

Jinxi, a pioneer soldier of the Chinese working class. The memorial hall was originally located at No.8 Jiefang Second Street, Daqing City. In February 2003, it was decided to relocate the Iron Man Wang Jinxi Memorial Hall through consultation between petroleum and petrochemical enterprises in the Daqing area, the Daqing Municipal CPC Committee and the Municipal Government, with the support of the Heilongjiang Provincial CPC Committee and Provincial Government, and with the approval of China National Petroleum Corporation. The new building was designed by the Architectural Design and Research Institute of Shenyang Jianzhu University and opened on September 26, 2006, the 47th anniversary of the discovery of the Daqing Oilfield. Wen Jiabao, then member of the Standing Committee of the Political Bureau of the CPC Central Committee and Premier of the State Council, inscribed the name of the memorial hall.

The Iron Man Wang Jinxi Memorial Hall covers an area of 11.6 hectares, with a main floor area of 21,500 square meters, a total exhibition hall area of 4,790 square meters and a total exhibition line length of 917 linear meters. The shape of the main building is a combination of two Chinese characters "工人", with a bird's eye view showing "工" and a side view showing "人", symbolizing that it is a memorial hall for workers. The main building is 47 meters high with 47 steps at the main entrance, which symbolizes the extraordinary 47-year life course of Iron Man Wang Jinxi. The top of the building is shaped like a drill bit, symbolizing Daqing people's pioneering and enterprising spirit.

"铁人"雕塑

油田旧影

钻机现状

忠烈祠外立面

国殇墓园

National Martyrs' Memorial Cemetery

腾冲国殇墓园位于云南腾冲西南 1 千米的叠水河畔小团坡下，占地 80 余亩，1945 年 7 月 7 日落成。1944 年夏，为了完成打通中缅公路的战略计划，策应密支那驻印军作战，中国远征军第二十集团军向占据腾冲达两年之久的侵华日军发起反攻，经历大小战斗 80 余次，于 9 月 14 日收复腾冲城，敌酋藏重康美大佐联队长及其下 6000 余人全部被歼，我军亦阵亡少将团长李颐、覃子斌等将士 8000 余人，地方武装阵亡官兵 1000 余人，盟军（美）阵亡将士 19 名。此役系第二次世界大战期间，中国远征军取得的重大战果。为纪念攻克腾冲而殉国的第二十集团军阵亡将士（含盟军阵亡官兵），辛亥革命元老李根源倡导修建此墓园，是全国建立最早的国军抗日烈士陵园之一。

国殇墓园整组建筑群由纪念广场、墓园大门、忠烈祠（抗日

The Tengchong National Martyrs' Memorial Cemetery, completed on July 7, 1945, is located at the foot of the Xiaotuan Slope along the Dieshui River, kilometers southwest of Tengchong, Yunnan Province, covering an area of more than 80 mu. In the summer of 1944, to complete the strategic plan of opening the China-Myanmar Highway and coordinate the operations of the Myitkyina Army in India, the 20th Army of the Chinese Expeditionary Force launched a counterattack against the Japanese invaders who occupied Tengchong for two years, and braved more than 80 battles, recovering Tengchong City on September 14. Kurashige Yasumi Osa, the captain of the Japanese troops and more than 6,000 soldiers were all killed. Our army also lost more than 8,000 generals and soldiers such as Major Generals Li Wei and Qin Zibin. The local armed forces lost over 1,000 officers and soldiers. The ally army (from the United States) lost 10 officers and soldiers. This campaign was a great success achieved by the Chinese Expeditionary Force during the Second World War. To commemorate the officers and soldiers of the 20th Army killed in the campaign (including the officers and soldiers of the Allied Forces killed) in the recovery of Tengchong, Li Genyuan, a veteran of the Revolution of 1911, advocated the construction of this cemetery, which was one of the cemeteries in memorial of martyrs of the national army

忠烈祠牌匾

入口围墙

英烈纪念堂）、英烈墓冢等几三个部分构成。以小团坡为起点，在其东北向的中轴线上，建有攻克腾冲阵亡将士纪念塔、腾冲战区抗日烈士墓、滇国殇墓园西抗战盟军阵亡将士公墓及纪念碑等。主体建筑为中轴对称、台阶递进形式，由大门经长通道循石级而上至第一台阶，再循石级而上，至嵌有蒋中正题李根源书之"碧血千秋"刻石的第二级台阶挡土墙，沿墙分两侧上至第二台阶，即忠烈祠所在地。

这座墓园在建筑风格上既沿袭了传统民居、殿堂的形式（如忠烈祠大殿、院墙等），也吸收了西式建筑影响（如方尖碑式的纪念碑），成功营造出庄严肃穆的纪念氛围。

1996年，国殇墓园被国务院列入全国重点文物保护单位。

resisting the Japanese aggression in China.

The whole complex of buildings in the National Martyrs' Memorial Cemetery consists of the memorial square, the cemetery gate, the martyrs' shrine (Memorial Hall of Martyrs in the War of Resistance against Japanese Aggression) and the heroes' tombs. With the Xiaotuan Slope as the starting point, on its northeast central axis, there are structures such as the Tower in Memorial of Generals and Soldiers Martyred in the Recovering of Tengchong, the Tombs of Martyrs Resisting Japanese Aggression in the Tengchong War Zone, the Tombs of Fallen Soldiers of Allied Force in the War against Japanese Aggression and the Monument, etc. The main structures are in the form of axial symmetry with progressive steps, from the gate to the first step through the long passage, and then to the second level of the retaining wall embedded with a carved stone of "Bi Xue Qian Qiu" (meaning "Loyal Martyrs Live Forever") composed by Chiang Kai-shek and inscribed by Li Genyuan. From two sides along the wall, visitors can go up to the second level, where the Martyrs' Temple is located.

In architectural style, this cemetery not only follows the forms of traditional dwellings and halls (such as the hall and walls of the Martyrs' Temple, etc.), but also absorbs the influence of Western architecture (such as the obelisk-style monument), thus successfully creating a solemn commemorative atmosphere.

In 1996, the National Martyrs' Memorial Cemetery was listed by the State Council as a key cultural heritage unit under state protection.

墓园局部（组图）

纪念碑

蒋氏故居

Chiang Kai-shek's Former Residence

工程地点	宁波市
建筑面积	600平方米
竣工时间	20世纪20年代末

蒋氏故居，位于浙江省宁波市奉化区溪口镇，昔日蒋氏家族在此地生活、工作、娱乐等。

蒋氏故居系群体建筑，主要包括玉泰盐铺、丰镐房、小洋房等建筑。玉泰盐铺乃蒋介石出生之地，前后两进，前进为三间一弄楼房，后进为为平房，占地716平方米，建筑面积600平方米。清末以来，玉泰盐铺曾两次失火，现存建筑系1946年所建。

丰镐房：镐（hào）为镐京，西周的国都，在今陕西省长安之西北。从上辈"周房"及蒋介石在宗谱中属"周字辈"，以及西周两位帝王的都城）——丰邑和镐京，推断出房名可能是在都城名中，各取第一字定为房名，而分给蒋介石的祖房故即名为"丰镐房"。丰镐两字，丰，代表蒋介石一房；镐，代表其亡弟蒋瑞青一房。瑞青早死，由蒋介石兼祧承袭，故称丰镐房。丰镐房，占地4800平方米，建筑面积共1850平方米，为清代建筑，其余都系蒋介石1929年扩建而成。

Chiang's Former Residence, located in Xikou Town, Fenghua District, Ningbo City, Zhejiang Province was the place where the Chiang's clan lived, worked and played.

Chiang's Former Residence consists of community buildings, mainly including the Yutai Salt Shop, the Feng Hao House, a small Western-styled house and other buildings. The Yutai Salt Shop, the place where Chiang Kai-shek was born, has two parts (front and back), with the front consisting of a three-jian building and one lane, while the back consisting of a bungalow, covering a land area of 716 square meters with a floor area of 600 square meters. Since the late Qing Dynasty, the Yutai Salt Shop has caught fire twice, and the extant structures were built in 1946.

Feng Hao House: Hao means Haojing, the capital of the Western Zhou Dynasty, located in the northwest of current Chang'an, Shaanxi Province. From a previous generation's house named "Zhou Fang" and Chiang Kai-shek's belonging to the "Zhou Zi" Generation in the genealogy, as well as Fengyi and Haojing, the capitals of the two emperors in the Western Zhou Dynasty, it can be inferred that the house might be named by taking the first character of the two capital names. The ancestral house assigned to Chiang Kai-shek is called "Feng Hao House". Of the two characters Feng and Hao, Feng represents Chiang Kai-shek's branch, while Hao represents his deceased brother Chiang Ruiqing's branch. As Ruiqing had died, Chiang Kai-shek inherited the whole

文昌阁外立面

丰镐房室内

建筑立面

小洋房为三间二层楼房，西式，前临剡溪，后依武山，占地240平方米，建筑面积共310平方米，建于1930年。蒋经国留苏回奉，偕妻蒋方良、儿子蒋孝文居此。建筑布局为传统的前厅后堂，两厢四堂格局。楼轩相接，廊庑回环，墨柱赭壁，富丽堂皇。前厅及左右有三个花园。厅堂廊庑内布满雕刻彩画。报本堂有蒋介石亲书的"报本尊亲是谓道德要道""光前裕后所望孝子顺孙"对联和"寓理帅气"匾额，是为私祝蒋经国40生日而立。

house, thus it was called Feng Hao House. The Feng Hao House, covering an area of 4,800 square meters and a total floor area of 1,850 square meters, was built in the Qing Dynasty. The rest were expanded by Chiang Kai-shek in 1929. The Little Western-Styled House built in 1930. is a three-jian two-story building in the Western style, facing the Shanxi Stream and backed by Mount Wushan, covering an area of 240 square meters with a total floor area of 310 square meters. Chiang Ching-kuo stayed here with his wife Chiang Fang-liang and son Chiang Hsiao-wen after he returned to Fenghua following his studies in the Soviet Union. The architectural layout is characterized by traditional front and back hall, with two wings and four halls. Buildings and pavilions are connected with looped corridors, black columns and reddish brown walls, resplendent and magnificent. There are three gardens by the front hall and on the left and right. The halls and corridors are full of colorful carved paintings. In the Baoben Hall there is a couplet and a stele written by Chiang Kai-shek and dedicated to his son Chiang Ching-kuo's 40th birthday. The couplet means, "The most important moral principle is to requite and follow one's ancestors and respect one's parents, and I wish that you would glorify your forefathers and enrich your posterity and have filial descendants." The stele means "Loft Ideal and Constant Self-improvement."

小洋房立面

小洋房室内

小洋房屋顶晒台

建筑组照

沿街立面

天津利顺德饭店旧址
The Former Site of Lishunde Hotel in Tianjin

利顺德饭店位于天津市和平区解放北路，一侧紧临美丽的海河，另一侧与中英花园式风格的维多利亚公园（后称市委公园），形成和谐的建筑环境，是原天津租界区现存极为珍贵的19世纪中叶建筑之一。该饭店始建于清同治二年（1863年），由英国基督教牧师殷森德创建，是中国历史上第一家外资饭店。建筑初为平房，被称之为"泥屋"，殷森德于1886年集资将"泥屋"改建成一座具有英式古典风格的三层豪华饭店，室内安装有中国最早的电灯、电话、电梯和数字统计机。他按照自己姓氏的中文译音，把饭店命名为"利顺德饭店"，并巧妙的把孟子名言"利顺以德"寓意其中，成为华北地区最负盛名的西式酒店。此饭店在1976年唐山大地震时有所损伤，2010年按文物修缮的严格规程予以修复。1996年被列为第四批全国重点文物保护单位。

Lishunde Hotel located in Jiefang North Road, Heping District, Tianjin, is one of the most precious buildings extant from the former Tianjin concession area in the middle of the 19th century. The hotel forms a harmonious architectural environment, with one side close to the beautiful Haihe River, and the other side adjacent to the Victoria Park featuring a Chinese and English integrated garden style (later called the Municipal Committee Park). Built in the second year of the Tongzhi Reign in the Qing Dynasty (1863), the hotel was founded by British Christian pastor John Innocent as the first foreign-funded hotel in Chinese history. The building was originally a bungalow, called the "Mud House". In 1886, John Innocent raised funds to transform the "Mud House" into a three-story luxury hotel with a classical English style, equipped with the earliest electric lights, telephones, elevators and digital counting machines in China. According to the Chinese transliteration of his surname Innocent, he named the hotel "Lishunde Hotel", while cleverly embedding Mencius' famous saying "Li Shun Yi De" (benefits are based on virtues), thus becoming the most famous Western-style hotel in North China. This hotel was slightly damaged during the Tangshan earthquake in 1976, and it was repaired in 2010 in strict accordance with the regulations on cultural relics repair. In 1996, it was listed in the fourth group of key cultural heritage sites

利顺德饭店建成后成为中国外交活动和政治活动的重要场所。英国、美国、加拿大、日本等国先后将各自的领事馆设在饭店内。《中国丹麦条约》《中国荷兰条约》《中葡天津通商条约》《中法简明条约》等在此签订。孙中山、黄兴、宋教仁、张学良、溥仪、蔡锷、梁启超、袁世凯、段祺瑞等均曾在此驻留，美国前总统胡佛年轻时也曾留宿于此。这里留下许多历史足迹，也留下了众多不同时期的文物，如宋庆龄弹奏过的钢琴、藏传佛教的稀世金佛、意大利文艺复兴时期的雕花长椅等。

under state protection.

After its completion, Lishunde Hotel became an important venue for China's diplomatic and political activities. The United Kingdom, the United States, Canada, Japan and other countries have set up their consulates in the hotel. The China-Denmark Treaty, the China-Netherlands Treaty, the China-Portugal Tianjin Trade Treaty and the China-France Concise Treaty were signed here. Sun Yat-sen, Huang Xing, Song Jiaoren, Zhang Xueliang, Puyi, Cai E, Liang Qichao, Yuan Shikai and Duan Qirui all stayed here once, and former US President Hoover also stayed here when he was young. Many historical footprints and cultural relics from different periods have been left here, such as the piano played by Soong Ching-ling, the rare golden Buddha of Tibetan Buddhism, and a carved bench of the Italian Renaissance.

建筑室内局部（组图）

沿街立面旧影

室内阶梯

建筑内老电梯

入口立面局部

室内走廊（组图）

建筑旧影（组图）

客房室内

餐厅室内

京师女子师范学堂
Jingshi Women's Normal School

工程地点 北京市 | 竣工时间 1909年 | 设计机构 黄瑞麟建议设立

京师女子师范学堂旧址位于北京市西城区新文化街45号，是清末民国初一处仿西方风格学校建筑，现为北京市鲁迅中学。清光绪三十四年（1908年），清廷学部决定在石驸马大街（今新文化街）斗公府旧址建筑校园，设立京师女子师范学堂，次年建成。是一组由四座楼组成的合院式校舍，教学楼面积4300平方米，礼堂建筑面积220平方米。

建筑整体坐北朝南，大门由四座砖砌立柱和铸铁栅栏组成，立柱为三段式结构，上下均为须弥座式，装饰砖雕花卉，上部装饰西式花球。南部楼房外立面由砖砌立柱和拱形门窗组成，正中为拱形大门，砖雕拱券，砖拱上部正中安放拱心石，立柱由青砖砌成凹凸的线条，窗户仍由拱形砖雕组成。楼房内立面为柱式连廊，上部为绿色挂檐板，二楼装有栏杆，廊内为教室，门窗仍为拱形。四面楼样式相同，门窗均为砖砌弧拱，内侧为连廊。礼堂为方形二层楼，铁皮瓦屋面，周围有廊柱。

中华民国成立后，京师女子师范学堂改建为北京女子师范学校，1924年又改为北京女子师范大学。鲁迅曾于1923—1926年在此执教。1926年"三一八惨案"后，于1931年在校内树立一座高约两米的"三一八遇难烈士刘和珍、杨德群纪念碑"。碑座由两块方形石基相叠而成，上为方尖碑形制。

京师女子师范学堂主体结构基本保持原貌，2006年5月被国务院定为第六批全国重点文物保护单位。

The former site of the Jingshi Women's Normal School located at No.45, Xinwenhua Street, Xicheng District, Beijing, is a campus imitating the Western style in the late Qing Dynasty and the early Republic of China, and is now the site of Lu Xun Middle School in Beijing. In the thirty-fourth year of the Guangxu Reign in the Qing Dynasty (1908), the Qing court decided to build a campus at the former site of Dougong Mansion in Shifuma Street (now Xinwenhua Street), and to establish the Jingshi Women's Normal School, which was completed the following year. It is a courtyard-style school architectural complex consisting of four buildings, including a teaching building with a floor area of 4,300 square meters and an auditorium with a floor area of 220 square meters.

The whole building complex faces south, and its gate is composed of four brick columns and cast-iron fences. The columns are of a three-stage structure, with a Sumeru seat at both the top and bottom, decorated with brick carved flowers and Western-style flower petal balls at the top. The facade of the southern building is composed of brick columns and arched doors and windows, with a gate in the middle topped by a carved brick arch. An arched stone is placed in the middle of the upper part of the brick arch. The columns are made of blue bricks with concave and convex lines, while the windows are also decorated with arched brick carvings. The inner facade of the building is a colonnade vestibule, the upper part being a green eave board, the second floor equipped with railings. Inside the vestibule are classrooms, with doors and windows still arched. The buildings on the four sides are of the same style, with brick arches for doors and windows and a vestibule inside. The auditorium is a square two-story building with a tin tile roof and colonnades around it.

After the founding of the Republic of China, the Jingshi Women's Normal School was transformed into Beijing Women's Normal School, and then changed into Beijing Women's Normal University in 1924. Lu Xun taught here from 1923 to 1926. After the "March 18th Massacre" in 1926, a "Monument to Liu Hezhen and Yang Dequn, Martyrs in the March 18th Massacre" about two meters high was erected in the school in 1931. The stele base is formed by stacking two square stone foundations, with the top shaped as an obelisk.

The original appearance of the main structure of Jingshi Women's Normal School has basically been maintained, which was listed by the State Council in the sixth group of key cultural heritage sites under state protection in May 2006.

入口立面

南开学校旧址

Former Site of Nankai School

南开学校位于天津市南开区四马路20—22号，始建于清光绪三十年（1904年），是我国近代教育家严范孙、张伯苓共同创办的私立中学。其前身为严氏家馆，1904年底易名为"私立敬业中学堂"；1905年终学校再次更名为"私立第一中学堂；1907年正月，学校由严宅迁入坐落在"南开洼"的校舍（欧式建筑二层灰砖楼房），同年秋，校名改称"私立南开中学堂"。次年增设高等师范科；1912年4月改校名为"私立南开学校"。

南开学校旧址占地3.3万平方米，建筑有东楼（1906年）、

Nankai School located at No.20-22 Sima Road, Nankai District, Tianjin, first built in the 30th year of the Guangxu Reign in the Qing Dynasty (1904), was a private middle school jointly founded by Yan Fansun and Zhang Boling, modern educators in China. Its predecessor was Yan's family school, which was renamed as "Private Jingye Middle School" at the end of 1904. At the end of 1905, the school was renamed "Private No.1 Middle School". In the first month of 1907 on the lunar calendar, the school moved from the Yan's Residence to the campus (a two-story grey brick building of the European architectural style) located in Nankai Wa. In the autumn of the same year, the school name was changed to "Private Nankai Middle School". The next year, the higher normal department was added. In April 1912, it was renamed "Private Nankai School".

The former site of Nankai School covers a land area of 33,000 square

伯苓楼沿街立面旧影

伯苓楼沿街立面效果图

伯苓楼门廊

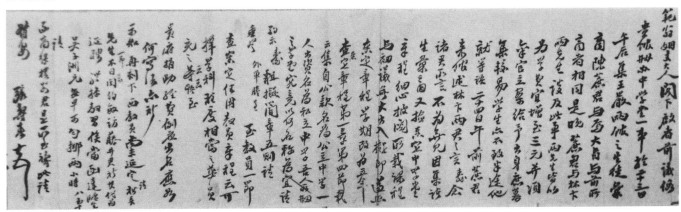

张伯苓先生手迹

范孙楼（1929年）、瑞廷礼堂（1936年）、物理实验楼等。东楼（即东教学楼，四马路20号）是其中建造最早也是最主要的建筑，总面积为952平方米。该建筑为二层砖木结构的楼房，坡屋顶，外立面以青砖镶嵌红砖饰面。建筑首层主入口突出并设有开方窗，建筑二层设有连续罗马式拱券窗和简化的爱奥尼克柱拱券。该建筑形体简单，但细部处理繁复，为典型的仿罗马式古典建筑风格，但又具有中国民族装饰的特点。南开学校其他的建筑也大体沿袭这一风格。

南开学校培养了许多在中国近现代史上有杰出贡献的各类精英，如周恩来同志即曾于1913—1917年在该校就读。

南开学校旧址于1996年被列为第四批全国重点文物保护单位。

meters. The buildings include the East Building (1906), the Fansun Building (1929), the Ruiting Auditorium (1936), and the Physics Laboratory Building, among others. The East Building (the East Teaching Building at No.20 Sima Road) is the earliest and most important building, with a total area of 952 square meters. The building is a two-story brick-and-wood structure with a sloping roof, and a facade decorated with blue bricks inlaid with red bricks. The main entrance of the first floor of the building protrudes and has square windows, while the second floor of the building has continuous Roman-style arched windows and simplified Ionic column arches. The building simple in shape, features complicated architectural details. It not only is of a typical Roman-like classical architectural style, but also has the characteristics of Chinese national decoration. Other buildings in Nankai School also generally follow this style.

Nankai School has cultivated many elites who have made outstanding contributions in modern Chinese history. For example, Comrade Zhou Enlai studied in the school from 1913 to 1917.

The former site of Nankai School was listed in the fourth group of key cultural heritage sites under state protection in 1996.

室内走廊与阶梯（组图）

伯苓楼窗饰及入口（组图）

伯苓楼侧立面

广州白云宾馆
Guangzhou Baiyun Hotel

工程地点	广州市
建筑面积	7800平方米
竣工时间	1976年
主要设计人	莫伯治

手绘剖透视图

白云宾馆1976年建成，高120米，共34层（包括地下室一层），有客房718间，为当时中国第一高楼。

设计遵循"适用、经济、在可能条件下注意美观"的建筑方针，广泛征集方案，多次审议，综合了各个方案的优点，不断改善设计，保证了设计质量。建筑师充分考虑自然环境与建筑的关系，为保留选址中的一座小山丘，特意将建筑后退至距道路200多米，山丘上的大树也被保留下来，成为建筑的自然景观。宾馆的中庭，利用原有的三颗古榕树，再通过瀑布、景石、水池，形成一个典雅的空间。

设计因地制宜，土洋结合，处理了高层建筑中一些具体问题。如，为了简化结构系统，在70米长的主楼中不设伸缩缝，解决了基础地质有软硬层的困难。在国内民用建筑中，首次采用钻（冲）孔桩基础；为了加强抗震的潜力，采用变截面钢筋混凝土剪力墙板结构；在外墙逐层挑出悬臂横板，使外墙的维修、清洁、防雨和遮阳等问题能够简易解决；采用外露光源或现成的代用品，避免使用昂贵的灯具。采用预制梯级饰面，简化多层梯级饰面的施工。

白云宾馆是岭南派建筑风格融入现代建筑的一个典范。

Built in 1976, Baiyun Hotel was the tallest building in China then, rising 120 meters high, with 34 floors (including the basement) and 718 guest rooms.
In the design process of the hotel, in accordance with the "applicable, economic, and aesthetic if possible" architectural principle, design schemes were widely collected and reviewed for many times, and the design quality was guaranteed by integrating the advantages of each scheme and continuously improving the design. The architect fully considered the relationship between the natural environment and the building. To preserve a small hill at the selected site, the building was retreated to more than 200 meters away from the road, and the big trees on the hill have also been preserved, becoming a natural landscape in front of the building. The atrium of the hotel forms an elegant space by using the original three ancient banyan trees, and adding a waterfall, landscape stones, and a pool.
The design is adapted to local conditions with an integrated local and foreign style. Some specific problems in high-rises have also been resolved. For example, to simplify the structural system, expansion joints are avoided in the 70-meter-long main building, addressing the difficulty of having soft and hard geological layers in the foundation. The drilled (punched) pile foundation was used for the first time in domestic civil buildings; to strengthen the anti-seismic potential, the variable cross-section reinforced concrete shear wall plate structure was adopted; cantilevered transverse plates were set on the external wall, so that the problems of maintenance, cleaning, rain protection and sunshade of the external wall could be easily solved; exposed light sources or ready-made substitutes were adopted to avoid using expensive lamps. Prefabricated cascade finishes were adopted to simplify the construction of multi-layer cascade finishes.
Baiyun Hotel is a paragon of blending the Lingnan architectural style into modern buildings.

建筑外立面

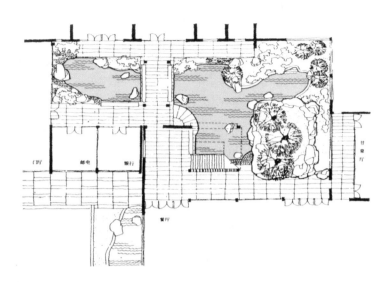

花园平面图

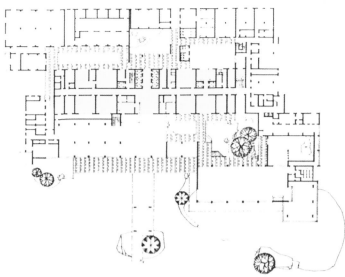

一层平面图

花园手绘速写

花园实景旧影

建筑外立面 2

旅顺火车站
Lvshun Railway Station

旅顺火车站位于辽宁省大连市旅顺口区，始建于1898年9月，是中俄联合修建的原东清铁路（中东铁路）南线终点站，也是东北地区最南端的铁路车站。1904年日俄战争后，建成现在的这个站舍，是中东铁路南支线南端的终点站，当时为三等站。中东铁路的修建，最初是军事上的需要，钢轨铺设一直延伸到旅顺的大连海湾附近，故火车站设在军港码头附近。满铁时期一度称之为"旅顺驿"。

这座建筑由俄藉工程师吉尔什曼负责施工。建筑为木质结构，砖石墙面，体量不大而比例适度，造型简洁明快，具有俄罗斯木构建筑特色。其平面呈"一"字形，立面在入口处加设

The Lvshun Railway Station, located in Lvshunkou District, Dalian City, Liaoning Province, was built in September 1898 as the southern terminal of the former Dongqing Railway (Chinese Eastern Railway) co-built by China and Russia and also the southernmost railway station in Northeast China. This station built after the Russo-Japanese War in 1904 was the terminal at the southern end of the South Branch of the Chinese Eastern Railway, and was a third-class station then. The Chinese Eastern Railway was constructed originally out of military need, and the steel rail was extended to the vicinity of the Dalian Bay in Lvshun. As a result, the railway station was located near the Army Harbor Wharf. During the South Manchuria Railway (SMR) period, the station was once called "Lvshun Post".

The building constructed with the supervision of Russian engineer Girshman. The building is a concise and bright wooden structure with masonry walls, small in size and moderate in proportion, characterized by the Russian

铁轨与火车站月台

门斗，中间顶部为绿色穹顶塔楼，加强了中央体积向上的动感。塔楼屋顶铺砌鱼鳞状铁皮瓦片，显然融合了俄罗斯式教堂的穹顶做法。站台上的木结构风雨棚构架合理，比例优美，与站舍浑然一体。

旅顺站是全国重点文物保护单位中东铁路的重要分项，同时被列为大连市第一批重点保护建筑物。

wooden architectural style. Its plane is " 一 "-shaped; the facade has a foyer at the entrance; and the top of the middle part is a green dome tower, which enhances the upward dynamism of the central volume. The roof of the tower is paved with fish-scale iron tiles, obviously drawing on the use of domes of Russian churches. The wooden canopy on the platform with a reasonable structure and elegant proportion is harmoniously integrated with the station building.

The Lvshun Station is an important sub-item of Chinese Eastern Railway, a key cultural heritage site under state protection, and meanwhile has been listed in the first group of key protected buildings in the city of Dalian.

铭牌

站台屋顶局部

火车站旧影

建筑铁路一侧立面

售票处入口

建筑立面局部

站台雨棚构造细部

屋顶细部

上海中山故居
Sun Yat-sen's Former Residence in Shanghai

工程地点 上海市 | 建筑面积 452平方米 | 竣工时间 20世纪初

上海中山故居位于香山路7号，现已辟为上海孙中山故居纪念馆。故居建于20世纪初，为两层欧式楼房，坐北朝南，占地面积1013平方米，建筑面积452平方米。外墙饰以灰色卵石，屋顶铺盖洋红色鸡心瓦。楼下是客厅、餐厅，楼上是书房、卧室、一个中型会客室和一个室内大阳台。楼的东面是汽车间。室内的陈设都按照宋庆龄的回忆布置，而且绝大部分是原物。

会客厅陈放着一只三人沙发和一对单人沙发。圆形红木茶几上放着镶银边的紫檀木雪茄烟盒和搪瓷烟灰缸。会客厅墙边放着六把红木椅子，四个墙角放着红木高茶几，茶几上陈列着铜质套盘。会客厅内陈列孙中山1912年就任中华民国临时大总统、孙中山夫妇与"永丰舰"官兵合影（1923年8月）等珍贵历史照片。

Sun Yat-sen's Former Residence in Shanghai at No.7 Xiangshan Road has been turned into the memorial hall of the former residence of Sun in Shanghai. Built in the early 20th century, the former residence is a two-story European-style building facing south, covering a land area of 1,013 square meters with a floor area of 452 square meters. The exterior wall is decorated with grey pebbles, while the roof is covered with magenta heart-shaped tiles. Downstairs there is a living room and a dining room, while upstairs there is a study, a bedroom, a medium-sized reception room and a large indoor balcony. To the east of the building is the garage. Indoor furnishings are arranged according to Soong Ching-ling's memories, and most of them are original objects.

There is a sofa for three and a pair of chair sofas in the living room. On the round mahogany tea table, there is a rosewood cigar box and an enamel ashtray. There are six mahogany chairs by the wall of the living room, and mahogany high tea tables with copper sets in four corners. Precious historic photos, such as Sun Yat-sen's appointment as the provisional president of the Republic of China in 1912, Sun Yat-sen and his wife taking a group photo with the officers and soldiers of Yongfeng Ship (August 1923), are displayed in the living room.

外立面

建筑转角处构造局部

屋面材质细部

建筑与环境的关系

餐厅与会客厅相通，中间可用两扇移门隔开。餐厅正中放着一张广东式样的红木餐桌椅。餐厅内陈列有孙中山就任中华民国军政府海陆军大元帅时使用的大元帅指挥刀等珍贵文物。

书房位于故居的二楼，书房的东、北、西三面都放着书橱，里面整齐地排列着各类书籍。墙壁上悬挂着孙中山为中国铁路建设所勾画的远景规划。

卧室陈设简单朴素。房间的北部正中放着一个由两个单人床拼成的对床。在对床的右侧，靠窗放着一张斜板写字台。

上海孙中山故居是孙中山和宋庆龄唯一共同的住所，于1918年入住于此。1925年3月孙中山逝世后，宋庆龄继续在此居住至1937年。抗日战争爆发后，宋庆龄移居香港、重庆。1945年底，宋庆龄将此寓所移赠给国民政府，作为孙中山的永久纪念地。

1961年3月，上海孙中山故居被国务院列为首批全国重点文物保护单位。1994年，故居被列为上海市爱国主义教育基地。

The dining room, connected with the living room, can be separated by two sliding doors. There is a mahogany dining table with chairs of the Guangdong style in the middle of the dining room. Precious cultural relics such as the Grand Marshal's command sword used by Sun Yat-sen when he took office as Grand Marshal of the army and navy of the military government of the Republic of China are exhibited here.

The study is on the second floor of the building. There are bookcases on the east, north and west sides of the study, in which books of various kinds are neatly arranged. On the wall hangs Sun Yat-sen's long-term plan for China's railway construction.

Bedroom furnishings are simple and plain. In the middle of the north of the room is a pair of single beds, on the right side of which there is a desk with a sloping panel by the window.

Sun Yat-sen's Former Residence in Shanghai was the only common residence of Sun Yat-sen and Soong Ching-ling, who moved here in 1918. After Sun Yat-sen passed away in March 1925, Soong Ching-ling continued to live here until 1937. After the outbreak of the War of Resistance against Japanese Aggression, Soong Ching-ling moved to Hong Kong and then to Chongqing. At the end of 1945, Soong Ching-ling donated this apartment to the National Government as a permanent memorial site for Dr. Sun Yat-sen.

In March 1961, Sun Yat-sen's Former Residence in Shanghai was listed by the State Council in the first group of key cultural heritage sites under state protection. In 1994, the former residence was listed as a patriotic education base in Shanghai.

光线照耀下的入口柱廊

西柏坡中共中央旧址
The Former Site of CPC Central Committee in Xibaipo

工程地点：石家庄市

竣工时间：1970年复原修建

设计机构：河北省建筑设计研究院等（复原设计）

西柏坡中共中央旧址坐落在河北省石家庄市平山县西柏坡镇西柏坡村，此处曾是中共中央和解放军总部所在地。

1948年5月26日至1949年3月23日，这里为中国共产党中央委员会和中国人民解放军总部驻地。

1949年3月，党中央离开西柏坡后，原中共中央旧址移交建屏县政府管理。

1955年，中央社会文化事业管理局与河北省文化局文管会、河北省博物馆派专人来西柏坡结合当地政府建立了西柏坡纪念馆筹备处。

1958年，因修建岗南水库，中共中央旧址拆迁。

1970年，西柏坡中共中央旧址复原时，采用了原来的木料门窗，墙屋顶采用了砖石混凝土结构，保存完好，对破损的门窗进行过修补。

The Former Site of CPC Central Committee in Xibaipo is located at Xibaipo Village, Xibaipo Town, Pingshan County, Shijiazhuang City, Hebei Province, where the CPC Central Committee and PLA headquarters were located.

From May 26, 1948 to March 23, 1949, this site was the headquarters of the Central Committee of the Communist Party of China and the People's Liberation Army.

In March 1949, after the CPC Central Committee left Xibaipo, the former site of the CPC Central Committee was handed over to the Jianping County Government.

In 1955, the Central Administration of Social and Cultural Undertakings, the Cultural Management Committee of the Hebei Provincial Cultural Bureau and the Hebei Provincial Museum dispatched people specially to Xibaipo to establish the preparatory office of the Xibaipo Memorial Hall in conjunction with the local government.

In 1958, the former site of the CPC Central Committee was relocated due to the construction of the Gangnan Reservoir.

In 1970, when the former site of the Central Committee of the Communist Party of China was restored in Xibaipo, the original wood doors and windows were used, and the roof of the wall was made of masonry concrete, which was well

纪念馆与环境

寝室与办公室

纪念碑

会议室

室外环境

1980年后，又先后复原了刘少奇同志旧居、中共中央接见国民党和平代表旧址、中央机关小学等。

现在的中共中央旧址大院是1958年修建岗南水库后，于1970年开始在原址北面山坡上按原布局、利用原房屋构件等复原修建的。大门向南，分前后两院，占地面积1.6万平方米，建筑为砖木结构平顶房。前院自东而西一座座小院依次为周恩来旧居、任弼时旧居、毛泽东旧居、军委作战室旧址、刘少奇旧居、董必武旧居、中共中央九月会议旧址。后院东北部三间窑洞式建筑为朱德办公室和居室旧址。大院西部前后院之间有中共七届二中全会旧址。

新中国成立前中共中央的许多重要会议在此召开，包括三大战役在内的一系列战役在此运筹和指挥，毛泽东主席的几十篇光辉著作在此诞生。1982年国务院公布西柏坡中共中央旧址为第二批全国重点文物保护单位。

preserved and the damaged doors and windows were repaired.
After 1980, Comrade Liu Shaoqi's Former Residence, the former site where the CPC Central Committee received the peace representatives of the Kuomintang and the primary school of the central government were restored successively. The former compound of the CPC Central Committee was rebuilt in 1970 after Gangnan Reservoir was built in 1958 on the northern hillside of the original site according to the original layout and using the original building components. With its gate facing south, the compound is divided into front and back courts, covering a land area of 16,000 square meters. The building is a flat-topped house with a brick and wood structure. From the east to the west, the small courtyards in the front yard are respectively Zhou Enlai's former residence, Ren Bishi's former residence, Mao Zedong's former residence, the former site of the military commission's war room, Liu Shaoqi's former residence, Dong Biwu's former residence and the former site of the September meeting of the CPC Central Committee. Three cave-style buildings in the northeast of the back court are the former site of the Zhu De's office and residence room. Between the front and back courtyards in the west of the compound is the former site of the Second Plenary Session of the Seventh CPC Central Committee.

Many important meetings of the CPC Central Committee before the founding of the People's Republic of China were held here, as well as the planning and command of a series of battles including the three major battles, and the writing of dozens of brilliant works of Chairman Mao Zedong. In 1982, the State Council announced the Former Site of the CPC Central Committee in Xibaipo be listed in the second group of key cultural heritage sites under state protection.

纪念馆室内局部

纪念馆入口立面图

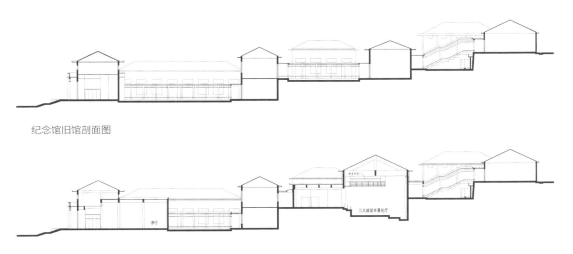

纪念馆旧馆剖面图

纪念馆改建后剖面图

纪念馆雕塑装饰

百万庄住宅区
Baiwanzhuang Residential Area

工程地点	北京市
建筑面积	120000平方米
竣工时间	1955年
设计机构	北京市建筑设计研究院
主要设计人	张开济

百万庄住宅区是为三里河地区"四部一会"及其他部委建设的干部职工住宅区，1953年开始建设，1955年陆续建成。建筑师张开济先生在设计中引用了"邻里单位"的概念，将整个用地分为12个区块，东西外侧安排为普通住宅，最北侧安排为办公楼，以此形成小区的门面和屏障。办公楼南面为干部住宅。中间的几个区块进行了错动，将商业、学校、绿地等公共设施安排在此，形成小区的居住中心。小区南面作为后续建设用地，暂以绿化相隔，为今后建设留下非常好的发展空间。小区以地支为名，分别以子、丑、寅、卯、辰、巳、午、未命名八个区，中间的干部住宅区为申区。小区规划整齐，设计严谨，是20世纪50年代初期国内住宅小区的代表作。

The Baiwanzhuang Residential Area is a residential area in Sanlihe area for cadres and workers of the "Four Ministries and One Commission". The construction of the area started in 1953 and was completed in 1955. Architect Zhang Kaiji introduced the concept of "neighborhood unit" in his design, dividing the whole land into 12 blocks. The east and west sides were arranged as ordinary residential houses, and the northernmost side was arranged as office buildings, forming the facade and barrier of the community. To the south of the office buildings is the cadre residential area. Several blocks in the middle were dislocated, and public facilities such as commerce, schools and green spaces were arranged here to form the residential center of the community. The south of the residential area is used as the follow-up construction land, temporarily separated by greening, leaving a very good development space for future construction. The community is divided into eight sections named after terrestrial branches: Zi, Chou, Yin, Mao, Chen, Si, Wu and Wei, and the cadre residential area in the middle is Shen Section. With neat planning and rigorous design, the Baiwanzhuang Residential Area is a representative work of residential quarters in the early 1950s in China.

住宅区现状

建筑与周边道路

建筑局部

建筑与环境

马勒住宅
Moller Residence

马勒住宅，又称马勒别墅，位于上海市陕西南路30号，占地5000余平方米，花园面积近2000平方米。1927年英籍犹太人马勒委托华盖建筑事务所设计建造该住宅，历时9年，于1936年竣工。

别墅主楼连接附楼，高高低低，屋顶陡翘，外形凹凸变化奇致。门窗上呈拱形，框架突出墙面。楼面陡峭，两座主塔高大、挺拔，像剑鞘一般，上开多层小窗。建筑物边梢楼角，都建有小的尖塔，以求与主塔呼应，造型绮丽。整座楼面呈现赭红色，一律由耐火砖建造，中嵌彩色瓷砖，望去像进入童话世界。室内划分各种大小用房、装修异常繁琐。楼梯、地板、壁板多为柚木，均呈红褐色，雕刻装饰带西方风味。花园四周设有耐火砖高围墙，琉璃瓦压顶。园内花房、葡萄房等都以磁砖铺地。入口处大门为红色，门楼采用红色木质拱券撑

Moller Residence, also known as Moller Villa, is located at No.30, Shaanxi South Road, Shanghai, covering a land area of more than 5,000 square meters and a garden area of nearly 2,000 square meters. In 1927, Moller, an English Jew, commissioned Huagai Architects to design and build the house, which was completed in 1936 after nine years' construction.
The main villa building is connected with the annex buildings, some high and some low. The roofs are steep with strange concave-convex changes in shape. The doors and windows are arched, with their frames protruding from the wall. The building surface is steep, and the two main towers are tall and straight, like scabbard, with multiple layers of small windows. There are small spires with beautiful shapes at the sides and corners of buildings to echo the main tower. The whole building surface is red ochre, all made of refractory bricks embedded with colorful ceramic tiles, making it a fairy tale world. Indoors there are rooms of various sizes with extremely elaborate decorations. Stairs, floors and sidings are mostly made of teak, all of which are reddish brown, and carved and decorated with a western flavor. The garden is enclosed with refractory brick walls topped by glazed tiles. The flower house and the grape house in the garden are paved with ceramic tiles. The entrance gate is red, and the gatehouse is supported by a red

托。门旁有一对石狮。沿街的围墙也为耐火砖建造，呈现赭红多彩色调，上覆绿色琉璃筒瓦。花园设在主楼南向，花园四周用彩色花砖铺地，并植有龙柏、雪松等名贵花木，中间草坪甚为宽阔。

马勒虽是英籍犹太人，但他发迹却在中国，所以楼房的外形虽是北欧挪威式，但花园和楼内装修的许多细部却颇有中国味道。大门口就像中国传统的豪门大宅一样，放置了一对中国式的石狮子，花园四周高大的围墙用耐火砖铺贴，以黄绿色中国琉璃瓦压顶，多少显示了他的中国情结。1941年，马勒为躲避战乱而离开中国。之后，马勒住宅多历战乱，几经转手，1949年后成为共青团上海团市委办公场所。

1989年，马勒住宅被列为上海市第一批优秀近代建筑、市级文物保护单位。

wooden arch. A pair of stone lions guard the gate. The fence along the street is also made of refractory bricks, showing auburn colors, covered with green glazed tiles. The garden located in the south of the main building is paved with colorful patterned tiles and planted with precious flowers and trees such as cypress and cedar, with a spacious meadow in the center.

An English Jew as he was, Moller made his fortune in China. Therefore, although the shape of the building is Nordic and Norwegian, many details of the garden and interior decoration are of the Chinese style. Just like traditional Chinese mansions, a pair of Chinese-style stone lions are placed at the gate. The tall walls around the garden are paved with refractory bricks topped with yellow-green Chinese glazed tiles, revealing his attachment to the Chinese culture to some extent. In 1941, Moller left China to escape the war. After that, Moller Residence went through many wars and changed hands several times. After 1949, it became the office building of the Shanghai Communist Youth League Committee.

In 1989, Moller Residence was listed in the first batch of outstanding modern buildings and cultural heritage sites under municipal protection in Shanghai.

阳台及屋顶装饰局部

北立面全景

南立面局部

东侧入口

门厅内景

楼梯间内景

盛宣怀住宅
Sheng Xuanhuai's Residence

工程地点：上海市
建筑面积：1775平方米
竣工时间：1900年
主要设计人：德籍建筑师

盛宣怀住宅位于淮海中路1517号，建于1900年。这座花园洋房原系德籍商人所建，占地面积12424平方米，总建筑面积1775平方米，采用欧式大草坪和中式庭院相结合手法，营造出宽敞、幽静、舒适的环境。后为清代洋务派主要人物、南洋公学（今上海交通大学）创办人盛宣怀购得，遂称为盛宣怀住宅。1912年秋，盛携家定居寓此。

住宅主楼前草坪面积达1668平方米，四周筑高围墙，设双扇镀金雕花大门，草坪东南角辟有网球场和休息室。主体建筑庭院平面呈长方形，横向布置于基地正中，佣人住房竖向布置于庭院的东西两侧，门房置于基地东北角。主屋平面将主人房与佣人房结合成整体，宅前花园植有冬青、雪松、龙柏等树木花卉，花园内砌有大理石喷水池，池中塑有希腊神

Sheng Xuanhuai's Residence at No.1517 Huaihai Middle Road was built in 1900. Originally built by a German businessman, this garden house of the Western style covers a land area of 12,424 square meters, with a total floor area of 1,775 square meters. By combining a European-styled lawn with a Chinese-styled courtyard, a spacious, quiet and comfortable environment has been created. Later, the house was purchased by Sheng Xuanhuai, a main figure of the Westernization School in the Qing Dynasty and the founder of Nanyang Public School (now Shanghai Jiaotong University). Thus it was called Sheng Xuanhuai's Residence. In the autumn of 1912, Sheng settled here with his family.

The lawn in front of the main residential building covers an area of 1,668 square meters, surrounded by high walls with double gold-plated carved gates. A tennis court with a lounge is in the southeast corner of the lawn. The courtyard plane of the main building is rectangular, which is horizontally arranged in the center of the base. The servants' houses are vertically arranged on the east and west sides of the courtyard, and the concierge is placed in the northeast corner of the base. The plane of the main house

建筑与周边环境

建筑立面旧影

俯瞰建筑全景

建筑入口

像（"文化大革命"中被毁）。

住宅的主楼为三层，建筑立面呈现新古典主义风格。出入口设在西边，有一个高大而宽敞的"∏"形门厅，左右两旁各立四根圆形厅柱，支撑着内藏式阳台，彩色玻璃天棚别有风韵。门厅两侧有衣帽间和会客室。主屋东面由柱廊和大玻璃窗组成的大休息厅，南面由柱廊、台阶组成宽敞的大门厅，北面由厨房和佣人房组成辅楼。楼上均设起居室、卧室和浴厕，生活设施齐全。室内墙饰有姿态各异的古希腊神像，壁面贴花绸和花纸，地板、扶梯、门窗均以柚木制成，外墙白色，与绿色草坪相映衬，构成一个优雅和谐的居住空间。

1916年盛宣怀病逝后该宅由其子盛重颐继承，后又相继成为陈调元宅、段祺瑞宅，抗日战争中曾被日本人占据。

1949年后，此宅归国家所有，中日复交后，日本驻上海领事馆设在这座花园洋房中。

combines the masters' houses with the servants' houses as a whole. The garden in front of the residence is planted with trees and flowers such as holly, cedar, dragon and cypress. There is a marble fountain in the garden, in which Greek statues were built (destroyed during the Cultural Revolution). The main building of the residence has three floors, and the facade of the building presents a neoclassical style. The entrance and exit are located in the west with a tall and spacious "∏"-shaped foyer with four circular hall columns on each of the left and right sides, which support the built-in balcony. The painted glass ceiling is uniquely charming. There are cloakrooms and reception rooms on both sides of the foyer. The main house consists of a large lounge with colonnade and large glass windows in the east, a spacious main hall with colonnade and steps in the south, and an auxiliary building with kitchens and servant rooms in the north. There are living rooms, bedrooms and toilets upstairs with complete living facilities. The interior walls are decorated with ancient Greek statues with different postures, and silk and flower wall paper. The floors, stairs, doors and windows are made of teak, and the exterior walls are painted white, setting off against the green lawn, forming an elegant and harmonious living space. After Sheng Xuanhuai passed away because of illness in 1916, the residence was inherited by his son Sheng Chongyi, and later became the residence of Chen Tiaoyuan and then the residence of Duan Qirui. It was occupied by Japanese during the War of Resistance against Japanese Aggression.

After 1949, this residence was owned by the state. After the reestablishment of diplomatic relations between China and Japan, the Japanese Consulate in Shanghai was located in this garden house.

沿街外景

院内水池及雕塑

四川大学早期建筑群
The Early Buildings of Sichuan University

工程地点：成都市
竣工时间：20世纪10年代至50年代
设计机构：美国、英国、加拿大、中国等国设计事务所等
主要设计人：弗雷德·荣杜义（英）、古平南 等

四川大学早期建筑包括包括华西校区的8栋文物建筑——华西校区办公楼（怀德堂）、华西校区第一教学楼（嘉德堂）、华西校区第二教学楼（苏道璞纪念堂）、华西校区第四教学楼（赫斐院）、华西校区第五教学楼（教育学院）、华西校区第六教学楼（万德门明堂学社）、华西校区老图书馆（懋德堂）、华西校区钟楼和望江校区的四川大学第一行政楼。四川大学早期建筑为全国重点文物保护单位。

华西校区办公楼（怀德堂）是其中最古老建筑，1915年动工，1919年建成。该楼为砖木结构，外观表现出中国古典建筑的特征，系由美国纽约罗恩甫氏为纪念白槐氏而捐建。该楼建成后设有校行政事务办公室、礼堂、文科教室和照相部等，二楼上的大讲演室还是星期日礼拜、聚会之所。怀德堂现为四川大学华西校区办公楼。

The early buildings of Sichuan University include eight cultural heritage buildings in West China Campus - West China Campus Office Building (Whiting Memorial Administration Building), West China Campus No.1 Teaching Building (Atherton Building for Biology and Preventive Medicine), West China Campus No.2 Teaching Building (Stubbs Memorial Building), West China Campus No.4 Teaching Building (Hart College), West China Campus No.5 Teaching Building (College of Education), West China Campus No.6 Teaching Building (Vandeman Memorial College), West China Campus Old Library (Lamont Library and Harvard-Yenching Museum), West China Campus Clock Tower, and No. 1 Administrative Building of Sichuan University, Wangjiang Campus. The early buildings of Sichuan University have been listed as key cultural heritage sites under state protection.

West China Campus Office Building (Whiting Memorial Administration Building) is one of the oldest buildings in Sichuan University. The construction of the building started in 1915 and was completed in 1919. Built with the donation from Mr. & Mrs. Joe Morrell in memory of Mr. Whiting, the building is a brick-wood structure with the characteristics of classical Chinese architecture. Upon the completion, the building had a university administration office, an auditorium,

华西校区老建筑群鸟瞰

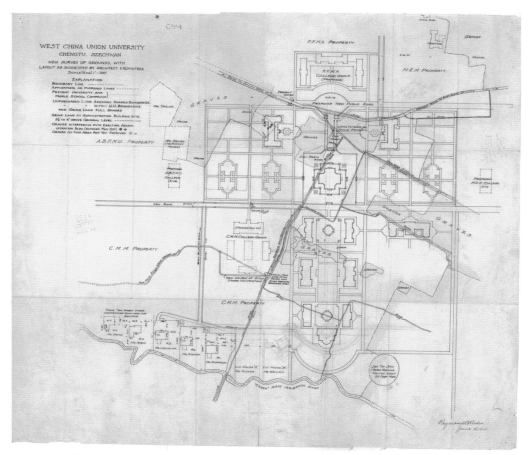

规划布局与现状图叠层

望江校区行政楼及牌坊

华西校区第四教学（合德堂）旧影

华西校区第四教学楼（合德堂）

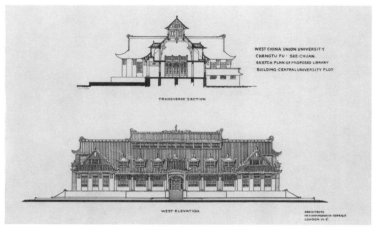
华西校区老图书馆（懋德堂）立面及剖面图

华西校区老图书馆（懋德堂）

华西校区办公楼（怀德堂）旧影1

华西校区办公楼（怀德堂）旧影2

华西校区办公楼（怀德堂）

278

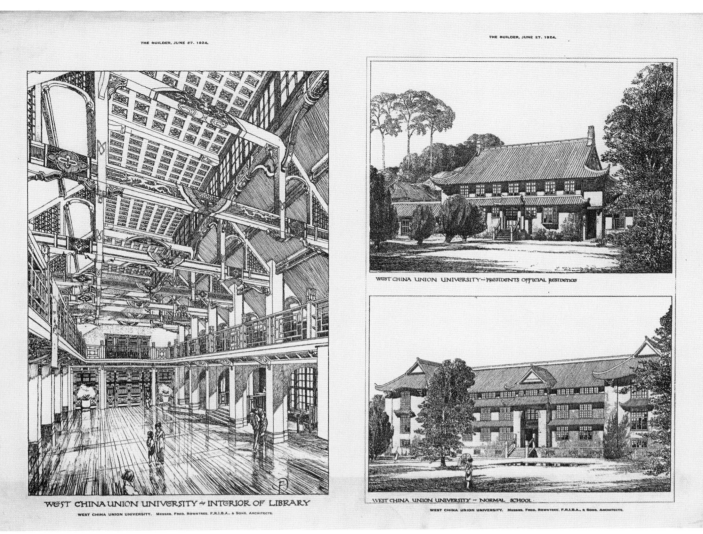

建筑透视效果图

华西校区钟楼，1925年动工，1926年竣工，系美国医生亚克门·柯里斯所捐建。其建筑形式采纳中国传统建筑的歇山式屋顶（50年代改为十字脊歇山顶），楼内大钟为美国梅尼利制钟公司于1924年专门铸造，早期由人工敲打，后改为电力，是今成都市现存唯一的机械式钟楼。

华西校区第一教学楼（嘉德堂），1924年竣工，为原华西协和大学生物楼，由美国夏威夷嘉热尔顿兄弟捐建，建成后为生物化学系、化学系、生物系、生理系教室，现为四川大学华西校区第一教学楼。

华西校区第二教学楼，1939年动工，1941年建成，为原华西大学苏道璞纪念堂，又名化学楼，由华西大学、内迁来蓉的金陵大学、齐鲁大学、金陵女子文理学院合资兴建。为纪念已故的来华英国化学家苏道璞博士（StubbsCM.），命名为苏道璞纪念堂。当时四校约定该楼由各校化学系及金陵大学的化工系合用，战后归华西大学所有，此楼沿用至今，

liberal arts classrooms and photography department, etc. The big lecture room on the second floor was also a place for Sunday worship and gathering. Whiting Memorial Administration Building is now the office building of West China Campus of Sichuan University.

The construction of the Clock Tower of West China Campus was commenced in 1925 and completed in 1926, with donations from an American physician Dr. Jonathan Ackerman Coles. The saddle roof of traditional Chinese architecture (changed to a saddle roof with a cross ridge in the 1950s) was adopted. The big clock in the building was specially cast in 1924 by Meneely Bell Company, the clock was originally knocked by a man and then changed to electricity-powered. It is the only extant mechanical clock tower in Chengdu today.

West China Campus No.1 Teaching Building (Atherton Building for Biology and Preventive Medicine) of West China Campus, completed in 1924, was the biology building of West China Union University, constructed with the donations from the Atherton brothers in Hawaii, USA. After completion, the building housed classrooms of the biochemistry department, the chemistry department, the biology department and the physiology department. Now, it is No.1 Teaching Building of West China Campus of Sichuan University.

No. 2 Teaching Building of West China Campus, which started in 1939 and was completed in 1941, is the former Stubbs Memorial Building of West China University, also known as Chemistry Building. It was jointly built by West China

现为四川大学华西校区第二教学楼。

华西校区第四教学（合德堂），1915年动工，1920年建成，由加拿大（英美）美道会为纪念最早到中国西南地区传教的赫斐氏所建，故又称赫斐院。建成时为物理系、数学系、农学院、宗教系教室，现为四川大学华西校区第四教学楼。

华西校区第五教学楼，1928年竣工，为原华西协和大学教育学院。英国嘉弟伯氏捐建该楼东端，为教育学院教学楼；1948年刘文辉捐建该楼西端。该楼现为四川大学华西校区第五教学楼。

华西校区第六教学楼（万德堂），建于1920年，又名万德门和明德学舍，由国浸礼会万德门夫妇捐建。建成时处于华西大学西部，即今人民南路上。由于人民南路工程的需要，1960年万德门被拆除，搬迁至现在的位置。一砖一瓦，均按原貌重建。由于迁建处地面有限，原万德门背后的侧楼去掉了，楼顶上的一座二层亭楼也除去。该楼建成后即为教学楼和学生宿舍，早年华西师范学校亦设于此。该楼现为四川大学华西校区第六教学楼。

华西校区老图书馆（懋德堂），1926年竣工，为原华西协和大学图书馆及博物馆，系美国赖孟德氏为纪念其子捐建，该楼共二层，建成后即为图书馆及博物馆，现为四川大学华

University, and universities moved to Chengdu, including Jinling University Cheeloo University and Jinling Women's College of Arts and Sciences. In memory of the late British chemist Stubbs CM., the building was named Stubbs Memorial Building. At that time, the four universities agreed that the building should be shared by the chemistry departments of each university and the chemical engineering department of Jinling University. After the War, the building was owned by West China University. This building is still in use today as No. 2 Teaching Building of West China Campus of Sichuan University.

West China Campus No.4 Teaching Building (Hart College), the construction of which started in 1915, was completed in 1920. It is also called Hart College because it was built in memory of the first missionary Hart in Southwest China by the Canadian (Anglo-American) Missionary Society of the Methodist Church. It housed classrooms of the physics department, the mathematics department, the agricultural college and the religion department, and now it is The No. 4 Teaching Building of West China Campus of Sichuan University.

West China Campus No. 5 Teaching Building, completed in 1928, is the former Education College of West China Union University. The eastern part of the building was built with donations from the Cadbury Brothers of the UK, used as the teaching building of the College of Education. In 1948, Liu Wenhui donated money to build the western part of the building. This building is now No. 5 Teaching Building of West China Campus of Sichuan University.

West China Campus No.6 Teaching Building (Vandeman Memorial) was built in 1920, also known as Wandemen and Mingde College, which was built with donations from Vandeman and his wife of the American Baptists. It was built in the west of West China University, that is, on Renmin South Road. Due to the need of the construction project on Renmin South Road, Vandeman Memorial was relocated in 1960 to its present place. Each detail of the building has been rebuilt according to their original appearance. Due to the limited land in the

华西校区钟楼

华西校区钟楼旧影

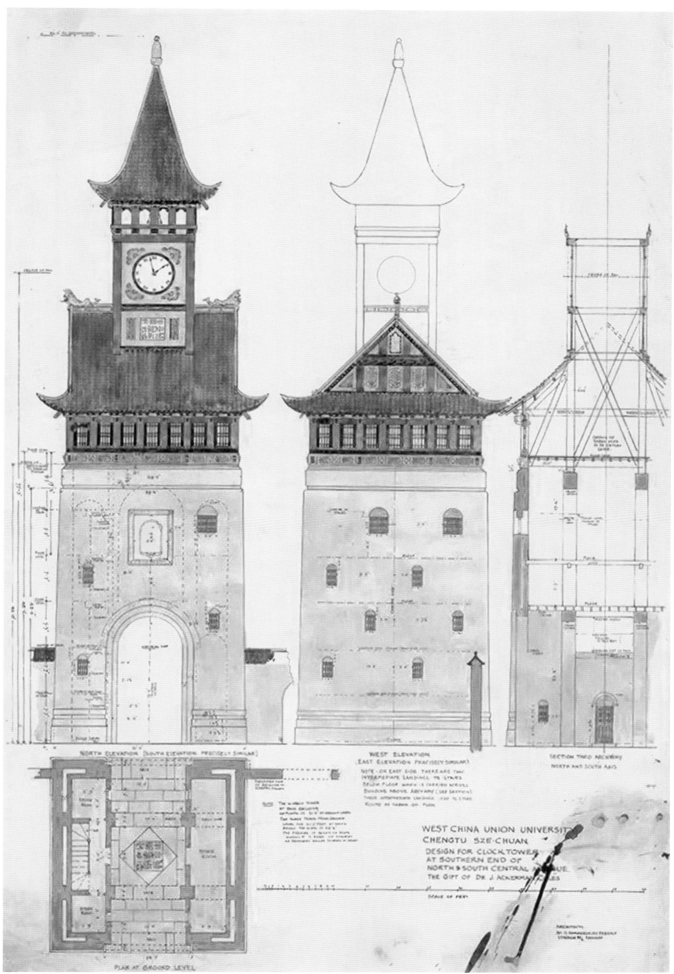

华西校区钟楼设计图纸

行政大楼门廊手绘效果图

西校区医学博物馆。

四川大学华西校区及其建筑，由英国设计师弗雷德·荣杜易（Fred Rowntree，1860 年出生于英国的斯卡布罗）规划设计。1900 年，华西协合大学在英国、美国和加拿大公开招标及举办了设计竞赛。荣杜易的设计所也受邀参与招标，并最终得标。荣杜易对整个华西坝进行了整体规划，为华西协合大学后期建设提供了框架。荣杜易在伦敦正式设计前考察了北京故宫、四川民居和日本建筑，建筑外观上着力突出中国特色的青砖黑瓦，间以大红柱、大红封檐板及清一色的歇山式大屋顶，在屋脊、飞檐上点缀以远古神兽、龙凤、怪鸟，檐下以斗拱为装饰，给人以神秘古朴的东方美感；内部则采用纯西式结构。

在修建过程中，大部分建筑材料都取自当地，例如在成都三

relocation area, the side building behind the original Vandeman Memorial was removed, so was a two-story pavilion on the top of the building. The building, upon its completion, became a teaching building and student dormitory. West China Normal School was also located here in its early years. This building is now No. 6 Teaching Building of West China Campus of Sichuan University.

West China Campus Old Library (Lamont Library and Harvard-Yenching Museum), completed in 1926, is the library and museum of former West China Union University, built with donations from the Lamont family in the United States in memory of their son. The two-story building turned into a library and museum after completion, and is now the Medical Museum of West China Campus of Sichuan University.

West China Campus of Sichuan University and its buildings were planned and designed by British designer Fred Rowntree (born in Scarborough, England in 1860). In 1900, West China Union University held open tenders and design competitions in the United Kingdom, the United States and Canada. Rowntree's design institute was also invited to participate in the bidding, and finally won the bid. Rowntree made an overall plan for the whole area of Huaxiba, providing a framework for the later construction of West China Union University. Then Rowntree investigated the Forbidden City in Beijing, Sichuan dwellings and Japanese buildings before starting the official design in London. The architectural appearance highlighted the blue bricks and black tiles with Chinese characteristics, featuring red columns, red gable boards and uniform saddle roofs. The roofs and cornices were decorated with statues of ancient beasts, dragons and phoenixes and strange birds, and the eaves were decorated with bucket arches, leaving an impression of a mysteriously simple oriental beauty. Internally, pure western structures were adopted.

瓦窑一带定制的砖瓦，来自四川西部高山的木材，从岷江上游河岸采集的石灰石等。建筑用的铁钉和玻璃等，主要是从汉口和上海甚至是国外购买而来。

嘉德堂、懋德堂、苏道璞纪念堂、赫斐院、教育学院、万德门明堂学社等，也皆为外籍设计师所设计，采用中西合璧砖木结构，外观融合中式古典韵味和西式建筑特征。

望江校区的四川大学第一行政楼建成于 1953 年，总建筑面积近 15000 平方米，造价为 160 万元，由四川省建筑工程局设计公司古平南（1914—1997 年）设计，成都市建筑工程局第一工程公司施工。大楼采用了纵横三段西方古典主义构图，外观运用中国传统建筑重檐歇山式屋顶、中国古代吉祥图案以及中国传统的红色梁柱等中国古典元素。此建筑造型浑厚凝重，是"中国固有式建筑"风格在共和国时代的延续和发展。1999 年，学校在保持该楼原貌的情况下，对大楼进行了彻底地加固和维修。

During the construction process, most of the building materials were taken from the local area, such as custom-made bricks and tiles in Sanwa Kiln, Chengdu, wood from the western mountains of Sichuan, limestone collected from the upper reaches of the Minjiang River, etc. Nails and glass for construction were mainly purchased from Hankou, Shanghai and even abroad.

Atherton Building for Biology and Preventive Medicine, Lamont Library and Harvard-Yenching Museum, Stubbs Memorial Building, Hart College, College of Education and Vandeman Memorial College were also designed by foreign designers with a brick and wood structure integrating both the Chinese and Western styles, presenting both classical Chinese appeal and Western architectural characteristics.

No. 1 Aadministrative Building of Sichuan University on Wangjiang Campus was built in 1953, with a total floor area of nearly 15,000 square meters at a cost of 1.6 million yuan. It was designed by Gu Pingnan (1914-1997) of The Design Company of Sichuan Construction Engineering Bureau, and constructed by No. 1 Engineering Company of Chengdu Construction Engineering Bureau. The building consists of three sections of Western classicism, and its appearance features Chinese classical elements such as double-eaves saddle roofs of traditional Chinese architecture, ancient Chinese auspicious patterns and traditional Chinese red beams and columns. With an immense and dignified appearance, this building is the continuation and development of the "Chinese Inherent Architecture" style in the PRC era. In 1999, the building was strengthened and repaired by the university while its original appearance was preserved.

室内阶梯

华西校区第六教学楼（万德堂）

中央银行、农民银行暨美丰银行旧址

The Former Site of Central Bank, Farmers' Bank and American-Oriental Bank of Fukien

工程地点	重庆市
建筑面积	3352.5平方米
竣工时间	1938年 1935年
设计机构	基泰工程司
主要设计人	杨廷宝

抗战时，今重庆市渝中区小什字、打铜街、道门口一带，是中国金融界大亨叱咤风云之地，曾被誉为"中国的华尔街"，其中位于道门口9号的重庆中央银行和位于新华路74号的美丰银行，均是当年重庆首屈一指建筑。因中央银行、农民银行都曾栖身于美丰银行大楼，故此次作为一个整体，以"中央银行、农民银行暨美丰银行旧址"的名目入选中国20世纪建筑遗产名单。

中央银行大楼由基泰工程公司1938年设计，主设计师为杨廷宝，建业营造厂施工。此建筑地下两层作为金库，为了防盗防爆，该库房混凝土墙身厚度达90厘米。中央银行是抗战时期国民政府最重要的金融机构之一，作为重庆抗战金融遗址群的重要组成部分，其旧址对于研究抗战时期国民政府的战时金融和重庆金融都有重要历史价值。

During the War of Resistance against Japanese Aggression, Xiaoshizi, Datong Street and Daomenkou, Yuzhong District of Chongqing, were arenas for Chinese financial tycoons, once known as "China's Wall Street". Chongqing Central Bank located at No.9 Daomenkou and American-Oriental Bank of Fukien located at No.74 Xinhua Road were the leading buildings in Chongqing then. As Central Bank and Farmers' Bank used to have domicile in the Building of American-Oriental Bank of Fukien, the building was selected as a whole in the list of architectural heritage in China in the 20th century under the name of "The Former Site of Central Bank, Farmers' Bank and American-Oriental Bank of Fukien".

The Central Bank Building was designed by Jitai Engineering Company in 1938 with Yang Tingbao as the main designer, and it was constructed by Jianye Construction Factory. The two floors underground of this building were used as the vault. In order to prevent theft and explosion, the concrete wall of this vault is 90 centimeters thick. The Central Bank was one of the most important financial institutions of the National Government during the War of Resistance against Japanese Aggression. The former site of the Central Bank as an important part of the financial sites of the War of Resistance against Japanese Aggression in Chongqing has significant

外立面现状

美丰银行

中央银行现状

美丰银行大厦为七层钢筋混凝土框架结构建筑，面阔37.8米，进深8.7米，建筑面积3352.5平方米，建筑占地面积407.5平方米。美丰银行大楼由上海基泰工程司建筑师杨廷宝先生设计，由馥记营造厂施工。大厦于1934年开工，1935年6月竣工，1935年8月正式落成剪彩，是当年重庆是首屈一指的建筑。该建筑立面第一至第二层用青岛黑色磨光花岗石贴面，第三层以上采用米黄色釉面薄瓷砖贴面，外观端庄简洁，底层入口门厅设钢卷门、板门、玻璃门三重，以保安全，其他都是钢窗硬木门。营业厅空间高达二层，顶部有气楼式天窗玻璃吊顶，增加了大厅的空间高度感。营业柜台用大理石贴面，第三至第六层均为写字间，由侧门厅电梯出入，地下室为银行金库。建筑外观像中国古代钱币的一种"布"币，是近代重庆较早出现的具有现代主义建筑风格的近代银行建筑，反映了设计者严肃的设计态度和纯熟的设计技巧，具有较高的建筑艺术、景观价值。抗战期间，内迁重庆的中央银行、中国农民银行在没有修建好新的办公地点之前，都曾在此租住办公。美丰大楼施工质量过硬，抗日战争初期经受了日机空袭的考验。

historical value for studying the wartime finance of the Government of the Republic of China and Chongqing finance during the War of Resistance against Japanese Aggression.

The Building of American-Oriental Bank of Fukien is a seven-story reinforced concrete frame structure with a width of 37.8 meters, a depth of 8.7 meters, a floor area of 3,352.5 square meters and a land area of 407.5 square meters. The Building was designed by Mr. Yang Tingbao, architect of Shanghai Jitai Engineering Department, and constructed by Fuji Construction Factory. The construction of the building was started in 1934, completed in June 1935, and officially celebrated for completion in August 1935, as one of the leading buildings of Chongqing at that time. The facade of the first and second floors are covered with black polished granit from Qingdao. The third floor and above are covered with beige thin glazed ceramic tiles, looking dignified yet unadorned. The entrance hall of the ground floor is decorated with a steel rolling door, a plate door and a glass door to ensure safety, and the rest are steel windows and hard wooden doors. The space of the business hall is as high as two floors, with a skylight glass ceiling at the top, which increases the space height of the hall.

The business counters are covered with marble, and the third to sixth floors are office rooms, which are accessed by an elevator in the side hall, and the basement is the bank vault. The architectural appearance is like a kind of "cloth" coin in ancient China. It is an early modern bank building with a modernist architectural style in Chongqing, reflecting the designer's serious attitude and skillful skills in design, and possessing both high architectural art and landscape value. During the War of Resistance against Japanese Aggression, the Central Bank and Farmers' Bank, which moved to Chongqing, rented offices here before they built new office buildings. The construction quality of the Building of American-Oriental Bank of Fukien was so excellent that it withstood the test of Japanese air strikes in the early days of the War of Resistance against Japanese Aggression.

井冈山革命遗址

Mount Jinggangshan Revolutionary Sites

井冈山革命遗址位于江西省吉安井冈山市境内,平均海拔1000米,为全国第一个农村革命根据地。井冈山革命遗址主要集中在江西省井冈山市的茨坪、茅坪、砻市、大小五井及五大哨口等地,保存完好的革命旧居旧址有几十处。旧址建筑多为砖木结构的祠堂、书院、店铺或民房。

茅坪位于原宁冈县境的东南面16公里处,井冈山黄洋界的北麓,曾是湘赣边界党、政、军最高领导机关所在地和革命斗争的指挥中心。位于湘赣边界的"一大"会址原为谢氏慎公祠,建于清末,砖木结构;毛泽东旧居为砖木结构二层楼房,有一个小厅堂,十间小房,楼上开有八角形天窗,故又称八角楼。中共井冈山前敌委员会旧址原为攀龙书院,始建于1867年,是一栋三层楼阁式的建筑,砖木结构,前后二厅,共十四间房。

茨坪于1928年7月后曾为革命根据地内党、政、军的最高指挥中心所在地,其中包括毛泽东旧居、红四军军部(含朱

Mount Jinggangshan Revolutionary Sites located in Jinggangshan City, Ji'an, Jiangxi Province, with an average elevation of 1,000 meters, formed the first rural revolutionary base in China. Mount Jinggangshan Revolutionary Sites are mainly concentrated in Ci Ping, Mao Ping, Longshi, Five Big and Small Wells and Five Great Outposts in Jinggangshan City, Jiangxi Province. There are dozens of well-preserved sites of the former revolutionary residences. Most of the old buildings are ancestral halls, academies, shops or private houses with a brick-and-wood structure.

Mao Ping located 16 kilometers southeast of former Ningkang County and at the northern foot of the Huangyang Border in Jinggangshan was once the seat of the highest leading organ of the party, government and army along the Hunan-Jiangxi border and the command center of revolutionary struggle. The site of the "First National Congress" located in the border of Hunan and Jiangxi was originally Xie's Shengong Temple, built in the late Qing Dynasty, with a brick-and-wood structure. Mao Zedong's former residence is a two-story building with a brick and wood structure, with a small hall, ten small rooms and octagonal skylights upstairs, so it is also called Octagonal Building. The former site of the CPC Jinggangshan Front Committee was originally Panlong Academy built in 1867. It is a three-story pavilion-style building with a brick and wood structure, front and back halls and fourteen rooms in total.

After July 1928, Ci Ping was the highest command center of the Party, government and army in the revolutionary base areas, including Mao Zedong's Former Residence, the Red Fourth Army Headquarters (including Zhu De's

中国工农革命军第一师师部旧址

茅坪八角楼

茅坪红军医院旧址

中国红军第四军军官教导队旧址

茅坪红四军军需处旧址

德旧居)、中共井冈山前委、湘赣边界特委、湘赣边界工农兵政府等,建筑均为青瓦土木结构的民房或店铺。茨坪周围的天险要隘五大哨口,当年均修筑有防御阵地、瞭望哨、营房等建筑,黄洋界哨口就是其中重要的一个哨口。

砻市是是著名的朱毛会师、红四军成立、红四军第二次党代会、成立红四军军官教导队的所在地。两军会师后,井冈山的斗争就进入全盛时期。这里的主要革命历史遗址有井冈山会师纪念馆、龙江书院、会师桥、井冈山会师纪念碑等。龙江书院建于清道光年间,是湘赣边界宁冈、酃县、茶陵三县的最高学府。建筑分三进,中间有天井,两边有厢房。后进是三层楼房,名文星阁,毛泽东曾在此举办过军官教导队。红四军建军广场,又名会师广场,位于宁岗县龙江河畔,会师桥位于广场前,桥头建立了井冈山会师纪念碑。

Former Residence), the Jinggangshan Front Committee of the Communist Party of China, the Hunan-Jiangxi Border Special Committee, the Hunan-Jiangxi Border Government of Workers, Peasants and Soldiers, and so on. The buildings are all private housing or shops with blue tiles and brick-and-wood structures. There are five outposts around Ci Ping, in which defensive positions, observation posts, barracks and other buildings were built in those years, among which the Huangyangjie Outpost was an important one.

Longshi is the place where the famous meeting of Zhu and Mao's armies took place, the Red Fourth Army was founded, the Second Party Congress of the Red Fourth Army was held, and the officers' training team of the Red Fourth Army was established. After the two armies joined forces, the struggle in Mount Jinggangshan culminated. The main revolutionary historical sites here include Mount Jinggangshan Joining Forces Memorial Hall, Longjiang Academy, Joint Forces Bridge and Mount Jinggangshan Joining Forces Monument. Longjiang Academy, built during the Daoguang Reign in the Qing Dynasty, is the highest educational institution of higher learning in Ninggang, Lingxian and Chaling counties on the Hunan-Jiangxi border. The building has three courtyards, with a patio in the middle and wing rooms on both sides. The back courtyard has a three-story building, named Wenxing Pavilion, where Mao Zedong once organized an officer training team. Red Fourth Army Jianjun Square, also known as Joining Forces Square, is located on the banks of the Longjiang River in Ninggang County. Joining Forces Bridge is located in front of the square, while the Monument to the Joining of Forces on Mt. Jinggang is built at the bridge head.

小井中国红军第四军医院旧址

红四军军部旧址

茨坪旧居朱德旧址

黄洋界营房旧址

中国红军第四军军械处旧址

大井毛泽东同志旧居

八角楼士兵委员会旧址

入口立面

青岛火车站
Qingdao Railway Station

青岛火车站位于山东省青岛市，胶济铁路和胶济客运专线的东端车站。

青岛站始建于清光绪二十六年（1900年）；1991年拆除重建；1994年8月8日投入运营。

青岛火车站主要由钟楼和候车大厅两大部分组成，北联一层办公用房，砖木钢混合结构，建筑具有德国文艺复兴建筑风格。车站候车大厅以高大的装饰山墙和三个大型券门突出面向市区的主入口，楼南角耸起一座造型优美的尖钟塔，钟塔的基座、窗边、门边及山墙和塔顶的装饰都用花岗石砌成，是仿半木构式样的公共建筑。屋顶为四坡顶，面覆中国杂色琉璃瓦。

The Qingdao Railway Station located in Qingdao, Shandong Province is at the east end of the Jiaoji Railway and the Jiaoji Passenger Dedicated Line.
The Qingdao Station was first built in the twenty-sixth year of the Guangxu Reign in the Qing Dynasty (1900); demolished and rebuilt in 1991; and was put into operation on August 8, 1994.
The Qingdao Railway Station is mainly composed of the Clock Tower and the Waiting Hall, connected to a one-story office building on the north, with a mixed structure of bricks, wood and steel. The building features a German Renaissance architectural style. The Waiting Hall of the station highlights the main entrance facing the urban area with a tall decorative gable and three large arched gates. A beautiful pointed clock tower is erected at the south corner of the building. The pedestal, window frames, gate frames, gable and the decoration on the top of the tower are all made of granite. This is a public building with an imitation semi-wooden structure. The roof has four slopes covered with Chinese variegated glazed tiles.
The new Qingdao Station takes a U shape as a whole, and its three station

新青岛站整体呈"U"字形,其三个站房均保持原车站红瓦黄墙的欧式风格,如老钟表楼和候车室,每个站房上均以隶书体写上了红色的"青岛"二字。西站房的装修风格与东站房的不相同,如其迎客广厅的屋顶不是列车车厢的形状,而是有六个圆形吊顶的平屋顶。

据悉,原青岛站由德国人魏尔勒和格德尔茨设计。

houses retain the European style of the original station with red tiles and yellow walls, such as the old clock tower and the waiting room, and on each house is written with "Qingdao" in red in the clerical script. The decoration style of the West Railway Station House is different from that of the East Railway Station House. For example, the roof of its welcoming hall is not the shape of a train carriage, but a flat roof with six circular ceilings.

It is reported that the original Qingdao Station was designed by two Germans, Weiler and Gedelz.

火车站钟楼

钟楼立面局部

立面窗饰

站内屋顶

站厅室内

屋顶局部

站厅室内全景

伪满皇宫及日伪军政机构旧址

The Former Sites of Puppet Manchurian Palace and Japanese Puppet Military Institutions

工程地点	长春市
竣工时间	1938年
设计机构	清水建设
主要设计人	相贺兼介（日）

伪满洲国是日本侵占中国东北后扶植的一个傀儡政权。首都设于新京（今长春）。行政区域辖东北三省全境（不含大连市），以及内蒙古东部、河北省承德市。伪满洲国在东北实行了14年之久的殖民统治，长春至今保留有大量的伪满军政统治机构的建筑遗存。

伪满皇宫旧址和日伪军政机构旧址包括伪满皇宫旧址（今伪满皇宫博物院）、伪满国务院旧址（今吉大基础楼）、伪满军事部旧址（今吉大第一临床医院）、伪满司法部旧址（今吉大新民校部）、伪满经济部旧址（今吉大中日联谊医院二部）、伪满交通部旧址（今吉大公共卫生学院）、伪满民生部旧址（今吉林省石油化工设计院）、伪满外交部旧址（今太阳会馆）、伪满综合法衙旧址（今解放军第

Puppet Manchukuo was a puppet regime supported by Japan after it invaded Northeast China. The capital of the puppet regime was located in Xinjing (present Changchun City). The administrative area covered the whole of the three northeastern provinces (excluding Dalian City), as well as eastern Inner Mongolia and Chengde City, Hebei Province. Puppet Manchukuo exercised colonial rule in Northeast China for 14 years, and Changchun still retains a large number of architectural remains of puppet Manchukuo military and political ruling institutions.

The former sites of Puppet Manchurian Palace and Japanese Puppet Military Institutions include the former sites of Puppet Manchurian Palace (now Puppet Manchurian Palace Museum), Puppet Manchurian State Council (now Jilin University Basic Building), Puppet Manchurian Military Department (now No. 1 Clinical Hospital of Jilin University), Puppet Manchurian Justice Department (now Jilin University Xinmin Campus), Puppet Manchurian Economic Department (now the Second Department of Jilin University Sino-Japanese Friendship Hospital), Puppet Manchurian Ministry of Communications (now Jilin University College of Public Health), and former site of the Puppet Manchurian Ministry of Foreign Affairs (now Taiyang Hall), Puppet Manchuria

同德殿

同德殿室内装饰

同德殿旧影

同德殿室内旧影

四六一医院）、日本关东军司令部旧址（今省委办公楼）、日本关东宪兵队司令部旧址（今省政府办公楼）、伪满洲国中央银行旧址（今工商银行）。

伪满皇宫建筑群形成于20世纪30年代前后，这里曾是清朝末代皇帝爱新觉罗·溥仪充当伪满洲国傀儡皇帝时居住的宫殿。其前身为建于1911年的吉黑榷运局官署。吉黑榷运局是民国时期管理吉林、黑龙江两省盐务的机构。

1932年3月，吉黑榷运局对其进行修缮，伪满洲国成立后，将这里的缉熙楼作为溥仪的寝宫，勤民楼作为他的办公楼。1936—1938年，同德殿建成。1938年，东御花园初步落成。1939年，同德殿东南30米处，修建防空地下室（御用防空洞），为钢筋混凝土结构。为增强防空地下室的防爆能力，在其上方加建了假山。1940年2月，建国神庙举行奠基仪式，5月28日竣工。1941年，嘉乐殿落成，这是伪满皇宫举行大型宴会的场所。1945年8月11日，溥仪仓皇出逃，建国神庙被日本关东军纵火烧毁，仅余基石。1945—1947年，伪满皇宫被国民党所办松北联中占用，后被国民党军队占用，内部受到严重破坏。2000年后，有关部门对伪满皇宫实施了全面保护恢复，2005年完成了勤民楼局部复原及展陈工程，恢复了二楼外廊、举架高度及楼梯间面积。

Comprehensive Law Office (now No. 461 Hospital of PLA), Japanese Kwantung Army Command (now the Office Building of Jilin Provincial Committee of the CPC), Japanese Kanto Gendarmerie Command (now the Provincial Government Office Building), and Puppet Manchukuo Central Bank (now ICBC).

The Puppet Manchurian Palace building complex was formed around the 1930s, where Aisin Gioro Puyi, the last emperor of the Qing Dynasty, lived when he acted as puppet emperor of Puppet Manchukuo. Its predecessor was the office of Jilin-Heilongjiang Transport Bureau, established in 1911 as an institution that managed salt affairs in Jilin and Heilongjiang provinces during the Republic of China.

In March, 1932, Jilin-Heilongjiang Transport Bureau had the building repaired. After the founding of the Puppet Manchukuo, Jixi Building here was used as Puyi's bedroom and Qinmin Building as his office building. From 1936 to 1938, Tongde Hall was built. In 1938, Dongyu Garden was initially completed. In 1939, an air defense basement (imperial air defense shelter) was built 30 meters southeast of Tongde Hall, which is a reinforced concrete structure. In order to enhance the explosion-proof ability of the air defense basement, a rockery was built above it. In February 1940, the foundation laying ceremony of Jianguo Temple was held, and the temple was completed on May 28th. In 1941, Jiale Hall was completed, which was the place where the Puppet Manchurian Palace held large banquets. On August 11, 1945, Puyi fled in a hurry, and the Jianguo Temple was set on fire by the Japanese Kwantung Army, leaving only the cornerstone. From 1945 to 1947, the Puppet Manchurian Palace was occupied by the Songbei Unified School run by the Kuomintang, and then occupied by the Kuomintang army, suffering serious internal damages. After 2000, the relevant departments implemented comprehensive protection and restoration of the Puppet Manchurian Palace. In 2005, the local restoration and exhibition project of Qinmin Building was completed, with the veranda, lifting height and stairwell area on the second floor restored.

原伪满国务院现状

原伪满国务院屋顶局部

伪满国务院正立面旧影

伪满国务院全景旧影

怀远楼始建于1933年7月3日，1934年10月30日竣工。1934年4月时，为满足溥仪称帝后"赐宴"的需求，利用勤民楼与怀远楼之间的空地，开始修建清晏堂。修建清晏堂时，将一楼东侧的车库和大食堂改造为办公室。2000年进行室内装修。

伪国务院办公楼，是伪满洲国政府机构建设形式的典型，即所谓"兴亚式"建筑式样。立面4层，中央另设塔楼，平面呈"H"形。以南、北、西3个入口为交通枢纽，由单面走廊连接通往各楼层房间，形成以主要入口（西入口）中央大厅为主轴的三轴线构图。立面采用中轴对称，基座、墙身、屋顶三段式技法。中轴线上塔楼突起，冠以重檐攒尖式琉璃瓦顶。入口外廊为欧洲古典陶立克柱式，与上部墙体形成强烈的虚实与明暗对比。重檐下再用一组附壁圆柱与门廊呼应。中轴线上，建筑外形下大上小，四面外墙逐渐收分，与重檐曲线屋顶协调一致，强调主体的宏伟效果。屋顶艺术是中国古建筑的一绝，但该楼屋顶的仙人、走兽、正吻、脊饰，连同制高点上的宝顶均非我国古建筑构件的形态特征。而西方古典柱式的齿石、额枋、檐壁等檐口构造也是一鳞半爪。设计者对众多建筑式样、建筑构件的杂糅、折中之术，意在赢得日本军国主义者的赏识。

注："兴亚式"建筑形式，是来源于日本的一种形式，是在某些近代建筑中运用古代建筑形式的一些特点，具有严重的形式主义、复古主义、拼凑集仿的特点。

The construction of Huaiyuan Building was started from July 3, 1933 and completed on October 30, 1934. In April 1934, in order to meet the need of giving a banquet after Puyi proclaimed himself emperor, Qingyan Hall was built by using the open space between Qinmin Building and Huaiyuan Building. During the construction of Qingyan Hall, the garage and dining hall on the east side of the first floor were transformed into offices. Interior decoration was carried out in 2000.

The office building of the Puppet State Council is a typical form of government department buildings in Puppet Manchukuo, namely the so-called "Xingya" architectural style. The facade has four floors, with another tower in the center, and the plane is "H"-shaped. The south entrance, the north entrance and the west entrance are the transportation hubs, which are connected by a single corridor to the rooms on each floor, forming a three-axis composition with the central hall of the main entrance (the west entrance) as the main axis. The three-stage technique was adopted with a symmetrical central in the facade, including axis, base, wall and roof. On the central axis, a tower rises, crowned with double eaves and pointed glazed tile roof. The entrance veranda is of the European classical Doric order, which, with the upper wall, forms a strong contrast between reality and light and shade. Under the double eaves, a group of attached columns are used to echo the porch. On the central axis, the shape of the building is big below and small above, and the four external walls are gradually narrowed, which is consistent with the curved roof with double eaves, emphasizing the magnificent effect of the main body. Roof art is a unique part of ancient Chinese architecture, but the immortals, beasts, front roof animals, and ridges on the roof of this building, together with the roof crown on the commanding height, are not the morphological characteristics of the components of ancient Chinese architecture. However, the eaves structures of the Western classical order, such as the tooth stones, the front columns and the eaves walls, are also fragmental. The designer's mixture and compromise of many architectural styles and components was intended to win the appreciation of Japanese militarists.

Note: The "Xingya-style" architectural form is a form originated from Japan, which uses some characteristics of ancient architectural form in some modern buildings, and is characterized by serious formalism, retro style and patchwork and imitation.

原伪满民生部现状

伪满民生部旧影

原伪满军政部现状

原伪满司法部入口及立面

伪满司法部旧影

伪满军政部旧影

原综合法衙现状正面

原伪满民生部现状

伪满综合法衙旧影

伪满交通部旧影

伪满经济部旧影

伪满兴农部旧影

梅园新村 30 号院

中国共产党代表团办事处旧址（梅园新村）

The Former Office Sites of the Delegation of the Communist Party of China (Meiyuan New Village)

中国共产党代表团办事处旧址（梅园新村）位于江苏省南京市，由梅园新村 30 号、梅园新村 35 号、梅园新村 17 号三处建筑组成。

梅园新村 30 号是周恩来、邓颖超办公和居住的地方，有二层楼房三幢，共 18 间，占地面积 431.75 平方米，建筑面积 361.1 平方米。主楼楼下有办公室、会客室、卧室、餐室等，楼上设有机要科等。为防止特务的监视和破坏，中共代表团将院墙加高了一倍，并在传达室和后边西晒台上各加盖了一层小楼。院内的翠柏、石榴、铁枝海棠、葡萄和蔷薇都是当年中共代表团留下的。整个院内依然保持着原来的风貌。

梅园新村 35 号是董必武、李维汉、廖承志、钱瑛等同志办公和居住的地方，有砖木二层楼房一幢，砖木平房二座，共 11 间，占地面积 155.25 平方米，建筑面积 192.1 平方米。楼房楼下是董必武、廖承志办公和居住的地方，楼上是 35

The former office sites of the Delegation of the Communist Party of China (CPC Delegation) (Meiyuan New Village) located in Nanjing, Jiangsu Province consist of three places: No.30, No.35 and No.17 Meiyuan New Village.

No.30 Meiyuan New Village, where Zhou Enlai and Deng Yingchao worked and lived, has three two-story buildings with 18 rooms, covering a land area of 431.75 square meters with a floor area of 361.1 square meters. There is an office, a sitting room, a bedroom and a dining room downstairs in the main building, as well as a confidential department upstairs. In order to prevent spy surveillance and destruction, the CPC Delegation doubled the height of the courtyard wall, and built a small building on the gatehouse and the west roof terrace behind it. The cypress, pomegranate, crown of thorns, grape vines and roses in the courtyard were all left by the CPC Delegation. The whole courtyard still maintains its original style.

No.35 Meiyuan New Village, where Dong Biwu, Li Weihan, Liao Chengzhi, Qian Ying and other comrades worked and lived, has one brick-wood two-story building and two brick-wood bungalows, with 11 rooms, covering a land area of 155.25 square meters with a floor area of 192.1 square meters. Downstairs of the main building is where Dong Biwu and Liao Chengzhi worked and lived, while upstairs is where Li Weihan and Qian Ying worked

号汉、钱瑛办公和居住的地方。中共代表团为了工作的方便与安全，将与梅园新村 31 号特务监视站相通的原有大门堵死，在东边开了一个小门，与梅园新村 30 号相通，并在小院内两边加盖了两座小平房，以此挡住 31 号特务的视线。东边平房是政策研究室办公室，西边平房是警卫室。

梅园新村 17 号是中共代表团办事机构的所在地，有砖木三层楼房一幢，二层楼房两幢，砖木平房两座，共 29 间，占地 502.13 平方米，建筑面积 725 平方米。其中，北边一幢楼房是中共代表团办事机构。楼下设有小会议室、新闻组、抄报室、第十八集团军驻京办事处处长办公室，楼上设有电讯组、外事组、军事组、党派组和妇女组办公室。南边楼房是中共代表团到南京后加盖的。楼上是工作人员的宿舍，楼下是饭厅，中共代表团常在这里举行记者招待会。

and lived. For the convenience and safety of work, the CPC Delegation blocked the original gate connected with No.31 spy monitoring station in Meiyuan New Village, opened a small gate in the east, which was connected with No.30 Meiyuan New Village, and built two small bungalows on both sides of the courtyard to block the sight of the spies from No.31. The bungalow in the east is a policy research office, and the bungalow in the west is a guard room.

No.17 Meiyuan New Village, the site of the office of the CPC Delegation, has a brick-wood three-story building, two two-story buildings and two brick-wood bungalows, covering a land area of 502.13 square meters and having 29 rooms with a floor area of 725 square meters. Among them, the building in the north is the office of the CPC Delegation, in which there is a small conference room, a newsgroup room, and a newspaper copying room, the office of the director of the 18th Army Beijing Office downstairs and there are offices of telecommunications, foreign affairs, military, political party group and women's group. The south building was built after the CPC Delegation arrived in Nanjing. Upstairs is the staff dormitory, and downstairs is the dining room, where the CPC Delegation often held press conferences.

入口局部

建筑夹道

入口及立面

室内展陈布置（组图）

室内展陈

沿街立面入口

百乐门舞厅
Paramount Hall

工程地点	上海市
建筑面积	2550平方米
竣工时间	1932年
设计机构	杨锡镠建筑师事务所
主要设计人	杨锡镠

百乐门舞厅位于愚园路218号与万航渡路（原极司菲尔路）转角处。中国商人顾联承投资70万两白银，购静安寺地营建Paramount Hall，并以谐音取名"百乐门"，全称"百乐门大饭店舞厅"。该建筑由杨锡镠建筑师设计，钢筋混凝土结构，建筑面积2550平方米，1931年开工，一年后完工。建筑共三层。底层为厨房和店面。二层为舞池和宴会厅，最大的舞池计500余平方米。

此建筑外观采用美国近代前卫的装饰艺术（Art-Deco）建筑风格，是当时30年代的中国乃至全世界建筑设计的新潮，号称"东方第一乐府"。舞池地板用汽车钢板支托，跳舞时会产生晃动的感觉，故称弹簧地板；大舞池周围有可以随意分割的小舞池；两层舞厅全部启用，可供千人同时跳舞，室内还装有冷暖空调，陈设豪华。三楼为旅馆，

The Paramount Hall is located at the corner of No.218 Yuyuan Road and Wanhangdu Road (formerly Jessfield Road). Gu Liancheng, a Chinese businessman, invested 700,000 silver dollars and bought land from the Jing'an Temple to build the Paramount Hall, and named it "Bailemen" homophonic with Paramount, and the full name was "Bailemen Hotel Ballroom". Designed by architect Yang Xiliu, the building is of a reinforced concrete structure with a floor area of 2,550 square meters. The construction was started in 1931 and completed one year later. The building has three floors. On the ground floor there are kitchens and stores. On the second floor there are ballrooms and banquet halls, with the largest ballroom covering more than 500 square meters.

The appearance of this building is of the modern avant-garde Art-Deco architectural style in the United States, a new trend of architectural design in China and even the whole world in the 1930s. The Paramount Hall was called "No. 1 Music Mansion in the East". The floor of the ballrooms was called "spring floor" because it was supported by automobile steel plate, which would produce a sense of shaking when dancing. Around the big dance floor, there were small dance floors dividable at will. When all the dance floors on the two

外立面现状

立面局部

前厅局部

顶层装有一个巨大的圆筒形玻璃钢塔，当舞客准备离场时，可以由服务生在塔上打出客人的汽车牌号或其他代号，车夫可以从远处看到，而将汽车开到舞厅门口。

百乐门是上海著名的综合性娱乐场所。1933年开张典礼上，时任国民党政府上海市长的吴铁城亲自出席发表祝词。当时百乐门的常客有张学良，陈香梅与陈纳德的订婚仪式也在此举行，卓别林夫妇访问上海时也曾慕名而来。

floors were opened, one thousand people could dance at the same time. The rooms were also equipped with air conditioning and luxurious furnishings. The third floor was a hotel, and the top floor was equipped with a huge cylindrical glass steel tower. Dancers ready to leave would ask a waiter to type the guest's car number or other codes on the tower, so that the driver could see it in the distance and drive the car to the gate of Paramount Hall.

Paramount is a famous comprehensive entertainment venue in Shanghai. At the opening ceremony in 1933, Wu Tiecheng, then mayor of the Kuomintang government in Shanghai, personally attended it and delivered a congratulatory message. Meanwhile, Zhang Xueliang was a frequent visitor to Paramount; the engagement ceremony between Chen Xiangmei and Claire Chennault was held here; and Chaplin and his wife also came to this place when they visited Shanghai.

建筑立面旧影

哈尔滨颐园街一号欧式建筑
European Building at No.1 Yiyuan Street, Harbin

工程地点：哈尔滨市
竣工时间：1919年
主要设计人：贝伦纳达（意）

颐园街一号位于哈尔滨市南岗区，地上三层，地下一层，砖木结构，建筑面积 2000 平方米。该建筑构图严谨，比例协调，主体不是对称形式，但巧妙地配以入口门廊，使主体显得均衡匀称，主次分明。主体为两层，上设阁楼层，中央部分做通高到顶的突出体，二楼设有别致的弧形阳台，辅以科林斯巨型壁柱。建筑采用孟莎式双折高坡屋顶，阁楼老虎窗冲破檐口形成断折山花，并以文艺复兴时期常用的花瓶栏杆联接，围成了女儿墙。

建筑主入口向北，西侧为主立面。室内客厅占据两层空间，成为建筑的中心，周围的一、二层分布安排工作室、书房、餐厅、卧室、阳光室等房间。向南的阳光室设置了喷泉。客厅采用巴洛克式木质装修，做工精致。副楼为两层带地下室建筑，与主楼通过门廊连接在一起。锅炉房设在主楼

The building at No.1 Yiyuan Street in Nangang District, Harbin is a building of a brick-wood structure with three floors above ground and one floor below ground, having a floor area of 2,000 square meters. The building features rigorous composition, with a coordinated proportion. The main body is not symmetrical, but it is cleverly matched with the entrance porch, which makes the main body appear balanced and symmetrical, with distinction of primary and secondary functions. The main body has two floors, with an attic on the top. The central part has a protrusion extending to the top, and the second floor has a unique curved balcony, supplemented by giant Corinthian pilasters. A Mansard-style double-folded steep-slope roof is used for the building, and the attic dormant windows break through the eaves to form broken pediments connected by vase railings commonly used in Renaissance to form a parapet. The main entrance of the building faces north and the west side is the main facade. The indoor living room occupies the space of two floors and becomes the center of the building. The surrounding first and second floors are distributed with a studio, a study room, a dining room, bedrooms, sun rooms and other rooms. A fountain is set up in the sun room facing the south. The living room adopts baroque wood decoration with exquisite workmanship. The

外立面

室内文物陈列

二层木饰

室内展陈布置

建筑立面及栏杆局部（组图）

背面的地下室，以确保地面环境的整洁；锅炉烟囱贴建筑后墙而上，避免了对建筑造型的干扰。

建筑整体造型简洁，手法娴熟，充分反映出法国古典复兴式建筑的韵味，灰白色水刷石墙面设有装饰线，深灰色鱼鳞瓦顶在蓝天白云的映衬下，显得格外宁静安逸。建筑位于院落中央，室外有圆形水池，周围布满植物，加以绿化环境和曲径通幽的小路，形成了一座美丽优雅的花园住宅。此处原为私人住宅，曾作为中共松江省委和哈尔滨市委接待处使用，1949年后，毛泽东、周恩来、刘少奇、朱德等领导人视察黑龙江时都曾住在这里，现为纪念馆。

auxiliary building is a two-story building with a basement, which is connected with the main building through the porch. The boiler room is located in the basement on the back of the main building to ensure the cleanliness of the ground environment. The boiler chimney is attached to the back wall of the building to prevent any interference to the building modeling.

The overall shape of the building is concisely and consummately designed, fully reflecting the beauty of French classical renaissance architecture. The gray-white whitewashed stone wall is decorated with lines, and the dark gray fish scale tile roof is particularly tranquil and serene against the blue sky and white clouds. The building is located in the center of a courtyard as a beautiful and elegant garden residence with a round pool outside, surrounded by plants, and winding paths.

Originally as a private residence, the building was used as the reception office of the Songjiang Provincial CPC Committee and the Harbin Municipal CPC Committee. After 1949, leaders such as Mao Zedong, Zhou Enlai, Liu Shaoqi and Zhu De all stayed here when they visited Heilongjiang. Currently the building has been transformed into a memorial hall.

室内屋顶木饰及吊灯

室内阶梯及双层采光窗

宣武门天主堂

Cathedral of the Immaculate Conception

工程地点 北京市 | **建筑面积** 1300平方米 | **竣工时间** 1904年

宣武门天主堂于1904年重修而成，因位于北京城南，故俗称"南堂"，是北京城内最早建成的教堂，现为北京主堂。"南堂"的大门采用中式三开间大门，共有三进院落，大门占据了教堂的第一进院落，其后的东跨院为教堂的主体建筑，西跨院为起居住房。教堂主体建筑为砖结构，面向南方，正面的建筑立面为典型的巴洛克风格，三个宏伟的砖雕拱门并列，将整个建筑立面装点得豪华而庄严。室内空间运用了穹顶设计，两侧配以五彩的玫瑰花窗，整体气氛庄严肃穆。屋顶建有高大的十字架。教堂内的柱子为砖砌而成。教堂内在北面设有圣台，南面建有乐楼。教堂还有一个西跨院，院子十分幽静，有两排房子，是主教、本堂神父及南堂所有神职人员居住的地方，俗称"神父院"。

The Cathedral of the Immaculate Conception, also known as the Xuanwumen Church and colloquially Nantang after its location in the south part of Beijing, is the earliest and the primary church in Beijing. The entrance to the church is a Chinese-style three-bay gate which forms part of the first of the three courtyards of the church compound. Behind it are the two other courtyards - the east one housing the primary church building and the west one forming the living quarters. The brick primary building faces south with a typical baroque facade into which three brick-carved archways open, giving the entire front an air of magnificence and solemnity. The interior is vaulted with colored rose windows on both sides, looking quite stately. The roof is topped with a massive cross. The columns inside are built with bricks. On the north side inside the cathedral is an altar, and on the south side a music tower. Inside the west courtyard, which is quiet and secluded, there are two rows of houses where the bishop, the priests and the clergy reside, colloquially known as the "Priests' House".

俯瞰建筑与周边环境

正立面

立面装饰局部（组图）

琉璃窗饰

室内空间

砖墙雕饰及院内雕像（组图）

内庭院

入口及门上雕饰细部

郑州二七罢工纪念塔和纪念堂
Zhengzhou February 7th Strike Memorial Tower and Memorial Hall

郑州二七大罢工纪念塔简称二七纪念塔，位于河南省郑州市二七广场，建于1971年，系为纪念发生于1923年2月7日的京汉铁路工人大罢工而修建的纪念建筑，由著名建筑师林乐义与杨国权、周培南等合作设计。

此纪念建筑为钢筋混凝土结构，外观为仿古联体双塔，通高63米，共14层，其中塔基座为3层，塔身为11层。塔顶建有钟楼，六面直径2.7米的大钟，整点报时演奏《东方红》乐曲。钟楼上高矗一枚红五星。塔平面为东西相连的两个五边形，从东西方向看为单塔，从南北方向看则为

Zhengzhou February 7 Strike Memorial Tower, which is called February 7 Memorial Tower for short, is located in February 7 Square, Zhengzhou City, Henan Province. Built in 1971, it is a memorial building built to commemorate the workers' strike on the Beijing-Han Railway on February 7, 1923. It was designed by famous architect Lin Leyi in cooperation with Yang Guoquan and Zhou Peinan.

This memorial building is of reinforced concrete structure, with the appearance of antique twin towers, with a height of 63 meters and 14 floors in total, of which the tower base has 3 floors and the tower itself has 11 floors. There is a bell tower on the top of the tower, with six big clocks with a diameter of 2.7 meters, playing the music of "Dongfanghong" on the hour. There is a red five-pointed star on the clock tower. The plane of the tower is two pentagons connected from east to west. It is a single tower from east to west and a

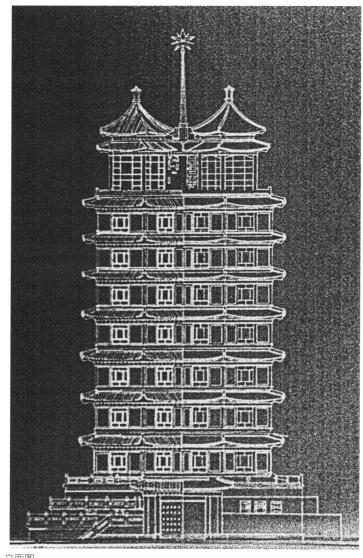

立面图

纪念塔旧影

入口细部

建筑主体与周边环境

317

双塔，塔底层有地道穿广场与道口相通。

二七纪念塔现为纪念馆，塔内二层以上辟为展厅，陈列有二七大罢工的各种历史文物、图片、文字资料。

此塔样式新颖、独特，雄伟壮观，建筑造型上具有中国民族建筑的特点，而室内空间则为展厅，具有合理的使用功能。

二七纪念塔是郑州市的重要标志，记入"郑州市近现代名优建筑名录"，是受到保护的闻名中外的纪念建筑，于2006年被列为全国重点文物保护单位。

double tower from north to south. The bottom of the tower has a tunnel through the square to connect with the crossing.

The February 7 Memorial Tower is now a memorial hall, and the second floor and above in the tower are turned into exhibition halls, displaying various historical relics, pictures and written materials of the February 7 Strike.

This tower is novel, unique and magnificent in style, with the architectural features of Chinese national architecture, while the indoor space is an exhibition hall, which has reasonable functions.

February 7 Memorial Tower is an important symbol of Zhengzhou City, which is recorded in the "List of Famous Modern Buildings in Zhengzhou City". It is a well-known memorial building under protection at home and abroad, and was listed as a national key cultural relics protection unit in 2006.

塔内局部

建筑与前广场喷泉

北海近代建筑
Beihai Modern Buildings

北海近代建筑位于广西壮族自治区北海市。1876年中英《烟台条约》签订后，北海被列为对外通商口岸，先后有英国、德国、法国、意大利等八个国家在北海设立领事馆和商务机构，同时建造了领事馆、教堂、学校、医院、海关等建筑。这些建筑均为欧式风格，其中有十几座建筑保存至今，多为一至两层，平面布置呈方正，设有回廊、地垄，地垄上铺着木地板。屋顶多为四面坡瓦顶，室内有壁炉，窗门多为拱券式。

这些近代建筑旧址，具有较高的历史价值，是中国近现代社会、经济、文化及对外开放等领域的历史见证，是中西文化交流的历史见证。

Beihai modern architecture is located in Beihai City, Guangxi Zhuang Autonomous Region. After the signing of Sino-British Yantai Treaty in 1876, Beihai was listed as a foreign trade port. Eight countries including Britain, Germany, France and Italy set up consulates and commercial organizations in Beihai, and built, churches, schools, hospitals, customs and other buildings. These buildings are of European style, among which more than a dozen buildings have been preserved up to now, mostly with one or two floors, square layout, cloisters and ridges, and wooden floors on the ridges. The roofs are mostly tiled on all sides, and there are fireplaces indoors, and the windows and doors are mostly arched.

These former sites of modern architecture have high historic value of witnessing China's modern society, economy, culture and opening to the outside world, as well as the cultural exchange between China and the West.

街景局部（组图）

街角雕塑

骑楼走廊

步行街入口

街景局部

街景局部（组图）

50年代航站楼

工程地点 北京市

建筑面积 10138平方米、58000平方米、32.6万平方米

竣工时间 1958年、1980年、1999年

设计机构 中国建筑设计研究院、北京市建筑设计研究院

主要设计人 许介三、刘国昭、马国馨 等

首都国际机场航站楼群
Terminal Buildings of Capital International Airport

首都机场航站楼最早于1955年开始建设，1958年正式投入使用，航站楼建筑面积仅为10000多平方米，现为中国国际航空公司办公楼。航站楼主体中部为瞭望塔，两翼逐步展开。停机坪一面设有高大的窗户，便于人们在室内观望飞机活动，两侧设有旅客出入口。航站广场一面，建筑立面采用花岗岩墙裙、钢门窗、米色墙面、汉白玉浮雕门楣，实现了朴实、庄重的效果。室内装修根据每一个部分不同的内容及要求，在形式、比例、色彩等方面相互协调，并配合不同的灯光效果、家具以及花卉，构成一个完美的整体。

首都国际机场T1航站楼1980年建成，建筑面积58000平方米。航站楼由主楼、卫星厅和输送廊道组成。主楼平面呈矩形，分为中央及东西两翼，首层中央为进港旅客大厅，二层为出港大厅，三层为迎送者休息廊。西翼首层为

The Capital Airport Terminal was built from 1955 and officially put into use in 1958, with a floor area of only over 10,000 square meters. It is now the office building of Air China. In the middle of the main body of the terminal building, there is an observation tower, with two wings spread. There is a tall window on the side facing the apron, to make it convenient for people to watch aircraft activities indoors. There are passenger entrances and exits on both sides. On the side facing the terminal square, granite dado, steel doors and windows, beige walls and white marble embossed lintels are adopted as the facade of the building, achieving a plain, solid and solemn effect. In interior decoration, according to the different contents and requirements of each part, all parts are coordinated with each other in form, proportion and color, and form a perfect whole with different lighting effects, furniture and flowers.

Terminal 1 of Capital International Airport was completed in 1980 with a floor area of 58,000 square meters. The terminal building consists of the main building, a satellite hall and a transmission corridor. The main building, rectangular in plane, is divided into the center, east and west wings. The center of the first floor is the inbound passenger hall, the second floor is the outbound hall, and the third floor is the welcoming lounge. The first floor of the west wing is the large, medium and small VIP lounges; on the second floor are civil aviation offices and delegation lounges; on the third floor are dining halls for transit

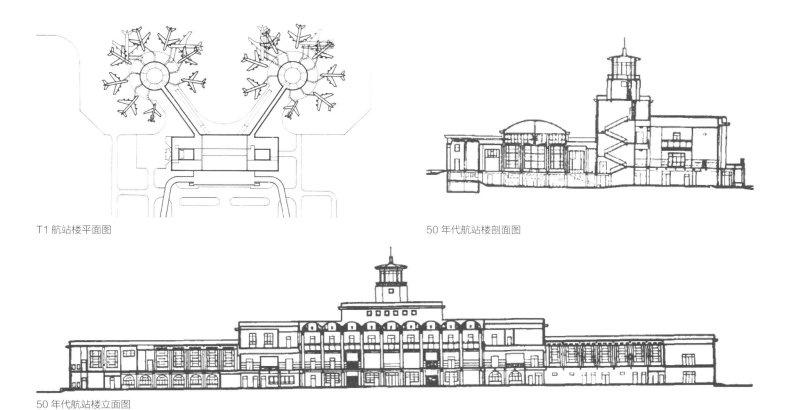

T1 航站楼平面图

50 年代航站楼剖面图

50 年代航站楼立面图

T1 航站楼鸟瞰

T1航站楼卫星厅室内

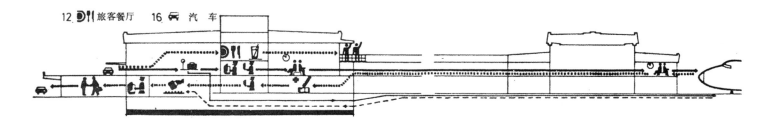

T1航站楼鸟瞰

T2 航站楼鸟瞰

贵宾用大、中、小休息厅，二层为民航办公及代表团休息厅，三层为旅客餐厅，地下室为厨房、通风机房和仓库。东翼首层为外国航空公司办事处、机组人员休息室及航运管理室，二层为航运调度中心，三层为过境旅客餐厅、机组人员餐厅等，地下室为厨房、通风机房和仓库。

主楼的东北和西北各有一座外径 50 米的卫星厅，每个卫星厅有 8 个登机门位和休息厅及服务设施。主楼与卫星厅间各设一条长 100 米的输送廊道，廊道内安装有双向自动步道，供旅客进出港使用。

航站楼在建筑造型及内外装修设计上，既体现现代化的建筑功能，又重点采用一些民族传统形式的纹样，外墙面为蛋清色面砖缀以折面几何图案，孔雀蓝釉面陶瓷板檐头额枋配以白色水刷石柱，色调明朗轻快。进出港大厅和过境餐厅分别以淡雅的色调、温暖的气氛和不同形式的沥粉彩画天花装饰。过境餐厅墙面是取材于哪吒闹海和巴山蜀水的大型重彩壁画，整个餐厅富有中国民族文化的特点。

结构为现浇钢筋混凝土框架结构，部分预制楼板，钢筋混凝土箱形基础，中部大厅为钢屋架，大型屋面板屋盖。

passengers and crews; and in the basement is the kitchen, ventilation rooms and warehouses. On the first floor of the east wing is the foreign airline office, the crew lounge and the shipping management room; on the second floor is the shipping dispatching center; on the third floor are dining halls for transit passengers and crews; and in the basement is the kitchen, ventilation rooms and warehouses.

There is a satellite hall with an outer diameter of 50 meters in both the northeast and northwest of the main building, and each satellite hall has 8 boarding gates, lounges and service facilities. There is a 100-meter-long transport corridor between the main building and each satellite hall. Two-way automatic trails are installed in the corridor for passengers to enter and leave the airport.

In architectural modeling and interior and exterior decoration design, the terminal building not only embodies modern architectural functions, but also highlights some national traditional patterns. The external wall is made of egg white bricks decorated with folded geometric patterns, and the eaves of peacock blue glazed ceramic plates are matched with white granitic stone pillars, with bright and light colors. The arrival and departure hall and the transit dining hall are decorated with quietly elegant colors, a warm atmosphere and different forms of embossed pastel ceiling. The walls of the transit restaurant are decorated with large-scale heavy-colored mural drawings about Prince Nezha's Triumph Against Dragon King and landscapes of Sichuan. The whole hall is immersed in Chinese national culture.

The building is of a cast-in-place reinforced concrete frame structure, with partially prefabricated floor slab, reinforced concrete box foundation, and steel roof truss and large roof plate in the middle hall.

This project was rated as one of the "Top Ten Buildings in Beijing in the 1980s".

该工程获北京市八十年代十大建筑称号。

首都国际机场T2航站楼1999年建成,为马国馨设计,总建筑面积32.6万平方米。航站楼地下一层,地上三层,其中二层为出港层,分国内干线和国际航线两部分,共36个固定机位,8个远机位。进港旅客经三层进港廊道到一层行李提取厅和进港大厅。在隔离区和公共区,设置了为旅客服务的餐厅、商业服务等设施,与原一号航站楼及新建停车楼之间均有通道相连。

建筑外立面以匀称流畅的曲线形金属屋面与采光天窗有机融合,配以虚实相间的弧形金属墙,形成富有时代特点和交通建筑特色的外观。入口处波浪形的雨棚使入口愈发突出。室内设计力求朴素、明朗、简洁,富有时代气息。主要大厅采用花岗石地面,铝合金复合金属墙面,吊顶除金属方板、条板外,还有浮云式曲线板材。候机厅与进港通廊空间流畅并且高低变化,再加上结构外露,同时新技术、新材料、新工艺的运用,给人深刻印象。

The Terminal 2 Building of Capital International Airport was completed in 1999 and designed by Ma Guoxin with a total floor area of 326,000 square meters. The terminal has one floor underground and three floors above ground, of which the second floor is the outbound floor, divided into domestic trunk lines and international routes, with 36 fixed bays and 8 remote bays in total. Inbound passengers go to the baggage claim hall and the inbound hall on the first floor via the third-floor inbound corridor. In isolated areas and public areas, restaurants and commercial services for passengers are available, connected with the original No. 1 Terminal and the newly-built parking building through passages.

The facade of the building is organically integrated with a symmetrical and smooth curved metal roof and a skylight, and is matched with an arc-shaped metal wall with both real and virtual appearances, thus forming a facade with the characteristics of the times and transportation architecture. The wavy canopy at the entrance highlights the entrance. In interior design, efforts have been made to make it plain, bright, concise and modern. Granite floor and aluminum alloy composite metal wall are used in the main hall. Besides metal square plates and slats, floating clouds curved plates are used in the ceiling. The building is very impressive for the smooth and varied space of the waiting hall and the entrance corridor, the exposed structure and the application of new technologies, new materials and new processes.

T2航站楼鸟瞰

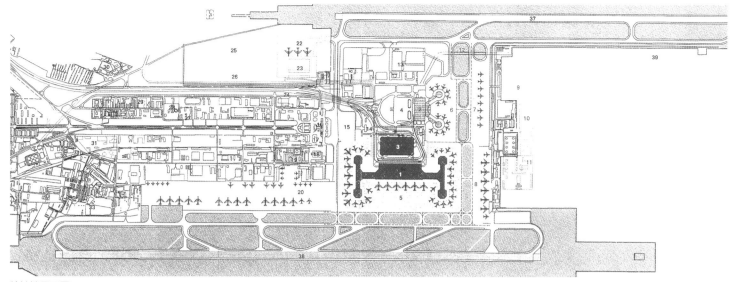

航站楼平面图

T2 航站楼路侧

T2 航站楼室内

天津市解放北路近代建筑群
Modern Building Complex on Jiefang North Road, Tianjin

工程地点 天津市

竣工时间 19世纪末至20世纪初

设计机构 景明工程司（英）/同和工程司（英）/华信工程司 等

主要设计人 沈理源/阎子亨 等

解放北路近现代建筑群位于天津市和平区，总面积近45公顷。解放北路建筑群为19世纪末20世纪初形成的英法租界银行办公区，是天津近代金融办公建筑最集中的地区。解放北路建筑群内的建筑主要有原美国海军俱乐部、原中央银行、原华俄道胜银行、原横滨正金银行、原汇丰银行、原中国实业银行、原久安商业银行、原泰兴洋行大楼、原中南银行、原花旗银行、原仁记洋行天津分行、原福利公司、原四行储蓄会天津分会、原麦加利银行、原怡和洋行大楼、原太古洋行大楼、原金城银行、原华比银行、利华大楼、原天津印字馆、原瑞隆洋行、原泰莱饭店、原屈臣氏大药房、

The Modern Building Complex on Jiefang North Road located in Heping District, Tianjin, with a total area of nearly 45 hectares, is the office area of British and French concession banks developed at the end of the 19th century and the beginning of the 20th century, which is the most concentrated area of modern financial office buildings in Tianjin.
The buildings in Jiefang North Road Building Complex mainly include the former US Navy Club, the former Central Bank, the former Russo-Chinese Bank, the former Yokohama Specie Bank, the former HSBC Bank, the former Industrial Bank of China, the former Jiu'an Commercial Bank, the former Tai Hing Co., Ltd. Building, the former Zhongnan Bank, the former Citibank, the former Gibb, Livingston & Co. Tianjin Branch, the former Welfare Company, the former Tianjin Branch of Four-bank Savings Association, the former Chartered Bank of India Australia and China, the former Jardine Matheson Building, the former Butterfield & Swire Co. Building, the former Kincheng

原天津印字馆

利顺德大饭店、原英国俱乐部等。

原美国海军俱乐部为二层砖木结构楼房，建筑外檐装饰丰富，入口处有醒目突出的阳台，拱券窗、扶壁柱等建筑元素交相辉映。建筑具有古典主义的建筑风格。

原中央银行由华信工程司建筑师沈理源设计，三层混合结构楼房，设有半地下室。外檐正立面用四根艾奥尼克巨柱和两根方柱构成对称立面，造型端正大方，布局严谨，室内装修华丽考究，具有古典复兴主义的建筑特征。

原华俄道胜银行为二层砖木结构楼房，建筑外檐墙面饰以黄色面砖，券形窗口、饰有人字形山花的平窗、弧形转角及盔顶，是一座具有浓郁俄罗斯风格的古典主义建筑。

原横滨正金银行建于1926年，三层混合结构楼房，外檐为石材墙面，造型稳重华丽，正立面的八根科林斯巨柱构成的开敞柱廊强调了对称构图，是典型的古典主义风格。

原久安商业银行是中国工程司阎子亨设计，四层钢筋混凝

Banking Corporation, the former Belgian Bank, Lihua Building, the former Tianjin Printing House, the former Building of Roy Davis, the former Tailai Hotel, the former Watsons Pharmacy, Lishunde Hotel and the former British Club, among others.

The former US Navy Club is a two-story brick-and-wood building, with rich decoration on the external eaves and a striking balcony at the entrance. Architectural elements such as arched windows and buttresses complement each other, presenting a classical architectural style.

The former Central Bank was designed by Shen Liyuan, architect of Huaxin Engineering Division. It is a three-story mixed structure building with a semi-basement. The facade of the external eaves is a symmetrical facade composed of four Ionic columns and two square columns. It has the characteristics of classical renaissance architecture, with its upright and generous shape, rigorous layout, gorgeous and elegant interior decoration.

The former Russo-Chinese Bank is a two-story brick-and-wood building, with yellow face bricks on the outer eaves and walls of the building, arched windows, flat windows decorated with herringbone-shaped pediments, curved corners and a hooded roof. It is a classical building with a prominent Russian style.

Built in 1926, the former Yokohama Specie Bank is a steady and splendid three-story mixed structure building with stone external walls. The open

原东方汇理银行

原大清邮政津局

解放北路街景 1

土框架结构楼房,局部五层。主入口位于建筑中部偏右,自二层砌筑竖向线条装饰,直至五层,使得建筑更加挺拔,建筑左侧另设拱券式入口。外檐为混水墙面,二层以上窗上挑出雨檐,同时作为横向装饰线条。建筑立面横竖线条对比明显,装饰简洁大方。

原泰兴洋行大楼建于 1937 年,建筑为四层钢混框架结构,外檐为水刷石断块饰面。首层以横线条为主,拱券形门窗,门上设有拱心石,二层以上作平窗,三层间隔设置由圆柱装饰的拱形阳台窗,配以水泥镂空护栏。建筑形体简洁大方。

原中南银行原为二层钢混结构建筑(有地下室),1938

colonnade composed of eight Corinthian columns on the facade emphasizes symmetrical composition, which is a typical classical style.

The former Jiu'an Commercial Bank is a four-story building of a reinforced concrete frame structure designed by Yan Ziheng of China Engineering Department, with five floors in part. The main entrance is located on the right to the middle of the building. It is decorated with vertical lines from the second floor to the fifth floor, which makes the building more upright. There is another arch entrance on the left side of the building. The external eaves are plaster walls, and the rain eaves are jetted out on the windows above the second floor, which are used as horizontal decorative lines at the same time. The vertical and horizontal lines of the building facade are obviously contrasted, with simple and elegant decoration.

The original Tai Hing Co., Ltd. Building, built in 1937, is a four-story steel-concrete frame structure, with the outer eaves decorated with granitic stone blocks. The first floor is mainly composed of horizontal lines, with arch-shaped

解放北路街景 2

解放北路街景 3

原太古洋行

年由华信工程司沈理源设计增建一层。建筑立面为石材装饰，规整统一。立面设计遵循古典主义的对称原则，设置贯通两层的半柱，柱身有艾奥尼克柱式花纹，但省略了柱头。入口门头细致婉约，与粗犷的主体建筑形成有趣的对比。室内装饰讲究，功能布局合理。

利华大楼建于1936—1938年，由法商永和营造管理公司设计，是一座兼具办公与住宅功能的公寓式大楼，也是天津最早的具有现代技术和功能的高层建筑之一。建筑主楼10层，高43米，钢筋混凝土框架结构，主楼外立面采用深棕色麻面砖，色彩稳重大方。内部设有两部电梯，布局合理。建筑构图采用不对称式，外部造型错落有致，典型的现代主义风格。

原泰莱饭店建于1929年，钢筋混凝土框架结构，首层为

doors and windows, with keystones on the doors. From the second floor, flat windows are used. Arched balcony windows decorated with columns at intervals and with cement hollow guardrails are found on the third floor. The architectural form is simple and elegant.

The former Zhongnan Bank was originally a two-story steel-concrete structure building (with a basement). In 1938, it was designed and built by Shen Liyuan of Huaxin Engineering Division. The facade of the building is neatly and uniformly decorated with stone. The facade has been designed according to the symmetry principle of classicism, with a half column which runs through two floors. The column body has Ionic column patterns, but with the column head omitted. The entrance gate head is meticulously and gracefully designed, which forms an interesting contrast with the rough main building. The interior decoration is exquisite and the functional layout is reasonable.

Lihua Building, built in 1936-1938 and designed by a French construction company Brossard, Mopin & Cie, is an apartment-style building with both office and residential functions, and is also one of the earliest high-rises with modern technology and functions in Tianjin. The main building of a reinforced concrete frame structure rises 43 meters high with 10 floors. The facade of the main building is made of dark brown spongy-surface bricks with a modest

原四行储蓄会天津分会

原法国俱乐部

商业用房,二层有写字间,三至六层为公寓,每套公寓均设有会客厅、餐厅、厨房、卫生间。外立面大面积使用黄褐色麻面砖,相间布置水刷石方壁柱。建筑造型典雅,色彩和谐,具有现代建筑特征。

原英国俱乐部为二层砖木结构楼房(设有地下室),建筑采用对称布局,窗间墙均匀布置的艾奥尼克巨柱强调了竖向构图,丰富的外立面装饰和拱券门窗,体现出典型的折中主义建筑风格。现存建筑为1904年重建。

解放北路历史文化街区经过近百年的历史积累,形成了大量保存较好、有价值的文物与历史风貌建筑。解放北路现存的文物与历史风貌建筑以银行办公、商业酒店建筑为主,相对正统的西洋古典主义风格建筑以罗马柱廊、三段式为主,建筑造价更高,建筑品质也更好。街区内现存大量银行、办公建筑,承载着众多真实的历史故事,保留着近代租界地的商贸金融街风貌,建成空间体现了英、法两国城市建设理念的异同,是租界时期的"城市博物馆",街区为近代历史的研究提供了史料。

and elegant color. There are two elevators inside, and the layout is reasonable. The building features a typical modern style, with asymmetrical architectural composition, and well-proportioned external shape.

The original Tailai Hotel was built in 1929, with a reinforced concrete frame structure. The first floor is for commercial use; the second floor has offices; and the third to sixth floors are apartments. Each apartment consists of a living room, a dining room, a kitchen and a bathroom. A large area of yellow-brown bricks with spongy surface are used in the facade, with granitic stone pilasters arranged alternately. The building has modern architectural characteristics, with an elegant architectural style, and harmonious colors.

The former British Club is a two-story brick-wood structure building (with a basement). The building is symmetrical and the evenly arranged Ionic columns between windows stress vertical composition. The rich facade decoration and arched doors and windows reflect a typical eclectic architectural style. The building extant was rebuilt in 1904.

Over nearly one century of historical vicissitudes, the historical and cultural block on Jiefang North Road has formed a large number of well-preserved and valuable cultural heritages and historical buildings. The existing cultural heritages and historical buildings on Jiefang North Road are mainly bank office buildings and commercial hotel buildings. The buildings with relatively orthodox Western classical style are mainly characterized by Roman colonnade and the ternary form, featuring higher construction cost and better construction quality. There are a large number of banks and office buildings in the block, bearing many real historical stories and retaining the style of commercial and financial street in the modern concession. The built-up space reflects the similarities and differences of urban construction concepts between the United Kingdom and France, and form an "Urban Museum" in the concession period, providing historical materials for the study of modern history.

正立面

西安易俗社
Xian Yisu Community Theater

西安易俗社剧场原为清朝末年修建的一座专供达官显贵听戏的室内剧场，创建于1917年6月。易俗社以重资购买剧场加以改造，并于同年投入使用，是西安城中最早、目前保护最为完整的秦腔剧场。全胜时期的易俗社，占地60余亩，几乎占据了西安市中心钟楼广场的东北区域。环绕其周围，还有露天剧场和电影院，以及经营手工艺品和地方小吃的店铺，使这里成为一个富有民间传统特色的文化娱乐中心。

易俗社剧场具有典型明清砖木结构的建筑风格，解放后曾多次维修，但仍保持着原有的风貌。入口为中国传统的垂花门，矩形观众厅为上下两层，共有955座。剧场剖面为高低跨坡屋顶，利用高侧窗自然采光与通风，具备了自然的音响效果。舞台正上方的传统彩绘记录了当时的艺术文

The Xi'an Yisu Community Theater was originally built in the late Qing Dynasty as an indoor opera theater specially for high-ranking officials. In June 1917, the Yisu Community purchased the theater with a heavy investment and put it into use in the same year. It is the earliest and most well-protected Shaanxi Opera theater in Xi'an. The Yisu Community in its heyday, covering more than 60 acres, almost occupied the northeast area of Zhonglou Square in the center of Xi'an. Surrounding it are open-air theaters and cinemas, as well as shops selling handicrafts and local snacks, making the area a cultural and entertainment center with rich folk traditional characteristics.

The Yisu Community Theater features an architectural style with a brick-wood structure typical in the Ming and Qing Dynasties. Despite the many repairs it has gone through since liberation, the theater still maintains its original style. The entrance is a traditional Chinese festoon gate, and the rectangular audience hall has two floors, with 955 seats in total. The section of the theater is a high and low cross-sloping roof, which uses high side windows for natural lighting and ventilation while assuming a natural sound effect. The traditional colorful paintings directly above the stage record the art and culture then. The whole theater was compact in space with plainly elegant decorations. The traditional form of beam-column and bucket-arch combination has been used

室内全景

化，整个剧场空间紧凑，装饰简朴淡雅，台口采用梁枋斗拱组合的传统形式，与两侧回廊柱子和木作纹样相协调，形成了统一和谐的氛围。平实简单的自然空间，使观众看得清、听得真。

1924年7月，鲁迅先生到西北大学讲学，曾多次在易俗社观看了秦腔演出，评价颇高，并亲笔以"古调独弹"四字与同来者联名题赠，还捐赠大洋50元以资赞誉和支持。易俗社剧场是一个拥有百年历史和西部文化特色的舞台，也是中国保存至今最早的室内新式剧场之一。

for the entrance of the stage, which coordinates with the cloister pillars and wooden patterns on both sides, presenting a sense of unity and harmony. The plain and simple natural space thus formed enables the audience to see clearly and hear authentically.

In July 1924, when Mr. Lu Xun delivered lectures in Northwest University, he watched Shaanxi Opera performances in the Yisu Community Theater many times, and highly praised the performances. He personally inscribed with his fellow guests the four characters "Gu Diao Du Tan" (Unique Performance of Ancient Tunes), and donated 50 silver dollars to show his appreciation and support.

The Yisu Community Theater, a centennial stage with Western cultural characteristics, is also one of the earliest indoor new theaters currently extant in China.

室内局部

从二层作为看舞台

创始人雕塑（组图）

室内装饰

地下工程主入口

工程地点｜重庆市
建筑面积｜10.4万平方米
竣工时间｜1984年

816 工程遗址
816 Project Site

位于重庆市涪陵区白涛镇的 816 工程遗址（816 地下核工程）是我国"三线建设"一个极具典型的案例。这里曾经是中国第二套核反应堆，现在以"世界上已知最大的人工洞体、全球解密的最大核军事工程"而著称。816 工程总建筑面积 10.4 万平方米，有大型洞室 18 个，道路、导洞、支洞、隧道及竖井等多达 130 条，所有洞体的轴向线长叠加达 20 余公里，其中，最大洞室高达 79.6 米，侧墙开挖跨度为 25.2 米，拱顶跨度为 31.2 米。这样"洞中有楼、楼中有洞，洞中有河"的工程设计在 1978 年曾获国家科技大会奖集体奖。1967 年 2 月，工程兵第 54 师所属三个团入川，承担起西南三线 816 工程的建设任务。国家核能工业部也陆续派出三个建筑公司，主要任务是架桥铺路、建厂房和住宅区。涪陵区县调配的民工有一万人左右，加

The 816 Project Site (816 Underground Nuclear Project), located in Baitao Town, Fuling District, Chongqing, is a typical case of the "Third-Front Movement" in China. This used to be the site of the second nuclear reactor in China, and is now known as "the largest artificial cave known in the world and the largest nuclear military project declassified in the world". The 816 Project has a total floor area of 104,000 square meters, with 18 large caverns, 130 roads, pilot tunnels, branch tunnels and shafts, etc. The axial line length of all caverns is more than 20 kilometers. The largest cavern is as high as 79.6 meters with a side wall excavation span of 25.2 meters, and a vault span of 31.2 meters. In 1978, such engineering design of "having a building in a cave, a cave in the building and a river in the cave" won the collective prize of the National Science and Technology Conference Award. In February 1967, three regiments of the 54th Division of Engineering Corps entered Sichuan to undertake the construction of 816 Project of Southwest China Third Front. The Ministry of Nuclear Energy Industry also dispatched three construction companies successively, whose main tasks were to build bridges, pave roads, and construct factories and residential areas. About 10,000 migrant workers were deployed in Fuling District and County, as well as nearly 10,000 talents recruited from the whole country and top technicians transferred from old

上从全国征调的人才和老厂转调的技术尖子也近一万人。1984年2月，由于国际形势的变化，"816工程"被封闭，始终没有正式投入过生产。此后，"816工程"中的极小一部分洞体被中国核工业建峰化工总厂作为物资仓库加以利用。1982年接到缓建指示时，816厂已完成85%的建筑工程、60%的安装工程、总投资达7.4亿元人民币。2002年4月8日，国防科工委以科工密办(2002)14号文同意对816工程解密。经过几年酝酿筹备，2010年4月底，作为世界第一大人工洞体、中国唯一解密核反应堆，816洞体工程的部分区域作为文旅项目向公众开放。

2015年6月《人民日报》对"816地下核工程"做了定位式报道：816，它既是一个历史名词，也是一种民族精神，一段共和国记忆，更是几代人的青春。这个历史名词

factories. In February 1984, due to the change of the international situation, the "816 Project" was closed and never officially put into production.

Since then, a very small part of the caves in the "816 Project" have been used as material warehouses by Jianfeng Chemical General Plant of China Nuclear Industry. In 1982, when the instruction of suspension of construction was received, 816 Plant had completed 85% of the construction works and 60% of the installation works with a total investment of 740 million yuan. On April 8, 2002, the Commission for Science, Technology and Industry for National Defense agreed to declassify the 816 Project with No.14 Document (2002). After several years of preparation, at the end of April 2010, as the largest artificial cave in the world and the only declassified nuclear reactor in China, part of the 816 Cave Project was opened to the public as a cultural tourism project.

In June 2015, People's Daily made a positioning report on "816 Underground Nuclear Project": 816 is not only a historical term, but also a national spirit, a memory of the PRC, and the youth of several generations. This historical term is called "Third-Front Movement"; this national spirit is called "selfless dedication"; this memory is known as "preparing for war and famine". In order to further explore and recognize the historical, social, technological,

白涛镇厂房及住宅（组图）

入口及山体环境

地下大空间

地下走廊（组图）

叫"三线建设";这种民族精神叫"无私奉献";这段共和国的记忆叫做"备战备荒"。为了进一步挖掘和认同816地下核工程的历史、社会、科技、经济和审美等诸多价值,实现816文化精神的重新梳理与弘扬,分析作为新中国重要"事件建筑"的"三线建设"的经典个案,2016年12月17日至18日,由中国文物学会20世纪建筑遗产委员会、中国"三线"建设研究会等单位联合主办,重庆市涪陵区人民政府、重庆建峰工业集团有限公司联合承办的"致敬中国'三线'建设的'符号'816暨20世纪工业建筑遗产传承与发展研讨会"在重庆市涪陵区举行。来自全国及重庆市的院士、大师与文博界、建筑界、旅游界知名专家百余人参加了考察和研讨活动,会议还形成了《为明天播种希望:中国重庆816共识——致敬中国"三线"建设的"符号"816暨20世纪工业建筑遗产传承与发展研讨会宣言》宣言,表达出对816项目保护与建设的六点建议:① 816地下核工程项目当属中国20世纪工业建筑遗产;② 816地下核工程应进入全国重点文物保护名录;③ 816地下核工程是中国"三线建设"的垂范典型;④ 816地下核工程应建成遗产变资源的深耕细作般的世界级创意园区;⑤ 816地下核工程是有着巨大影响力的传播空间;⑥ 816地下核工程保护与"活态"使用需要顶层设计。据悉,2009年"816工程遗址"授牌为重庆市重点文物保护单位,2018年该项目等42项入选工业和信息化部公布的"第二批国家工业遗产名单"。

economic and aesthetic values of the 816 underground nuclear project, to reorganize and carry forward the spirit of the 816 culture, and to analyze the classic case of the "Third-Front Movement", which is a significant "Event Architecture" of the People's Republic of China, the " 'Symbol' 816 Dedicated to China's 'Third-Front Movement' and the Seminar on the Inheritance and Development of Industrial Architecture Heritage in the 20th Century" were held in Fuling District from December 17 to 18, 2016, jointly sponsored by the 20th Century Architectural Heritage Committee under the Chinese Society of Cultural Relics and the China Third-Front Movement Research Association, and jointly undertaken by Chongqing Fuling District People's Government and Chongqing Jianfeng Industrial Group Co., Ltd.

Academicians and masters from all over the country and Chongqing, as well as over 100 well-known experts from cultural and museum circles, architectural circles and tourism circles, participated in the investigation and discussion activities. At the meeting, a declaration was also formed, "Sowing Hope for Tomorrow: Chongqing 816 Consensus of China: 'Symbol' 816 Paying tribute to China's Third-Front Movement and the Declaration of the Seminar on Inheritance and Development of Industrial Architecture Heritage in the 20th Century", proposing six suggestions for the protection and construction of the 816 Project: ① the 816 underground nuclear engineering project is an industrial architectural heritage of China in the 20th century; ② the 816 underground nuclear project should be listed in the key cultural heritages under state protection; ③ the 816 underground nuclear project is a typical example of China's "Third-Front Movement"; ④ the 816 underground nuclear project should be built into a world-class creative park with deep cultivation of turning the heritage into resources; ⑤ the 816 underground nuclear engineering is a communication space with great influence; ⑥ the protection and "dynamic" use of the 816 underground nuclear project call for top-level design.

It is reported that in 2009 the "816 Project Site" was listed as a key cultural heritage protection unit in Chongqing, and in 2018, 42 projects including this project were selected into the "Second List of National Industrial Heritages" published by the Ministry of Industry and Information Technology.

遍布裸露钢筋的地下通道

未完成施工的巨大洞体

地下印有标语的墙壁

洞体局部（组图）

大雁塔风景区三唐工程
Three Tang Projects in Big Wild Goose Pagoda Scenic Area

工程地点	西安市
建筑面积	27240平方米
竣工时间	1988年
设计机构	中建西北建筑设计研究院公司
主要设计人	张锦秋 等

三唐工程位于国家文物保护单位西安市南郊慈恩寺唐大雁塔东侧,是改革开放初期由中国建筑师主持设计的外资旅游项目之一。三唐工程包括唐华宾馆、唐歌舞餐厅和唐代艺术展览馆。

设计充分运用我国传统的空间理论并将其与现代生活结合,形成以雁塔高耸、"三唐"奔趋、雁塔刚健、唐华幽深为特色的刚柔相济、虚实相生的园林化格局;又借景古塔,以景寓情,把塔影组织在各组建筑的主景之中,创造意境,从而使这组时差一千多年的建筑群和谐统一而又气韵生动。整组建筑群充满诗情画意,步移景异,创造了充满历史文化情趣、舒适文明的旅游休闲环境,是传统与现代结合、探索和谐建筑的重要项目。

The Three Tang Projects, located on the east side of the Big Wild Goose Pagoda, a cultural heritage site under state protection in the Daci'en Temple in the southern suburbs of Xi'an, are a part of the foreign-funded tourism projects designed by Chinese architects in the early stage of reform and opening up. The Three Tang Projects include the Tang Hua Hotel, the Tang Song and Dance Restaurant and the Tang Art Exhibition Hall.

In the design, traditional Chinese space theories have been made full use of and combined with modern life to develop a garden landscape integrating strength and grace, virtuality and reality, characterized by the towering Wild Goose Pagoda, the spreading "Three Tang Projects", the vigorous Wild Goose Pagoda and the tranquil Tanghua Hotel. Also, the ancient pagoda is fully integrated into the main scenery of each building complex to create artistic conception, thus harmonizing, unifying and invigorating this group of buildings with a time difference of more than 1000 years. The whole group of buildings is idyllic, with different sceneries emerging from different directions, creating a comfortable and civilized tourism and leisure environment full of historical and cultural interests. This is an important project of combining tradition with modernity in the exploration of harmonious architecture.

唐华宾馆全景

总平面图

唐华宾馆主庭院东部廊道

唐华宾馆主庭院局部

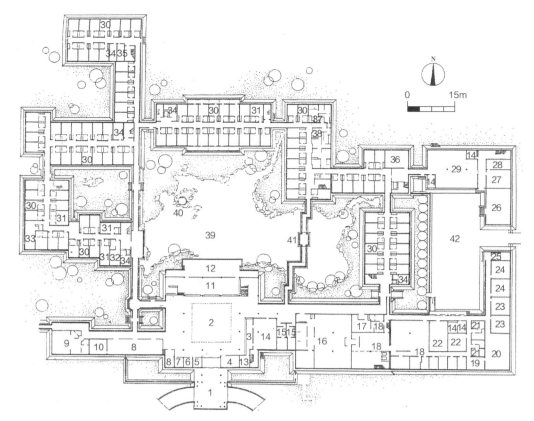

唐华宾馆首层平面图

从唐华宾馆主庭院向东望

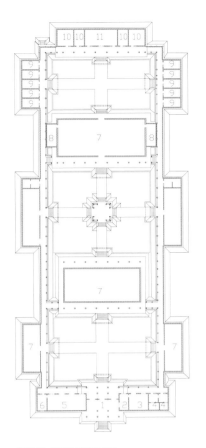

唐代艺术展览馆平面图

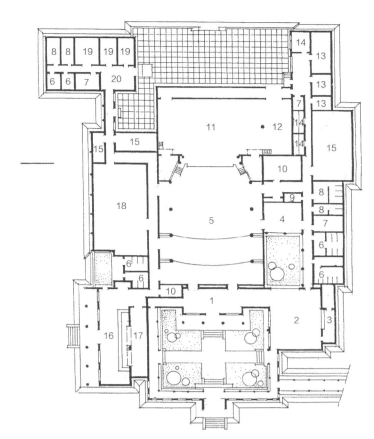

唐歌舞餐厅平面图

唐华宾馆雪景

唐歌舞餐厅内景

唐华宾馆大堂

自大雁塔上鸟瞰唐华宾馆

联接唐华宾馆和歌舞餐厅长廊

唐代艺术展览馆第二进庭院

唐代艺术展览馆改成特色酒店后原大门外景

唐代艺术展览馆改成特色酒店后第二进庭院

唐华宾馆暂时停业进行维修提升

唐华宾馆至唐歌舞餐厅长廊上开设遗址公园大门

唐代艺术展览馆改成特色酒店后第二进庭院看大雁塔

茅台酒酿酒工业遗产群
Maotai Liquor Industry Heritage Group

工程地点：怀仁市
设计年代：清代、民国及新中国后
设计机构：贵州省建筑设计研究院（改扩建）

茅台酒酿酒工业遗产群始建于明代，因战乱几次被毁，清同治元年（1862年）重建。遗产群包括现存的清同治元年（1862年）、光绪五年（1879年）、民国十八年（1929年）先后建成的"成义酒坊""荣和酒坊"和"衡昌酒坊"旧址，1951年贵州省人民政府统一接管后的"贵州省专卖事业公司仁怀茅台酒厂"时期陆续扩建和新修的各类代表性酿酒厂房等基础设施共10处。该10处不同时期的设施占地20余亩，以工业建筑及附属物为载体，含踩曲房、粮仓、曲药房、石磨房、酒库、窖池、烤酒房、古井等一套完备的酿酒工业体系。其中，部分设施至今仍发挥着重要的生产作用，整体保存完好。代表性工业建筑如下几项：

① "成义"烧坊烤酒房旧址。"成义"烧坊始建于1862年，现保存的旧址原为"成义"烧酒坊生产房。1985年茅台酒厂在原址上改建成制酒一车间生产房，该建筑体为砖石结构台梁式小青瓦顶仿古建筑，大门上刻有"茅酒之源"四个大字，面积约450平方米，保存完好。现使用的窖池就是原烧房窖池扩建，在生产空地上还有填埋的原烧坊窖池。生产房后面是原"成义"烧坊生产茅台酒取水的杨柳湾水井。

The Maotai Liquor Industry Heritage Group was founded in the Ming Dynasty, which was destroyed several times due to war and turmoil and rebuilt in the first year of the Tongzhi Reign in the Qing Dynasty (1862). The heritage group includes the extant sites of "Chengyi Distillery", "Ronghe Distillery" and "Hengchang Distillery", which were successively built in the first year of the Tongzhi Reign in Qing (1862), the fifth year of the Guangxu Reign in Qing (1879) and the eighteenth year of the Republic of China (1929), and ten basic facilities including various representative distillery factories that were successively expanded and newly built during the period of "Renhuai Maotai Distillery of Guizhou Exclusive Sales Company" after Maotai was taken over by the Guizhou Provincial People's Government in 1951. The 10 facilities from different periods cover an area of more than 20 mu, and with industrial buildings and appendages as carriers, include a complete liquor-making industrial system, such as yeast treading house, granary, yeast house, stone mill, wine depot, pit, liquor bakery and ancient well. Among them, some facilities still play an important role in production, well preserved as a whole. Representative industrial buildings are as follows:

The former site of the "Chengyi" Distillery. The "Chengyi" Distillery was built in 1862, and the former site extant now was originally the production room of "Chengyi" Distillery. In 1985, The site was transformed into a workshop of liquor making by Maotai Distillery. The building is an antique house with a brick and stone structure, and its gate is engraved with the four characters "Mao Jiu Zhi Yuan" (The Source of Maotai Liquor), covering an area of about 450 square meters, which is preserved intact. The pit used now is an expansion of the pit of the original distillery, and there is also the original distillery pit in the open space of production. Behind the production room is the Yangliuwan well where the

酒窖外立面

厂区旧影

历史遗迹

"荣和"烧坊干曲仓旧址。"荣和"烧坊始建于清光绪五年（1879年），保存至今的"荣和"烧坊干曲仓（现茅台酒厂制曲一片区干曲仓）为全木结构的仓储式建筑，台梁式小青瓦顶，面积约200平方米，是专门存放酒曲的建筑。1935年红军长征途经茅台时，曾在此驻留。"荣和"烧坊干曲仓为茅台酒厂现存最老的生产厂房，该建筑前面的道路修建在当年"荣和"烧房窖池遗址之上；

② "荣和"烧坊踩曲房旧址。1954年，将原"荣和"烧坊踩曲房旧址改建为茅台酒厂制酒一片区踩曲房，该建筑体为砖木结构，台梁式小青瓦顶，面积约150平方米。

"荣和"烧坊烤酒房旧址。"荣和"烧坊烤酒房旧址上现存有1954年改建的茅台酒厂制酒一车间烤酒房。该建筑为砖木结构，台梁式小青瓦顶，面积约30平方米，保存状况较好，内存大量原始生产工具，用于展示茅台酒工艺。

③ "衡昌"烧坊烤酒房旧址。"衡昌"烧坊始建于1929年，保存至今的烤酒房旧址（茅台酒厂制酒一车间2号生产房"茅酒古窖"）是原"衡昌"烧坊烤酒房。1941年，"衡昌"烧坊更名为"恒兴"烧坊，1985年改建为茅台酒厂一车间2号生产房，建筑体为砖石结构台梁式小青瓦顶仿古建筑，大门上刻有"茅酒古窖"四个大字，面积约400平方米，保存完好，内存大量生产工具，现仍在进行生产。现使用的窖池就是原烧坊窖池扩建，在生产空地上还留有填埋的原烧坊窖池。

④ 制曲一片区发酵仓。制曲一片区发酵仓是1954年茅台酒厂为提高产量而修建的专门用于发酵酒曲的建筑，为砖木结构小青瓦顶，面积约1000平方米，保存完好，为现存茅台酒厂最早的发酵仓，内存大量的生产工具。

original "Chengyi" distillery workshop drew water for Maotai production.

The former site of "Ronghe" Distillery Dry Yeast Warehouse. The "Ronghe" Distillery was built in the fifth year of the Guangxu Reign in the Qing Dynasty (1879). The extant dry yeast warehouse of the "Ronghe" Distillery (now the dry yeast warehouse of Maotai Distillery) is a warehouse-type building all made of wood, covered by a small blue tile roof with a floor area of about 200 square meters, and it is a building dedicated to the storage of liquor yeast. The Red Army stayed there when it passed through Maotai in 1935. The "Ronghe" Distillery Dry Yeast Warehouse is the oldest extant production plant of Maotai Distillery, and the road in front of the building was built on the former site of "Ronghe" Distillery Fermentation Pit.

The former site of the "Ronghe" Distillery Yeast Treading House. In 1954, the former site of the "Ronghe" Distillery Yeast Treading House was transformed into the Yeast Treading Room of No. 1 Area in Maotai Distillery. The building is of a brick-wood structure with a small blue tile roof covering an area of about 150 square meters.

The former site of the Liquor Distilling House of the "Ronghe" Distillery. On the former site of the Liquor Distilling House of the "Ronghe" Distillery, there is a distilling house of a workshop of Maotai Distillery, which was rebuilt in 1954. The building is a brick-wood structure with a small blue tile roof, covering an area of about 30 square meters. It is well preserved with a large number of original production tools, which are used to display Maotai liquor techniques.

The former site the Distilling House of the "Hengchang" Distillery. The "Hengchang" Distillery was founded in 1929, and the old site of the Distillery (Maojiu Ancient Pit, No.2 Production Romm of Maotai Distillery No.1 Liquor Workshop) is the original distilling house of "Hengchang" Distillery. In 1941, "Hengchang" Distillery was renamed as "Hengxing" Distillery. In 1985, it was transformed into No.2 production room of Maotai Distillery No. 1 Workshop. The building is an antique house with a brick-and-stone structure, having four characters "Mao Jiu Gu Jiao" (Ancient Pit of Maotai Distillery) engraved on the gate, covering an area of about 400 square meters. It is preserved intact with a large number of production tools, and is still in production. The pit used now is an expansion of the original distillery pit, and there is still a buried original pit in the open production space.

Fermentation Warehouse in the First Section of Yeast Making. The fermentation warehouse in the first section of yeast making is a building constructed in 1954 specially for fermenting yeast by Maotai Distillery. It is a brick-wood structure with a small green tile roof, covering an area of about 1,000 square meters,

复建后的门楼

⑤制曲二片区踩曲房、发酵仓。制曲二片区踩曲房、发酵仓是20世纪70年代茅台酒厂在1954年改建的基础上修建的第二批厂房之一，专门用于踩制酒曲。建筑体为砖木结构台梁式小青瓦顶建筑，面积约850平方米，保存完好，是现在仍在使用的最老的踩曲房。

⑥制曲二片区石磨房、干曲仓。制曲二片区石磨房、干曲仓是20世纪70年代茅台酒厂在1954年改建的基础上修建的第二批厂房之一，专门用于磨制酒曲。建筑体为砖木结构台梁式小青瓦顶建筑，面积约800平方米，保存完好，为现存茅台酒厂最早的石磨房，现仍在使用，内存大量的生产工具。

茅台酒酿酒工业遗产群见证了茅台酒由籍籍无名到享誉世界，以及由手工作坊向工业化、由民营向国营的转变。这里既是茅台酒及其文化的实物载体，也是我国民族工业自清代以来艰难前行，不断发展壮大，并创造辉煌的历史见证，是一个时代变迁的缩影。

2013年，茅台酒酿酒工业遗产群被国务院批准为"第七批全国重点文物保护单位"。

which is preserved intact. It is the earliest fermentation warehouse extant in the Maotai Distillery and has a large number of production tools.

Yeast Treading Room and Fermentation Warehouse in the Second Section of Yeast Making. The yeast treading room and fermentation warehouse in the second section of yeast making are one of the second group of factory workshops built by Maotai Distillery in the 1970s after the reconstruction in 1954, which are specially used for treading and making yeast. The building is preserved intact, having a small green tile roof with a brick-wood structure, covering an area of about 850 square meters. It is the oldest yeast treading house still in use.

Stone Mill Room and Dry Yeast Warehouse in the Second Section of Yeast Making. The stone mill and dry yeast warehouse in the second section of yeast making are one of the second factory buildings built by Maotai Distillery in the 1970s after the reconstruction in 1954, which are specially used to grind yeast. The building is a well preserved brick-and-wood-structure house with a platform-beam-type small green tile roof, covering an area of about 800 square meters. It is the earliest stone mill room in the Maotai Distillery, which is still in use now, with a large number of production tools.

The Maotai Liquor-making Industry Heritage Group has witnessed the transformation of Maotai from an unknown brand to a world-famous brand, from a manual workshop to industrialization and from a private business to a state-owned company. This place is not only the physical carrier of Maotai and its culture, but also the epitome of the vicissitudes of the times, a historical witness to the difficult struggle, gradual development and brilliant achievements of China's national industry since the Qing Dynasty.

In 2013, the Maotai Liquor Industry Heritage Group was approved by the State Council to be listed in the *Seventh Group of Key Cultural Heritage Sites under State Protection*.

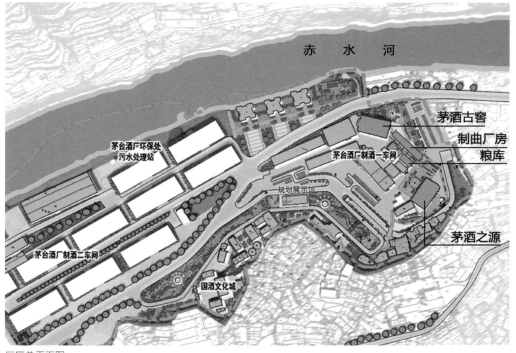

厂区总平面图

粮库外观

铭牌

酒窖室内

厂区内景

茂新面粉厂旧址
Former Site of Maoxin Flour Mill

茂新面粉厂位于江苏省无锡市，是民族工商业先驱荣宗敬、荣德生等于1900年筹资创办的，是荣氏家族创办得最早的企业，原名保兴面粉厂，后改称茂新面粉厂。抗战期间厂房被炸，设备受损。1945年重建该厂，并由荣德生之子荣毅仁先生出任厂长。其生产的"兵船牌"面粉当时享誉全国，还曾远销英、法等国及南洋各地。

现存旧址有灰色办公楼和红色厂房两栋建筑。红色厂房现在是无锡中国民族工商业博物馆以及无锡市城市规划展示馆。据悉，现存建筑均为1946年重建，设计者为著名建筑师童寯。

Maoxin Flour Mill, located in Wuxi City, Jiangsu Province, was founded in 1900 by Rong Zongjing and Rong Desheng, pioneers of national industry and commerce. It was the earliest enterprise founded by the Rong family, formerly known as Baoxing Flour Mill, and later renamed Maoxin Flour Mill. During the War of Resistance against Japanese Aggression, factory workshops were bombed and the equipment was damaged. In 1945, the factory was rebuilt, and Mr. Rong Yiren, the son of Rong Desheng, became the factory director. The "Warship Brand" flour produced by the mill was famous all over the country at that time, also exported to countries like the UK, France and other countries and all parts of South Asia.

The extant site has two buildings: the gray office building and the red factory building. The red factory building is now the site of the Wuxi Museum of Chinese National Industry and Commerce and the Wuxi Urban Planning Exhibition Hall. It is reported that the buildings extant were rebuilt in 1946, and the designer was Tong Jun, a famous architect.

工程地点	竣工时间	设计机构	主要设计人
无锡市	1946年重建	华盖建筑师事务所	童寯

厂区建筑现状

厂区旧有设施陈列

建筑立面局部

主入口立面

于田艾提卡清真寺
Aitika Mosque of Yutian County

工程地点｜和田市
建筑面积｜6666.7平方米
竣工时间｜20世纪末

于田县艾提卡清真寺位于新疆和田地区于田县老城巴扎（集市）内，又称"艾提卡尔清真寺""大清真寺"，建于1200年，至今已有800余年历史。于田艾提卡清真寺总面积为6666.7平方米，建筑为砖木结构，对研究和田地区古建筑文化具有重要价值。

清真寺由大殿、南门、北门、净身房组成，自建成至今先后历经7次扩建和维修。其中，大殿分内殿和外殿，共有11个开间，长71.68米，宽41.98米，建筑面积3009.7平方米，殿内林立着153根木柱，平顶结构，外殿前设有棚檐。南门建筑面积115.4平方米，两端各有一根高16.24米的装饰性塔柱，门侧设螺旋式楼梯可通往二层平台，平台上建有一座5平方米的两层小阁楼，高4.43米。北门重建于1997—1998年，门塔总面积227平方米，宽25米，

Aitika Mosque of Yutian County located in the Bazaar of the old town of Yutian County, Hetian District, Xinjiang, is also called "Aitikaer Mosque" and "Great Mosque". Built in 1200, it has a history of more than 800 years. The Aitika Mosque of Yutian County covering a total area of 6,666.7 square meters, with a brick-wood structure, is of great value to the study of ancient architecture culture in the Hetian area.

The mosque consisting of a main hall, a south gate, a north gate and a clean room has been expanded and repaired seven times since its establishment. The main hall is divided into inner hall and outer hall, with a width of 11 jian totaling 41.98 meters, a length of 71.68 meters, and a floor area of 3,009.7 square meters. There are 153 wooden columns in the hall, which has flat roof structure, and ridge eaves in front of the outer hall. The south gate structure has a floor area of 115.4 square meters, with a decorative tower column 16.24 meters high at each end. A spiral staircase is set at one side of the gate leading to the second floor platform, on which there is a small attic with a height of 4.43 meters and an area of 5 square meters. The north gate was rebuilt in 1997-1998. The total area of the gate tower is 227 square meters, with a width of 25 meters and a height of 15 meters. On the gate there is the Xuanli Tower and the Wangyue House. The structure has

北立面

室内柱式及空间

高 15 米，大门上建有宣礼塔、望月楼。房屋为三层，屋顶有 9 个观望塔和一个圆形顶棚。净身房位于清真寺的东南角，砖木结构，布局简洁，净身房正中布置矩形的洗手池，洗手池四角设有四根木柱以支撑上方的屋顶天窗。

建筑装饰体现了维吾尔传统建筑艺术特点。大殿木柱柱头由木雕的钟乳拱龛层层堆叠，玲珑典雅；大殿台明装饰以席纹、人字纹、波浪纹、菱形格等图案的砖雕；门侧、木柱、顶棚等部位饰以精美彩绘。

three floors, with nine watchtowers and a circular ceiling on the roof. The purification room is located in the southeast corner of the mosque, with a brick-wood structure and simple layout. A rectangular hand-washing sink is arranged in the middle of the purification room, and four wooden columns are set at the four corners of the sink to support the roof skylight above.

The architectural decoration of the mosque embodies the characteristics of Uygur traditional architectural art. The pillar heads of the main hall are stacked layer by layer by wood carving stalactite niches, which are exquisite and elegant; the main hall is decorated with brick carvings with mat patterns, herringbone patterns, wavy patterns and diamond grids, among others; gate sides, wooden columns, ceilings and other parts are decorated with exquisite color paintings.

屋顶构造

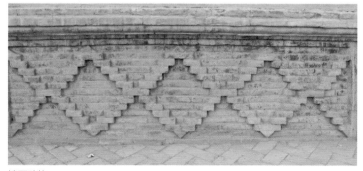

墙面砖饰

大殿俯瞰

柱头彩绘细部

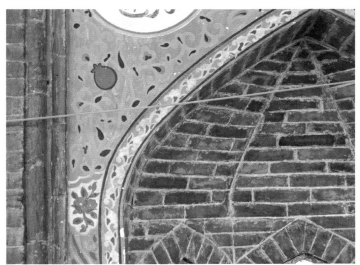

彩绘细部

大连中山广场近代建筑群
Modern Building Complex in Zhongshan Square, Dalian

工程地点 大连市

竣工时间 多为1908年至1936年

主要设计人 田秋韵（日）等

大连中山广场的建设最初始于1898年沙俄侵略中国时，当时仿照巴黎进行城市规划，在市中心围绕圆形广场设计放射状道路，周边设置银行、邮局、警察局、旅馆等机构。日俄战争后日本占领大连，对已建成区域进行大规模改造，同时进行城市扩建。今日中山广场已成为大连市的标志性景观。大连中山广场近代建筑群围绕在一个直径213米、向外辐射10条道路的圆形广场周围，建造时间多为1908年至1936年间，其中包括朝鲜银行（现中国工商银行中山广场支行）、大连民政署（现辽阳银行大连分行）、英国驻大连领事馆（现广东发展银行、上海浦东发展银行）、大和旅馆（现大连宾馆）、大连市役所（现中国工商银行大连市分行）、东洋拓殖株式会社（现交通银行大连市分行）、中国银行支店（现中信银行中山支行）、横滨正金银行支店（现中国银行辽宁省分行）、关东递信局（大连市邮政局）。广场上的大连人民文化俱乐部建于1950年。英国驻大连领事馆于1995年拆除，2000年在原址重建。

The construction of the Zhongshan Square in Dalian was started when Russia invaded China in 1898, modeled after Paris, with radial roads designed around the circular square in the center of the city, around which institutions such as banks, post offices, police stations and hotels were set up. After the Russo-Japanese War, Japan occupied Dalian, and carried out a large-scale transformation of the built area, while expanding the city. Today, the Zhongshan Square has become a symbolic landscape of Dalian. The modern architectural complex of the Zhongshan Square in Dalian was built around a circular square with a diameter of 213 meters and 10 radial roads. The buildings in the complex were mostly built from 1908 to 1936, including the Bank of Korea (now the Zhongshan Plaza Branch of the Industrial and Commercial Bank of China), the Dalian Civil Affairs Department (now the Dalian Branch of the Liaoyang Bank), the British Consulate in Dalian (now Guangdong Development Bank and Shanghai Pudong Development Bank), the Yamato Hotel (now the Dalian Hotel), the Dalian Municipal Service Office (now the Dalian Branch of the Industrial and Commercial Bank of China), Toyo Takushoku Co., Ltd. (now the Dalian Branch of the Bank of Communications), the Bank of China Branch (now the Zhongshan Branch of China CITIC Bank), the Yokohama Specie Bank Branch (now the Bank of China, Liaoning Branch), and the Kanto Mailing Bureau (now the Dalian Post Office). The Dalian People's Culture Club in the square was built in 1950. The British Consulate in Dalian was dismantled in 1995 and rebuilt in 2000.

大和旅馆旧址

英国驻大连领事馆旧址

大连市民政署旧址

大连人民文化俱乐部

大连市役所旧址

朝鲜银行旧址

大和宾馆旧址局部

东洋拓殖株式会社旧址

旅顺监狱旧址

The Former Site of Lvshun Prison

工程地点：大连市

占地面积：226000平方米

竣工时间：1907年

主要设计人：俄罗斯建筑师、日本建筑师松室重光 等

旅顺日俄监狱是俄国和日本在中国修建的一座法西斯监狱，旅顺监狱旧址是一个坐北朝南的平面略呈矩形的高墙院落，其北院墙外为丘陵地带，是原监狱窑厂、菜地、林场等旧址，南院墙偏西位置的原监狱办公楼是监狱的正门。监狱围墙长725米，高4米，墙内面积26000平方米，设置有普通牢房253间，病牢18间，暗牢4间，工厂15间，另有检身室、刑讯室、绞刑场、食堂、浴室、仓库等，加上院外的林场、窑厂、菜地，总占地面积226000平方米。这座监狱可同时关押2000多人，是当时东北地区最大的一座监狱。

旅顺日俄监狱始建于1902年，当时修建了监狱的办公楼和85间牢房。俄国人修建的监狱办公楼是一座典型的俄式古典样式建筑，为二层砖木混合结构，底边长约34米，宽约15米，通高17米，以砖墙承重，木构桁架支撑四面坡屋顶。

The Lvshun Japan-Russia Prison is a fascist prison built by Russia and Japan in China. The former site of the Lvshun Prison is a high-wall courtyard facing south with a slightly rectangular plane. Outside the north courtyard wall are hills, the former sites of the kiln factory, vegetable fields, and the forest farms of the prison. To the west of the south courtyard wall of the former prison office building is the main entrance of the prison. The prison wall is 725 meters long and 4 meters high, and the prison has an area of 26,000 square meters. There are 253 ordinary cells, 18 cells for the sick, 4 secret cells and 15 factory workshops. There are also other function areas like examination rooms, torture rooms, hanging grounds, canteens, bathrooms, warehouses, etc. Together with forest farms, the kiln factory and vegetable fields outside, the prison covered a total area of 226,000 square meters. This prison could hold more than 2,000 people at the same time, which was the largest prison in Northeast China at that time.

The Lvshun Japan-Russia Prison was founded in 1902, when the office building and 85 cells were built. The prison office building built by the Russians is a two-story building of the typical Russian classical style of a brick-wood mixed structure, with a bottom edge of about 34 meters long, a width of about 15 meters, and a height of 17 meters. The building

办公楼旧址

旅顺刑务所土地范围图

牢房内天窗

牢房立面图

其中部首层辟拱形门洞,并以此为中心左右对称排列开四个矩形窗的主体墙面,再向两侧缩进一步为开两个矩形窗的墙面;二层中部及两侧均为圆券形窗,中部两圆券窗之上设城堡式装饰墙面,而翼楼窗形则与首层翼楼一致;屋顶四个坡面均设葱头顶的老虎窗;圆券窗楣上套尖券窗楣,四面墙体均采用凹凸线脚装饰。

监狱房位于东、中、西三面,呈放射状分布,在三面牢房交会处设有中心看守台。这种牢房的分布格局反映了19世纪初叶,英国狱政学家杰里米·边沁的辐射式监狱构想理念。在上层牢房走廊中间铺设铁栅栏,一方面便于上下层牢房间的通风采光,另一方面便于看守管理,有助于上下楼之间传递管理信息。

1907年11月,日本殖民当局开始扩建和使用旅顺监狱。关东都督府民政部土木课主持了对旅顺监狱改扩建的设计工作,主要设计师是松室重光。日本秉承了俄国合理的构图手法,采用跨度宽、体量大、层位高的方式,体现厚重牢固的特点。监狱的牢房部分,将西部和中部保留,延长东部并将其加到两层,屋顶由灰色瓦棱铁制起脊屋顶转变为灰瓦起脊屋顶,监狱整体材料则由沙俄的灰砖变为红砖。

is supported by brick walls and the sloping roofs on the four sides are supported by wooden trusses. The first floor of the central part has an arch-shaped door opening and the main wall surfaces with four rectangular windows are bilaterally symmetrical with the opening as the center, and then further reduced to the wall surfaces with two rectangular windows on both sides. The middle and both sides of the second floor are round arched windows, and castle-style decorative walls are set on the two round arched windows in the middle. The window shape of the wing building is consistent with that of the first floor wing building. The four slopes of the roof are opened with dormant windows topped onion heads. Round arched window lintels are covered with sharp arched window lintels, and all four walls are decorated with concave and convex architraves.

Prison rooms are located in the east, middle and west, distributed radially, and there are central lookout points at the intersection of the cells. This distribution pattern of cells reflects the radiation prison conception of British prison scientist Jeremy Bentham in the early 19th century. Iron fences are laid in the middle of the corridor of the upper prison cell for ventilation and lighting of the upper and lower prison rooms on the one hand, and for guarding management on the other hand, which is helpful for transferring management information between the upper and lower floors.

In November 1907, the Japanese colonial authorities began to expand and use the Lvshun Prison. The Civil Engineering Department of the Ministry of Civil Affairs of Kanto Governor's Office supervised the design work of the reconstruction and expansion of the Lvshun Prison, and the main designer was Matsumuro Shigemitsu. Japan followed the reasonable composition technique of Russia, and adopted the method of wide span, large volume and high horizon, embodying the characteristics of thick and firm structures. In the prison cell part, the western and central parts were retained, and the eastern part was extended and added to two floors. The roof was changed from a grey tiled iron ridged roof to a grey tiled ridged roof, and the whole prison material was changed from grey bricks of Russia to red bricks.

建筑旧影

全景俯瞰旧影

牢房外立面（俄罗斯修建部分）

牢房外立面（日本修建部分）

暗牢通道

绞刑室

唐山抗震纪念碑
Tangshan Earthquake Monument

1984年，唐山抗震纪念碑在唐山中心城区落成。纪念碑广场由南北、东西两条轴线建立起纪念性的空间氛围。唐山抗震纪念碑蕴含着人们对这场灾难的痛彻反思，对逝去亲人的哀思与缅怀，对未来美好生活的企盼与憧憬，因而获得了社会各界的认可。

建筑师选取了"裂缝"这一客观现象加以运用，将主碑设计为单元构件重复组成，构件间留出缝隙的形式。其目的在于用隐喻的手法，使人们产生裂缝—地震的联想，来表现地震这一特定的主题。

主碑为交叉排列的四个30米高的梯形变截面柱，柱间形成缝隙。下面八组浮雕，相互之间也留有缝隙，形成断断续续的正方形环状平面。纪念碑为双重台基，为了丰富表现层次，组成观赏的空间序列，将主副碑分别设计在两层台基上。拾级而上，第一层台基上是一座象征震害的残垣断壁形式的副

In 1984, the Tangshan Earthquake Monument was completed in the city proper of Tangshan. A commemorative space atmosphere has been developed in the Monument Square by the north-south axis and the east-west axis. The Tangshan Earthquake Monument has won recognition from all walks of life because it holds people's deep reflection on the disaster, the grief and memory of their deceased relatives, and their expectation and longing for a better future life.

The architect has applied the objective phenomenon of "crack" to design the main monument as a repeated composition of unit components, leaving gaps between the components, for the purpose of using a metaphorical technique to associate cracks with earthquakes, so as to express the specific theme of earthquake.

The main monument consists of four 30-meter-high trapezoidal columns with variable cross-sections and gaps formed between the columns. The eight groups of reliefs below also have gaps between them, forming intermittent square annular planes. The monument is a double platform base. To enrich expressive levels and form an ornamental spatial sequence, the main and auxiliary monuments have been designed on two platforms respectively. Climbing up the stairs, we can see that on the first platform is a secondary monument in the form of ruins symbolizing earthquake damages.

纪念碑及基座全景

全景俯瞰

浮雕局部

隔水眺望纪念碑

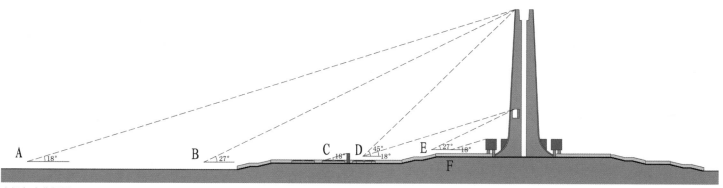

人视角度分析图

碑。碑芯镶砌花岗石镌刻的碑文，记述了地震的发生以及抗震救灾的宏伟斗争。

更上一层台阶便是那突兀而起的主碑，四片碑身与地面以圆滑曲线相交，恰似从地面自然升起，象征新唐山鳞次栉比的楼房从废墟中拔地而起，预示着新唐山的欣欣向荣。顶端为抽象的手的造型，犹如伸向天际的四只巨手，宣示着人定胜天的真理。下部由八块浮雕组成的正四边形的环，环绕碑身，象征全国四面八方的支援。碑下台基四面各设四段台阶，每段七步，共计28步，象征7月28日，这个灾难发生的日子。正面台阶设计为垂带台阶，垂带部分兼可作为坡道供伤残者轮椅通过。

主碑下部八块大型浮雕由地震灾害、抗震救灾、恢复生产、重建新唐山四组以及蒙难、救援、重建、崛起四个转角部分组成，浮雕不设边框，自然围成方形平面，与碑体紧密地、有机地结合。

The inscription on granite in the core of the monument describes the occurrence of the earthquake and the great earthquake relief efforts.

One level higher is the main monument that rises abruptly. The four pieces of monuments intersect with the ground in smooth curves, as if rising naturally from the ground, symbolizing the rise of rows of buildings in the new Tangshan from the ruins, and indicating the prosperity of the new Tangshan. The shape of abstract hands at the top is just like four giant hands reaching out to the sky, declaring the truth that man can conquer the rage of nature. The lower part is a regular quadrilateral ring composed of eight reliefs, which surrounds the monument to symbolize the support of the whole country from all directions. There are four terraces on the four sides of the foundation of the monument, each with seven steps, totaling 28 steps, symbolizing July 28th, the day when the disaster occurred.

The front terrace is designed as a vertical belt terrace, and the vertical belt part can also be used as a ramp for wheelchairs of the disabled to pass through.

The eight large reliefs at the lower part of the main monument are composed of four groups, namely earthquake disaster, earthquake relief, recovery of production and reconstruction of the new Tangshan, and four corner parts, namely disaster, rescue, reconstruction and rise. The frameless reliefs naturally form a square plane, which is closely and organically integrated with the monument.

纪念碑夜景灯光

手绘渲染图 1

手绘渲染图 2

周边环境（组图）

中苏友谊纪念塔
Sino-Soviet Friendship Memorial Tower

中苏友谊纪念塔位于辽宁省大连市旅顺口，建于1956年。塔身用汉白玉、大理石、花岗岩建成，塔高22.2米，基座呈正方形，长宽各22米，四周围以汉白玉栏杆，四面正中皆出台阶，双重月台。

中苏友谊纪念塔基座呈方形，砌在第二层月台中心，四面各为一幅浮雕图象：东为鞍钢高炉，西为中苏友谊农场，南为天安门和克里姆林宫，北为旅顺口胜利塔。

中苏友谊纪念塔座之上为多棱面柱状体的汉白玉塔身，断面为十二角形，高1.2米，象征中苏人民友谊的人物群雕像环绕塔身下端。中苏友谊纪念塔顶饰有莲花瓣，其中镶嵌着中苏友谊徽。塔身雕刻有中苏两国人民群像，塔顶雕刻有莲花、中苏友谊徽以及和平鸽。

中苏友谊纪念塔于1961年被国务院公布为第一批全国重点文物保护单位。

The Sino-Soviet Friendship Memorial Tower, located in Lvshunkou, Dalian, Liaoning Province, was built in 1956. The tower made of white marbles, marbles and granites rises 22.2 meters high, with a square base of 22 meters in both length and width. It is surrounded by white marble railings, having double platforms with steps in the middle of all sides.

The foundation of Sino-Soviet Friendship Memorial Tower is square, built in the center of the second platform, with a relief image on each of its four sides: an image of the Angang Blast Furnace in the east, an image of the Sino-Soviet Friendship Farm in the west, an image of the Tiananmen Square and Kremlin in the south, and an image of the Lvshunkou Victory Tower in the north.

The upper part of the Sino-Soviet Friendship Memorial Tower is a prismatic white marble tower with a cross-section of dodecagon and a height of 1.2 meters. The statues of figures symbolizing the friendship between the Chinese and Soviet peoples surround the lower end of the tower. The top of the Sino-Soviet Friendship Memorial Tower is decorated with lotus petals, inlaid with the Sino-Soviet Friendship Emblem. The tower is engraved with the group statues of the Chinese and Soviet peoples, and the top of the tower is engraved with lotus flowers, the Sino-Soviet friendship emblem and the doves of peace.

The Sino-Soviet Friendship Memorial Tower was announced by the State Council in 1961 as one of the first group of key cultural heritage sites under state protection.

纪念塔及基座全景

装饰细部

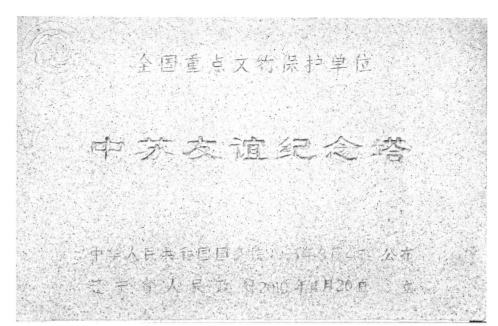
铭牌

纪念塔及基座旧影

纪念碑浮雕（组图）

细部设计（组图）

纪念碑与周边环境

金陵兵工厂
Jinling Arsenal

金陵兵工厂的前身金陵机器制造局，始创于1865年，位于南京市秦淮区中华门外，与上海江南制造局、福州船政局、天津机器制造局齐名，是我国洋务运动期间创办的四大兵工企业之一。据《续纂江宁府志》记载：其规模为"委员住房一所十二间，机器汽炉房等八十余间，廊五十余间"，"屋宇形式皆仿照西洋建筑"。

1929年6月，金陵机器制造局改隶兵工署直辖，称为金陵兵工厂。1934年，决定将厂房翻新，增购机器设备。1935年6月初发出通告，6月20日开标。扩建工程包括改扩建厂房、职工住宅、浴室、物料库、试验室等，还购置了机器设备。1937年7月抗战爆发，金陵兵工厂被迫西迁重庆，原址被日军占领。

The Jinling Arsenal, preceded by the Jinling Machinery Manufacturing Bureau founded in 1865, and located outside Zhonghuamen, Qinhuai District, Nanjing, is one of the four major ordnance enterprises founded during the Westernization Movement in China, enjoying the same reputation as the Shanghai Jiangnan Manufacturing Bureau, the Fuzhou Shipping Bureau and the Tianjin Machinery Manufacturing Bureau. According to the record in A Continuation of the Jiangning Chronicles, its scale was "12 rooms for Committee members of, more than 80 rooms for machinery and steam furnaces, and more than 50 corridors", and "the forms of buildings are all modeled after Western architecture".

In June 1929, the Jinling Machinery Manufacturing Bureau was transferred to the Ministry of Ordnance Industry and was renamed the Jinling Arsenal. In 1934, it was decided to renovate the factory buildings and purchase additional machinery and equipment. A notice was issued in early June 1935, and a bid was opened on June 20. The expansion project included the renovation and expansion of the factory buildings, staff residences, bathrooms, material warehouses, laboratories, etc., and the purchase of machinery and equipment. When the War of Resistance against Japanese Aggression broke out in July

入口牌坊

厂区运作旧影

厂区建筑旧影（组图）

厂房锯齿状侧立面

厂区建筑夹道

1945年8月，日本宣布无条件投降后，工厂重新迁回首都南京。1949年4月南京解放，同月解放军接管工厂。1950年8月，工厂改属华东军区。1952年11月改属地方，由原二机部领导。1957年4月，采用国营晨光机器厂作为第二厂名，对外公开使用。1980年3月，改称南京晨光机器厂。1996年更名为南京晨光集团有限责任公司。2007年，变身"1865创意产业园"。

现保存一些清朝和民国时期的建筑，早期的机器正厂、机器右厂和机器左厂虽已拆迁，但建厂标牌仍保存在拆迁后的厂房门额上。这些厂房具有西洋风格，人字形屋顶，三角析架，门窗上部为拱形青砖清水墙，坚固宽敞。除清朝时期遗留下的厂房建筑外，还有民国时期的厂房及办公用房，至今仍在使用。

现存民国时期建造的厂房，主要有以下几类类。一是建于1935年的二层西式厂房楼群，向南，呈凹形分布，厂房之间东西间隔约10米，南北间隔约8米，架空的过街楼将整个楼群串联在一起，从任何一座厂房的楼梯进入二楼后，不需要下楼就可以通过过街楼进入楼内任何地方。这组楼平均高16米，东面一组的第一座厂房二楼东面开有大门，与依山而筑的马路相连，电瓶车等车辆可以从此门进入厂房。过街楼的建立，为人员流动、零件转换等提供了便利，是一种既经济有科学的设计方法。有的厂房内还开有用于垂直运输的方口，并装有吊斗等设备，可以方便地在楼层之间进行转

1937, the Jinling Arsenal was forced to move westward to Chongqing, while the original site was occupied by Japanese troops.

In August 1945, after Japan announced its unconditional surrender, the arsenal moved back to Nanjing, which was liberated in April 1949, and the PLA took over the arsenal in the same month. In August 1950, the arsenal was transferred to the jurisdiction under the East China Military Zone. In November 1952, it became a unit under local jurisdiction, led by the former Second Machinery Department. In April 1957, the State-owned Chenguang Machinery Plant was adopted as the second plant name, open to the public. In March 1980, it was renamed the Nanjing Chenguang Machinery Plant. In 1996, it was renamed Nanjing Chenguang Group Co., Ltd. In 2007, it was transformed into the "1865 Creative Industry Park".

Some buildings from the Qing Dynasty and the Republic of China are now extant. Although the early central machine plant, the right machine plant and left machine plant have been demolished, the plant building signs are still kept on the front of the demolished plants. These workshops are solid and spacious, featuring a Western style, with a herringbone roof, a triangular frame, doors and windows arched with blue plain brick walls. In addition to the plant buildings left over from the Qing Dynasty, there are also plant buildings and office buildings from the Republic of China, which are still in use today.

The existing workshops built during the Republic of China mainly fall into the following types. First, the two-story Western-style plant buildings built in 1935 are distributed in a concave shape to the south, with an east-west interval of about 10 meters and a north-south interval of about 8 meters. Overhead street-crossing buildings connect the whole buildings together. After entering the second floor from the stairs of any plant building, you can enter any place in a building without going downstairs. The average height of this group of buildings is 16 meters, and the east side of the second floor of the first plant building in the east group has a gate connected with the road built along a mountain, and vehicles such as battery cars can enter a plant building through this gate. The establishment of street-crossing buildings provides convenience for people flow and parts conversion, which is an economical and scientific design method. Some workshops also have square openings for vertical transportation, and

建筑与广场陈列

厂区现状

运。同时在厂房靠墙的位置设有明沟，便于污水外泄。二是建于 1936 年的多跨连续厂房，厂房顶上设有锯齿形天窗。天窗开向北侧，这样就避开了太阳光的直射，使厂房内的光线柔和。屋顶上设置白铁明沟，用于雨水和雪水的排放。室内高 8.68 米，水泥地面，屋顶铺瓦，木屋架与瓦之间铺有油毛毡。厂房四面钢窗呈连续性分布，四面均开有大门。内部采用钢架结构，工字钢支柱，支柱内槽设有涡杆，与开启天窗的涡轮相连。开天窗时，只要转动涡杆下端的旋转手柄，就可带动涡轮涡杆系统工作，打开离地面 5 米多高的天窗。在金陵兵工厂遗存建筑中，还有一种带"戴帽儿"式的建筑，是生产机关枪的厂房，厂房屋顶的"戴帽儿"保证了厂房内的通风和采光，使得加工过程中产生的油烟灰尘，通过空气对流，有效地从厂房内排出，从而保证了厂房内的空气流通。保留至今的金陵兵工厂厂房建筑，反映了中国近代工业建筑发展的特点和历程，具有一定的历史研究价值，并保持着中国近现代历史文化的风貌。

are equipped with cableway buckets and other equipment, which can be conveniently transported between floors.

Meanwhile, an open ditch is set near the walls of the workshop to facilitate the outlet of sewage. Second, a multi-span continuous plant buildings built in 1936 have zigzag skylights on their top. The skylight opens to the north to avoid direct sunlight and soften the light in the workshop. A white iron open trench is set on the roof to discharge rainwater and snow water. The interior is 8.68 meters high, with concrete floor and tiled roof. Oil felt is laid between the wooden frame and the tiles. Steel windows are continuously distributed on the four sides of the workshop, and gates are opened on four sides. The interior is of a steel frame structure, with "I"-shaped steel struts. The inner groove of a strut is provided with a vortex rod connected with a turbine for opening the skylight. When opening the skylight, one just has to turn a rotary handle at the lower end of the vortex rod, and the turbine vortex rod system can be driven to work, thus opening the skylight more than 5 meters above the ground.

Among the remaining buildings of the Jinling Arsenal, there is also a kind of buildings "wearing a hat", which were for producing machine guns. The "hat" on the roof of the plant building ensured ventilation and daylighting in the factory building, so that the oil fume and dust produced during the processing was effectively discharged from the factory building through air convection, thus ensuring air circulation in the plant building.

The buildings of the Jinling Arsenal, which have been preserved up to now, reflect the characteristics and course of the development of modern industrial buildings in China, have certain value for historical research, while maintaining the style and features of modern Chinese history and culture.

建筑与环境

原机器大厂入口

立面窗饰局部及铭牌

天津广东会馆

Guangdong Guild Hall in Tianjin

广东会馆位于天津市南开区南门里街，建于1907年，由天津海关道唐绍仪等倡议集资兴建，是旅津广东人议事集会的场所。该建筑群是天津市现存规模最大、装饰最精美的会馆。会馆内的戏楼装饰精美，音响效果好。孙中山、邓颖超等人曾经在这里发表演说，组织活动；孙菊仙、杨小楼、梅兰芳、荀慧生等京剧大师都曾在此演出。整个建筑群采用中国传统的合院建筑形式，具有广东潮州四合院的建筑特征：砖木结构，青砖饰面，重要的房间号门窗均有精美的砖雕、木雕，工艺精湛；会馆内的戏楼采用穹窿状密集旋转斗拱藻井，装饰华美。

The Guangdong Guild Hall located in Nanmenli Street, Nankai District, Tianjin was built in 1907, with funds raised by an initiative proposed by Tang Shaoyi and others of Tianjin Customs. It is a place where Guangdong people living in Tianjin gathered for discussions. This complex is the largest and most beautifully decorated club extant in Tianjin. The theater in the club is beautifully decorated and has good sound effects. Sun Yat-sen, Deng Yingchao and others once gave speeches and organized activities here; Sun Juxian, Yang Xiaolou, Mei Lanfang, Xun Huisheng and other Peking Opera masters have performed here. The whole building complex assumes the traditional Chinese courtyard building form, featuring the architectural characteristics of Chaozhou quadrangle dwellings in Guangdong Province: brick-wood structure, blue brick veneer, exquisite brick carvings and wood carvings on important room doors and windows, created with consummate craftsmanship. The theater in the club is splendidly decorated, having a dome-shaped dense rotating bucket arch caisson.

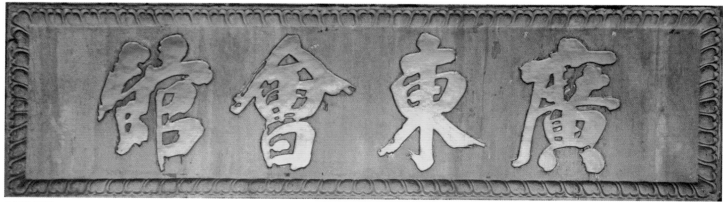

入口处牌匾

正立面

立面装饰局部（组图）

木雕装饰及牌匾

展陈局部 1

室内空间及构造

展陈局部 2

戏台局部

张謇塑像

办公用房

南通大生纱厂
Nantong Dasheng Yarn Factory

南通大生纱厂是中国近代早期的民族资本企业之一，是民族实业先驱张謇创办的近代股份制纺织企业，1895年开始筹办，次年开始购地建厂，1898年开始建设，后逐步扩大生产规模。历经社会动荡、战争以及发展变化，原南通大生纱厂的钟楼、公事厅、专家楼、清花间厂房、南通纺织专门学校旧址、唐闸实业小学教学楼等建筑、设施，仍基本保持着原有的历史面貌和格局，门类较全。

南通大生纱厂纱锭数量曾占全国纱锭总数的11.9%；张謇在南通兴办一系列工厂、学校、博物苑和地方事业对南通的城市建设和社会进步起到了巨大的推动作用。当时的纺织染传习所是我国最早创办的纺织高等院校。

Nantong Dasheng Yarn Factory, one of the enterprises funded by national capital in early modern China, is a modern joint-stock textile enterprise founded by Zhang Jian, a pioneer of national industry. The preparations for the factory started in 1895, and the land was purchased starting from the following year, construction started in 1898. Then the production scale was gradually expanded. Going through social unrest, war and development and changes, the buildings and facilities of former Nantong Dasheng Yarn Factory, such as the bell tower, the official hall, the expert building, the Qinghuajian Workshop, the former site of Nantong Textile Vocational School, and the teaching building of Tangzha Industrial Primary School, have basically maintained their original historical appearance and pattern, with relatively complete categories.

The number of spindles in Nantong Dasheng Yarn Factory once accounted for 11.9% of the total number of spindles in China. Zhang Jian set up a series of factories, schools, museums and local undertakings in Nantong, which greatly promoted the urban construction and social progress of the city. The Textile and Dyeing Training Center he established was the earliest textile institution of higher learning in China.

办公用房立面

仓库

仓库入口

入口处装饰及厂名

钟楼立面

大青楼正立面局部

工程地点 沈阳市
建筑面积 27600平方米
竣工时间 1914年至1930年
设计机构 基泰工程司
主要设计人 杨廷宝、张作霖、张学良

张学良旧居

Zhang Xueliang's Former Residence

张学良旧居又称"大帅府"或"少帅府",位于辽宁省沈阳市,是张作霖及其长子张学良的官邸和私宅。

张学良旧居始建于1914年9月,总占地3.6万平方米,总建筑面积为2.76万平方米。1916年张作霖正式入住后又不断扩建,逐步形成了由东院、中院、西院和院外建筑等四个部分组成的的建筑体系,主要有大青楼、小青楼、西院红楼群及赵四小姐楼等建筑。

张学良旧居既是张作霖官邸,也是张氏家族的私宅。张作霖将自己的府邸比作皇宫宝殿,以期通过建筑的表现形式达到张氏政权的至高无上。他头脑中的权贵思想在这所建筑中得以表现。其建筑风格也是中外皆有,各不相同。

张学良旧居是仿王府式建筑。整个建筑以东、中、西三路南北纵向排列布局,营造"府"的氛围。整体建筑坐北朝南,

Zhang Xueliang's Former Residence located in Shenyang, Liaoning Province, also known as "Dashuai Mansion" or "Shaoshuai Mansion" (Commander Zhang's Mansion) was the official domicile and private residence of Zhang Zuolin and his eldest son Zhang Xueliang.

Zhang Xueliang's Former Residence, built in September 1914, covered a total land area of 36,000 square meters with a total floor area of 27,600 square meters. After Zhang Zuolin moved in officially in 1916, the mansion continued to expand, gradually developing an architectural complex consisting of four parts, namely, the East Court, the Middle Court, the West Court and the buildings outside the courts, mainly including the Daqing Building, the Xiaoqing Building, the Red Mansions in the West Court and the Miss Zhao's Building.

Zhang Xueliang's Former residence served not only as the official domicile of Zhang Zuolin, but also as the private residence of the Zhang's clan. Zhang Zuolin had his residence built in the style of the royal palace to display the supremacy of Zhang's regime through architectural expression. His ambition to be a powerful dignitary was expressed in this mansion, which features a great diversity of architectural styles, both at home and abroad.

Zhang Xueliang's Former Residence is a palace-like architectural complex.

大青楼室内

大青楼立面窗饰

大青楼侧立面

主入口

入口处影壁

室内陈列布置

呈"目"字形结构，共有 11 栋 57 间。 四合院正门南侧有一座起脊挑檐的影壁，刻有"鸿禧"大字的汉白玉板镶嵌在影壁正中。

旧居内有侧门回廊相连，并有角门通往东面的帅府花园和大小青楼。四合院体现出典型的中国传统建筑形式，门廊柱的油饰彩绘独具特色，窗下墙身的砚石浮雕堪称一绝。旧居花园建有假山、花坛、雨路、亭台水榭、荷池和隧道。西院的七座红楼建筑群，是 1930 年由张学良规划并筑好地基，"九一八"事变后建成的。

The whole complex is vertically arranged from north to south along the east, middle and west roads, creating a "palace" style. The whole complex faces south, with a "目"-shaped structure, including 11 buildings and 57 rooms. On the south side of the main entrance of the courtyard house, there is a screen wall with ridges and overhangs, and a white marble plate engraved with the Chinese characters "Hong Xi" (meaning great bliss) are embedded in the center of the screen wall.

The buildings in the former residence are connected by side doors and corridors, with corner doors leading to the Shuaifu Garden, Daqing and Xiaoqing Buildings in the east. The quadrangle courtyard embodies the typical traditional Chinese architectural form, with unique oil painting on the porch posts and inkstone reliefs on the wall under the window. There are rockeries, flower beds, rain roads, pavilions, ponds and tunnels in the old garden. The seven red chamber buildings in the West Court were planned Zhang Xueliang, with its foundation laid in 1930, and built after the September 18th Incident.

垂花门细部

室外环境

中华民国临时参议院旧址

The Former Site of Provisional Senate of the Republic of China

旧址位于南京市鼓楼区湖南路10号，现为江苏省军区司令部。清光绪三十四年（1908年），清政府颁布《各省咨议局章程》。宣统元年（1909年）端方奏请建筑江苏省咨议局，由孙支厦仿法国文艺复兴建筑式样设计。1911年辛亥革命后，宣布起义的17个省的代表共45人，于12月10日聚集于此商讨组织临时中央政府。12月29日在此宣布成立中华民国，确定1912年为中华民国元年，推选孙中山为临时大总统。孙中山就任临时大总统后，以原江苏省咨议局院为中华民国临时政府参议院院址。参议院在此地通过《中华民国临时约法》。孙中山于1912年4月1日辞去临时大总统职务之后，此建筑又移用为国民党中央党部办公地。1937年至1945年一度由汪伪政府所用。1949年后由江苏省军区司令部接管。

此建筑群占地面积6300平方米，主要建筑仍为当年江苏省咨议局之遗存，为前后两进及东西厢房组成的四合院，占地面积4600平方米。主楼仿法国文艺复兴建筑式样，由孙支

The former site of Provisional Senate of the Republic of China located at No.10 Hunan Road, Gulou District, Nanjing, is now the headquarters of the Jiangsu Military Zone. In the thirty-fourth year of the Guangxu Reign in the Qing Dynasty (1908), the Qing government promulgated the Articles of Association of Provincial Consultative Bureaus. In the first year of the Xuantong Reign (1909), Duan Fang requested the construction of the Jiangsu Provincial Consultative Bureau, and Sun Zhisha was assigned to design with the French Renaissance architectural style. After the Revolution of 1911, a total of 45 representatives from 17 provinces that announced the uprising gathered here on December 10th to discuss the organization of the interim central government. On December 29th, the founding of the Republic of China was announced. 1912 was confirmed as the first year of the Republic of China, and Sun Yat-sen was elected as Interim President. After Sun Yat-sen took office as interim president, he took the former Jiangsu Provincial Consultative Bureau as the Senate seat of the provisional government of the Republic of China. The Senate passed the Provisional Constitution of the Republic of China here. After Sun Yat-sen resigned as Interim President on April 1, 1912, the building was transferred to the office of the Kuomintang Central Party Department. It was once used by the Wang's Puppet Government from 1937 to 1945. After 1949, the building was taken over by the Jiangsu Military Command. The building complex covers an area of 6,300 square meters, and the main buildings are still the remains of the Jiangsu Provincial Consultative Bureau, which is a quadrangle composed of front and back rooms and east and west wing rooms, covering an area of 4,600 square meters. The main building was modeled

建筑旧影 1

厦设计。正立面面阔十间,共 73.6 米,中间入口有突出的门厅,蒙莎式屋顶,中间耸起钟塔楼,室内进深 10.5 米,前后有廊,廊深约 2.90 米。后进面阔十间,共 57 米,室内进深 8 米,前后走廊 2.90 米。四合院中间原有大厅,1929 年奉安大典时,孙中山遗体自北京迁来南京,曾停柩厅内举行公祭。该大厅在"文革"期间拆去。

此建筑群于 1991 年被国家建设部、国家文物局评为近代优秀建筑,现为全国重点文物保护单位。

after French Renaissance architecture and designed by Sun Zhisha. The façade is 73.6 meters wide, with ten jian. There is a prominent entrance hall, a mansard roof in the middle and a bell tower in the middle. The indoor depth is 10.5 meters, and there are corridors in front and back, which are about 2.90 meters deep. The back courtyard has a total width of 57 meters, with ten jian. The indoor depth is 8 meters, and the front and back corridors are 2.90 meters in depth. There was originally a big hall in the middle of the quadrangle. When the Feng'an ceremony was held in 1929, Sun Yat-sen's body moved from Beijing to Nanjing, where a public sacrifice was held. The hall was demolished during the Cultural Revolution. In 1991, this architectural complex was appraised as an excellent modern building complex by the Ministry of Construction and the State Administration of Cultural Heritage of China, and is now a key cultural heritage site under state protection.

建筑旧影 2

汉冶萍煤铁厂矿旧址

The Former Site of Hanyeping Coal and Iron Plant

工程地点 黄石市

竣工时间 1908年

设计机构 盛宣怀建议组织兴建

汉冶萍煤铁厂矿旧址位于湖北省黄石市西塞山区、黄石港区，是中国历史上最早的钢铁煤联企业所在地。

光绪十六年（1890年），为修建芦汉铁路，湖广总督张之洞创建汉阳铁厂，1908年，在汉阳铁厂、大冶铁矿、萍乡煤矿的基础上，成立了汉冶萍煤铁厂矿有限公司（简称汉冶萍公司），它集勘探、冶炼、销售于一身，是中国历史上第一家用新式机械设备进行大规模生产的规模最大的钢铁煤联合企业。据《汉冶萍公司志》记载，1915年前的一段时间内，该企业的钢产量占中国钢铁产量的100%。1948年，汉冶萍公司停产。该公司先后经历了官办、官督商办、商办三个时期，几乎见证了中国近代资本主义发展的全部历程，堪称中国近代钢铁工业发展的缩影。

汉冶萍煤铁厂矿旧址现完整保留有汉冶萍时期的冶炼铁炉、

The former site of the Hanyeping Coal and Iron Plant located in Xisaishan District and Huangshigang District, Huangshi City, Hubei Province is the earliest iron, steel and coal joint enterprise in Chinese history.

In the 16th year of the Guangxu Reign (1890), to build the Luhan Railway, Governor-General Zhang Zhidong of Huguang founded the Hanyang Iron Works. In 1908, based on Hanyang Iron Works, Daye Iron Mine and Pingxiang Coal Mine, Hanyeping Coal and Iron Plant and Mine Co., Ltd. (referred to as Hanyeping Company for short) was established. Integrating exploration, smelting and sales, it is the first large-scale steel produced by new machinery and equipment in Chinese history. According to Records of Hanyeping Company, the steel output of this enterprise accounted for 100% of China's steel output in a period of time before 1915. In 1948, Hanyeping Company stopped production. Having experienced three periods: government-run, government-supervised and private business-run, and private business-run, the company witnessed almost the whole course of the development of modern Chinese capitalism, and thus could be called an epitome of the development of modern Chinese iron and steel industry.

The former site of Hanyeping Coal and Iron Plant has completely preserved the smelting furnaces, a blast furnace trestle, a Japanese-European building

厂区现状

厂区现状（组图）

厂区鸟瞰旧影

曾发行的股票

高炉栈桥、日欧式建筑群、瞭望塔、卸矿机及天主教堂（又称小红楼）等。

冶炼铁炉、高炉栈桥位于黄石市西塞山区，建于1919年，现存1、2号化铁炉，建筑面积1452平方米。冶炼铁炉于1919年12月8日动工兴建，1921年5月，1号化铁炉身建成，6月1日，汉冶萍公司在大冶钢铁厂设置开炉筹备处，筹备开炉炼铁事宜。7月24日，储备化铁炉生产用水1500吨的水塔在抽水试压时崩塌，致使开炉日期后推。1922年6月24日，1号化铁炉正式举火开炼，26日晨出铁。7月5日，因炉盖失灵，下部铁水凝固而停炼，仅出铁千余吨。1号化铁炉停炉后，汉冶萍公司令大冶钢铁厂抓紧对2号化铁炉的施工。1923年4月4日，2号化铁炉举火开炼，于5日晨出铁152吨。从此，大冶钢铁厂进入生产时期。平均日出铁335吨，如焦炭产量达400吨以上（每日需焦炭500吨）。1924年底，2号化铁炉终因焦炭供应不济而停炉，

complex, a watchtower, ore unloaders and a Catholic church (also known as Xiaohonglou) from the Hanyeping period.

The iron smelting furnaces and the blast furnace trestle, located in Xisaishan District of Huangshi City, was built in 1919. There are No.1 and No.2 iron smelting furnaces with a floor area of 1,452 square meters. The construction of smelting furnaces started on December 8, 1919, and the No.1 smelting furnace body was completed in May 1921. On June 1, Hanyeping Company set up the furnace opening preparation office in Daye Iron and Steel Plant to prepare for the furnace opening and ironmaking. On July 24th, the water tower with 1,500 tons of production water for cupola collapsed during a pumping and pressure test, which caused the delay of the opening date. On June 24, 1922, No.1 cupola was put on fire, and iron was tapped on the morning of June 26. On July 5th, due to the failure of the furnace cover, the lower molten iron solidified and the refining was stopped, and only over 1,000 tons of iron was tapped. After the shutdown of No.1 cupola, Hanyeping Company ordered Daye Iron and Steel Plant to pay close attention to the construction of No.2 cupola. On April 4th, 1923, No.2 cupola started smelting, and 152 tons of iron was produced on the morning of the 5th. From then on, Daye Iron and Steel Plant entered the production period. The average daily iron tapped was 335 tons, if the coke output was over 400 tons (500 tons of coke was needed every day). By the end of 1924, No.2 cupola was shut down due to poor coke supply. It was only in operation for

仅生产了一年零八个月，共出铁 204350 吨。

日欧式建筑群位于冶钢集团厂区西总门外行政区主干道南北两侧，是厂方为解决当时工程技术及管理人员办公和生活起居，于1917年以后相继动工兴建并陆续竣工的。该建筑群中的日式建筑，又称大字号楼，现存四栋，是厂部的高级管理及工程技术人员使用的宿舍楼，位于行政区主干道北侧，每栋两层，占地面积192平方米，建筑面积2366平方米；其中的欧式建筑仅存公事房一栋，三层砖木结构，是当时大冶铁厂厂部办公楼，为典型欧式建筑，占地面积240平方米，建筑面积720平方米。

瞭望塔旧址位于冶钢集团公安处后侧江堤上，建于1918年，主要功能为警戒、报警。塔身平面呈等边六角形，全高约13米，塔身高约11.1米，最大直径2.6米。塔分上下两层：上层为敞开式，下层在六个不同方位设12个椭圆形瞭望孔，塔门朝东，设旋转楼梯上下。该塔采用带有"铁锤钢钳"交叉标识的红砖和米灰色砖（无标识）错缝平砌。整个建筑坚固结实、美观，为欧式建筑风格。

卸矿机位于黄石市港务局11号码头，于1939年开始挖基打桩，1940年开始安装，1941年5月竣工投产。建筑面积8447平方米。

该卸矿机由两排贮矿仓和皮带运输机两大部分组成。共有8个矿仓。矿仓上部铺设铁轨，以供火车运矿、卸矿。卸矿口下，架设皮带运输机，通过卸矿口直接将铁矿装船。

小红楼位于黄石市黄石港区，建于1880年左右，为砖木结构，欧式风格。调查资料证实，该建筑原为英国人在黄石修建的天主教堂，建筑面积260平方米。现保存有一幢主楼、一幢附属设施，新中国成立后改为办公楼。

汉冶萍煤铁厂矿旧址是中国近（现）代工业发展和日本帝国主义疯狂掠夺我国资源的真实写照，是我国现存最珍贵的工业遗产。

2006年，汉冶萍煤铁厂矿旧址由国务院公布为第六批全国重点文物保护单位。

one year and eight months, with a total output of 204,350 tons of iron.

The European-style building complex is located on the north and south sides of the main road outside the West General Gate of Yegang Group Plant. The construction of the buildings was started and completed by the Plant after 1917 to provide office and living space for engineering, technical and management staff members then. There are four Japanese-style buildings extant in the complex, which were dormitory buildings used by senior management, engineering and technical personnel, located on the north side of the main road of the administrative district. Each building has two floors, covering a land area of 192 square meters with a floor area of 2,366 square meters. Among them, only one European-style building, which was a three-story office building of a brick-wood structure. The office building of Daye Iron Works at that time was a typical European building with a land area of 240 square meters with a floor area of 720 square meters.

The former site of the watchtower located on the river bank behind the Public Security Department of Yegang Group was built in 1918, mainly functioning as alarm reporting. The plane of the tower is an equilateral hexagon, with a total height of about 13 meters, a tower height of about 11.1 meters and a maximum diameter of 2.6 meters. The tower is divided into upper and lower floors: the upper floor is open, and the lower floor is provided with 12 oval observation holes in six different directions, with the tower door facing east and spiral staircase up and down. The tower is made of red bricks and beige bricks (without marks) with cross marks of "Hammer Gang Qian". The whole building is solid, beautiful and of Europear style.

The unloader is located at No.11 Pier of Huangshi Port Authority. It started to dig foundation and pile in 1939, started to install in 1940, and was completed and put into production in May 1941. The floor area is 8,447 square meters.

The ore unloader consists of two rows of ore storage bins and belt conveyor. There are 8 ore silos in total. Rails are laid on the upper part of the ore bin for ore transportation and unloading by train. Under the ore unloading port, a belt conveyor is erected to directly load the iron ore through the ore unloading port.

Xiaohonglou (Little Red Mansion), located in Huangshigang District, Huangshi City, was built around 1880, with a brick-wood structure and European style. Investigation data have confirmed that the building was originally a Catholic church built by the British in Huangshi, with a floor area of 260 square meters. Now a main building and an ancillary facility are extant, changed into an office building after the founding of the People's Republic of China.

The former site of Hanyeping Coal and Iron Plant as a true portrayal of China's recent (present) industrial development and Japanese imperialists' crazy plundering of China's resources is the most precious industrial heritage extant in China.

In 2006, the former site of Hanyeping Coal and Iron Plant was announced by the State Council as one of the sixth group of key cultural heritages sites under state protection.

水塔

甲午海战馆
Jiawu Naval Battle Memorial Hall

工程地点：威海市
竣工时间：1995年
设计机构：天津大学建筑学院
主要设计人：彭一刚

建筑位于山东省威海市刘公岛南岸，与威海市中心隔海相望。作为纪念性建筑，除保证展出功能要求外，还必须与特定的环境相融合，以形成独特的场所精神。为再现这一重大历史事件，设计师以象征主义手法使建筑形象犹如相互穿插、撞击的船体，悬浮于海滩之上，每当风起云涌，掀起阵阵狂浪，便形成一种悲壮气氛。为纪念以丁汝昌、邓世昌等人为代表的英雄人物，在建筑物上设置一尊巨大雕像，巍然屹立于"船首"，手持望远镜，怒目凝视万里海疆；随风扬起的头蓬，预示着一场恶战风暴即将来临。

The building is located on the south bank of Liugong Island, Weihai City, Shandong Province, facing the center of Weihai City across the sea. As a memorial building, it must not only meet the functional requirements of exhibition, but also integrate with the specific environment to form a unique spirit of place. To re-present the important historical event, the designers have adopted symbolism to make the architectural image resemble hulls which are interspersed, colliding with each other, suspended on the beach. Whenever the wind is surging and clouds are scudding, wild waves surge, developing a tragic atmosphere. To commemorate the heroes represented by Ding Ruchang, Deng Shichang and others, a huge statue has been set up on the building, erecting tall at the bow of a ship, holding a telescope and staring at the vast sea with anger. The hood raised in the wind indicates an imminent fierce battle.

立面全景

外立面浮雕

沿海阶梯

外立面局部

室内展陈(组图)

室内展陈

国润茶业祁门红茶旧厂房
Old Factory Buildings of Qimen Black Tea of Guorun Tea Industry

工程地点｜池州市
竣工时间｜20世纪50年代初
设计机构｜中贸部基建处华东分处

1950年，中国茶叶公司皖南分公司在安徽省贵池县池口村（今池州市池口路33号）筹建大型新式机制茶厂——贵池茶厂，之后陆续扩建，形成规模。工厂至今还留有一些老建筑。其中，制茶车间是老厂区内占地面积最大的建筑，六个一面坡屋顶并联组成大车间屋顶，立面呈锯齿状，室内由巨大的水泥柱形成无隔断墙的柱网空间。其一面坡顶为垂直的玻璃墙与斜坡瓦面组成，水泥柱中空，实为排水管道。建筑形式受当年苏联工业建筑风格的影响，外观简洁朴素，以保证实用功能为主。室内的机械设备由原贵池茶厂人自主设计、自主建造的木质联装生产线，设备优良，布局科学合理，国内仅此一条，于1961年4月施工，1962年5月竣工，至今仍可使用。整体建筑规整，风格简约，较多地采用了新技术、

In 1950, the Wannan Branch of China Tea Company established a large-scale new mechanical tea factory-Guichi Tea Factory in Chikou Village, Guichi County, Anhui Province (now at No.33 Chikou Road, Chizhou City), and then expanded it successively to form a scale. Some old buildings are still extant in the factory. Among them, the tea-making workshop is the largest building in the old factory area. Six roofs with one slope each are connected in parallel to form the roof of the large workshop, and the facade is serrated. The indoor column grid space without partition walls has been formed by huge cement columns. The top of the one-slope roofs is composed of vertical glass walls and slope tile surface, and the cement columns are hollow, which are actually drainage pipes. The architectural form was influenced by the industrial architectural style of the Soviet Union back then, with a plain appearance, mainly to ensure practical functions. The indoor mechanical equipment was independently designed and built by the former Guichi Tea Factory. The equipment is excellent and the layout is scientific and reasonable, being the only one in China. Construction was started in April 1961 and completed in May 1962, and the building can

制茶车间锯齿状立面

制茶车间室内局部

制茶车间室内局部（组图）

老毛茶仓库过廊

老毛茶仓库储藏间

新结构、新材料，代表了当时制茶产业的最高水平。

毛茶老木仓建于1952年，是外部为青砖砌墙、内部为木板包裹的库房建筑。由三个"人"字形坡顶并联组成屋顶，外观正立面呈"山"字形，整体风格简洁明快。其室内长廊居中，屋顶设亮瓦自然采光，长廊内现存同时期的大型地磅，两侧分列10个尺度统一的内库房，室内地板与四面壁板均为大兴安岭红松板材。这些与建筑同龄的红松木板，散发着让人难以忘怀的茶香。

手工拣厂是"精制工艺"的生产场地，为两栋二层建筑，分别建于20世纪50年代初和60年代初。一层室内空间宽敞，四壁有玻璃窗，采光明亮。居中有一列略显细瘦的木柱上承横梁，横梁是由上下两层木板夹蜂巢状木格组成的，具有简洁内敛又不失装饰性的美感。经分析，原设计应为平房，后因生产扩大而加建上层，为安全起见而增加立柱。

现存原贵池茶厂老厂区的建筑，自1951年后陆续建成，是新中国成立初期引进苏联等国外建筑技术而建造的工业建筑佳作，其简洁的外观和实用功能相结合，为存世较少的现代工业建筑实例，客观记录了新中国成立初期的经济状况，具有浓郁的现代工业气息和现代建筑艺术风格。

still be used today. The whole building is regular with a concise style. New technologies, new structures and new materials were adopted, representing the highest level of the tea industry at that time.

The Old Wooden Warehouse for Primary Tea, built in 1952, is a warehouse building with blue bricks on the outside and wooden boards on the inside. The roof is composed of three "herringbone"-shaped slopes connected in parallel, and its facade is "E"-shaped. The overall style is simple and bright. Its indoor long corridor is in the middle, with bright tiles on the roof for natural lighting. There are large scales in the long corridor left from the same time, with 10 inner storerooms with uniform sizes on both sides. The indoor floor and four wall boards are made of Korean pines in Daxing'anling. These Korean pine boards, which are the same age as the building, give out unforgettable tea fragrance.

The Manual Picking Factory is the production site of "refined techniques", which consists of two two-story buildings built in the early 1950s and 1960s respectively. The first floor is spacious, with glass windows on the four walls and bright lighting. In the center, there is a slightly thin wooden column bearing the cross beam which is composed of upper and lower layers of wooden boards sandwiched with honeycomb-shaped wooden lattices, simple and restrained without losing its function as an ornament. After analysis, the original design of the building should be a bungalow. Later, due to the expansion of production, an upper floor was added, and the columns were added for safety reasons.

The extant buildings in the old factory area of Guichi Tea Factory, built successively since 1951, are excellent industrial buildings built with foreign architectural technologies introduced from places such as the Soviet Union in the early days of the founding of the People's Republic of China. With their simple appearance and practical functions, these buildings are rare examples of modern industrial buildings, objectively recording the economic situation in the early days of the People's Republic of China, while assuming a strong modern industrial aura and modern architectural art style.

职工宿舍旧址

手工拣厂结构

手工拣厂屋内墙面

手工拣厂内部陈设

手工拣厂立面

手工拣厂室内陈设与结构

本溪湖工业遗产群
Lake Benxi Industrial Heritage Group

本溪湖工业遗产群经历了清朝末年、中华民国、伪满洲国和新中国四种不同的社会制度。本溪湖工业遗产群主要包括本钢一铁厂旧址（一号炼铁高炉、焦炉、洗煤楼）、本钢二电冷却塔及发电车间、本溪湖煤铁有限公司及事务所旧址（大白楼小红楼）、本溪煤矿中央大斜井（含肉丘坟）、彩屯煤矿竖井、本溪湖火车站、张作霖别墅等。

其中，本钢一铁厂旧址前身为1905年中日合办的本溪湖煤铁有限公司制铁工场，是中国最早的中外合资企业。始建于1914年的本钢一铁厂旧址1号炼铁高炉，其炉体当时从英国匹亚逊诺尔斯工厂引进，装料卷扬机、锅炉、鼓风机、发电机等设备从德国购进，为当时全亚洲最大高炉，在技术上首屈一指，是我国使用现代高炉炼铁的开端。本钢一铁厂旧

The Lake Benxi Industrial Heritage Group has gone through four different social systems: the late Qing Dynasty, the Republic of China, the Puppet Manchukuo and the People's Republic of China. The Industrial Heritage Group mainly includes the former site of No.1 Iron Works of Benxi Iron and Steel Co., Ltd. (No.1 Ironmaking Blast Furnace, Coke Oven and Coal Washing Building), the former site of Lake Benxi Coal and Iron Co., Ltd. and its office (Big White Building and Little Red Building), the central inclined shaft of Benxi Coal Mine (including Rouqiu Tomb), Caitun Coal Mine Shaft, Lake Benxi Railway Station and Zhang Zuolin's Villa.

Among them, the former site of No.1 Iron Works of Benxi Iron and Steel Co., Ltd., formerly known as the iron making workshop of Lake Benxi Coal and Iron Co., Ltd. jointly established by China and Japan in 1905, was the earliest Sino-foreign joint venture in China. Built in 1914, the No.1 ironmaking blast furnace at the former site of No.1 Iron Works of Benxi Iron and Steel Co., Ltd. was imported from the British Pyason Knowles Plant at that time, and equipment such as loading winches, boilers, blowers and generators were purchased from Germany. As the largest blast furnace in Asia at that

一号高炉

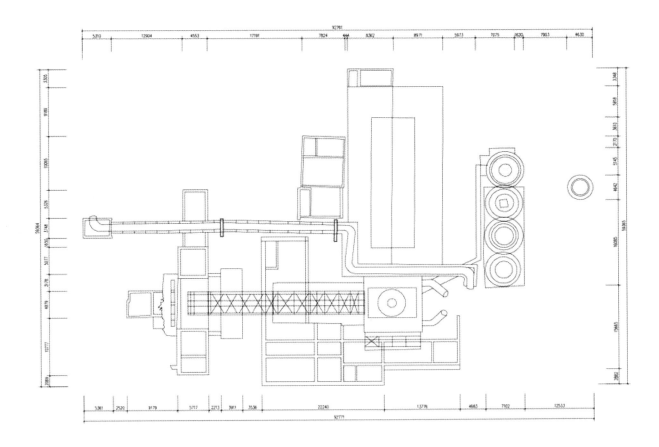

一号高炉平面图

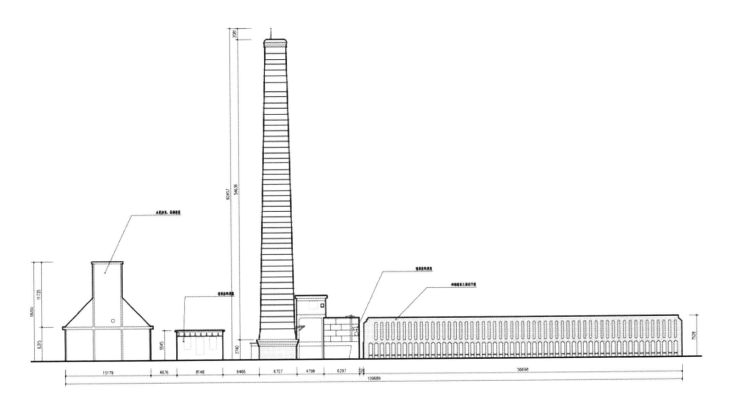

焦炉东立面图

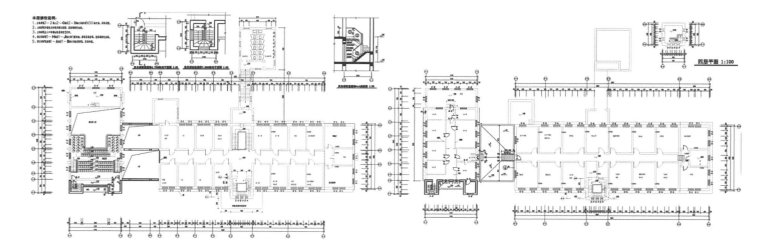

张作霖别墅二层及四层平面图

张作霖别墅现状

址，始建于1905年，为一个整体炼铁工业遗址，包括一条完整的炼铁生产线，厂区内仍保留着炼焦炉、热风炉和洗煤车间等一系列设施设备，是中国现存最早的钢铁企业。

本溪湖火车站，始建于1905年，是日本帝国主义掠夺中国东北经济物资的重要运输线，现存的候车室基本保持原有形制，仍在使用。

time, it was second to none in technology, witnessing the beginning of using modern blast furnaces to ironmaking in China. The former site of No.1 Iron Works of Benxi Iron and Steel Co., Ltd., which was built in 1905, is an integrated iron-making industrial site, including a complete iron-making production line. A series of facilities and equipment such as coke ovens, hot-blast stoves and coal washing workshops remain in the plant area. This was the earliest extant iron and steel enterprise in China.

Benxihu Railway Station, built in 1905, is an important transportation line for Japanese imperialists to plunder economic materials in Northeast China. The extant waiting rooms basically in their original shape are still in use.

高炉及制铁工厂旧影

第二发电厂冷却塔

安奉铁路本溪太子河铁路甲线桥

焦炉旧址

本溪湖煤铁公司事务所旧影

本溪湖煤铁公司旧影

洗煤楼和焦炉旧影

本溪湖火车站旧影

哈尔滨防洪纪念塔
Harbin Flood Control Memorial Tower

防洪纪念塔位于中央大街北侧尽头、松花江江岸，为纪念哈尔滨市人民战胜1957年特大洪水所建。该建筑为苏联设计师巴吉斯·兹耶列夫与哈尔滨工业大学李光耀教授于联合设计，并于1958年落成，现已成为这座城市的象征之一。

纪念塔高22.5米，由基座塔身、喷泉、围廊和广场等四部分组成一个完整的建筑景观组群。

塔的基座呈方形，上窄下宽，带有收分，由深绿色花岗碎石砌成。基座上方采用波浪式水泥杆，镶嵌着与真人大小一样的24位古铜色人物浮雕，集中描述了防洪筑堤大军从宣誓上堤、运土打夯、抢险斗争到胜利庆功等场面，表现出哈尔滨人民在防洪斗争中所表现的英雄气概。

塔身为椭圆形，每周由20块带有弧形凹槽的花岗岩组成。塔身的顶部设高为3.5米的工、农、兵、知识分子的全身立像雕塑，其中一人为苏联友人形象。塔下阶表示海拔标高119.72米，标志1932年洪水淹没哈尔滨时的最高水位；上阶表示海拔标高120.30米，标志1957年战胜大洪水时的最高水位。

塔基前的喷泉，象征勇敢智慧的哈尔滨市民正把惊涛骇浪的江水驯服成细水长流，兴利除患，造福人民。

纪念塔周围，以塔后身为中心设有20根科林式圆柱，顶端用一条宽带将圆柱两端的画壁连结在一起，组成一个35米长的半圆形罗马式回廊。

The Flood Control Memorial Tower located at the end of the north side of Central Street and on the bank of Songhua River was built to commemorate the triumph of Harbin people over the catastrophic flood in 1957. The building was jointly designed by Soviet Designer Bagis Zyelev and Professor Li Guangyao of the Harbin Institute of Technology. Completed in 1958, it has become one of the symbols of the city.

The memorial tower, standing 22.5 meters high, consists of four parts: the base and body of the tower, fountain, veranda and square.

The base of the tower is square, narrow at the top and wide at the bottom, and is made of dark green granite gravels. Wave-type concrete poles are used above the base, inlaid with 24 bronze figures with life-size reliefs, which focus on describing the scenes of the army engaged in flood control and dike building from swearing to dike building, transporting soil and tamping, fighting for emergency rescue and celebrating victory, presenting the heroic spirit of Harbin people in the struggle for flood control.

The body of the tower is oval, each circle consisting of 20 granite blocks with arc grooves. At the top of the tower is a 3.5-meter-high statue of workers, peasants, soldiers and intellectuals, one of whom is a Soviet friend. The lower stage of the tower represents the elevation of 119.72 meters above sea level, marking the highest water level when Harbin was flooded in 1932. The upper stage represents an elevation of 120.30 meters above sea level, marking the highest water level when the great flood was conquered in 1957.

The fountain in front of the tower base symbolizes that the brave and intelligent Harbin citizens are taming the stormy river into a long flowing stream, which can benefit the people and remove troubles.

Around the memorial tower, with the back of the tower as the center, there are 20 Colin columns, the top of which connects the painted walls at both ends of the columns with a wide band to form a 35-meter-long semicircular Roman cloister.

纪念塔全景

塔身

回廊细部（组图）

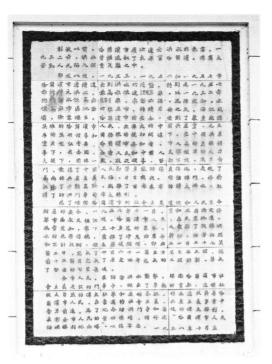

塔身铭文

回廊全景

哈尔滨犹太人活动旧址群
The Former Sites of Jewish Activities in Harbin

哈尔滨犹太人活动旧址群分散于南岗区秋林商业区和道里区中央大街一带，由20多处不同建筑风格、不同用途的单体建筑构成，其造型独特，特色鲜明，虽饱经百年沧桑仍典雅壮观。其代表性建筑遗存有1909年建成的哈尔滨犹太总会堂、1921年竣工的哈尔滨犹太新会堂、1923年兴建的哈尔滨犹太人国民银行、1914年建成的犹太巨商斯基德尔斯基欧式私邸、建于1920年的犹太穆棱煤矿公司和最早进入哈尔滨经商的犹太商人索斯金的故居等。

哈尔滨犹太人总会堂位于中央大街西侧的通江街，始建于1907年并于1909年落成，设计师为H.A.卡兹·吉列。建

The former sites of Jewish activities in Harbin are scattered in Lin Qiu Business District of Nangang District and Central Street of Daoli District, consisting of more than 20 single buildings with different architectural styles and different uses. These buildings are unique in shape and distinctive in characteristics, elegant and spectacular despite a hundred years of vicissitudes. The representative architectural remains include the Harbin Jewish General Hall built in 1909, the Harbin Jewish New Synagogue built in 1921, the Harbin Jewish National Bank built in 1923, the European-style private residence of Jewish baron Skidelski built in 1914, the Jewish Muling Coal Mine Company built in 1920, and the former residence of Jewish businessman Soschin who was the first Jew entering Harbin to do business. The Harbin Jewish General Hall located in Tongjiang Street on the west side of Central Street was built from 1907 and completed in 1909. The engineering designer was H.A. Kaz Gillette. The building suffered a

犹太国民银行旧址

秋林洋行道里分行旧址

远东银行旧址

协和银行旧址

颐园街三号立面

颐园街三号室内

穆棱煤矿公司旧址

筑曾于 1931 年遭受过严重的火灾，后由哈尔滨犹太宗教公会进行了修复，入口两侧的旋转阶梯即在此次修复中加建。2004 年及 2013 年，哈尔滨政府又先后对建筑进行了修复、改造，使其满足室内乐欣赏等功能，满足了现代人们对于历史建筑的功能需求。建筑为二层砖混结构，门窗设计由尖拱、圆拱组成，建筑顶部设有塔楼穹顶，尖顶处设置了六角圣星标识。建筑内部面积不大，其音乐厅最多时可满足百余人，室内装饰风格独特，造型精美，韵味十足，结合暖色调的光线设计和中轴对称的柱廊，更凸显了庄严、典雅的空间感受。据悉在建筑建成长达半个多世纪的时间里，这里始终是哈尔滨犹太人宗教、文化、经济和社会的管理中心，在世界犹太文化研究中占特殊地位。昔日的哈尔滨犹太人总会堂成为犹太民族与中国文化相交集的见证者，今日的老会堂音乐厅更将成为中外艺术交流的传播之所。

犹太新会堂坐落于在道里区经纬街与经纬五道街交口处，是一座以红白两色为主色调的老建筑，楼顶有巨大的金色圆形穹顶，中心立着一个六角星。据《哈尔滨市志》记载，犹太新会堂是东北地区最大的犹太教堂，可容纳 800 人同时做礼拜。

severe fire in 1931, and was later repaired by the Harbin Jewish Religious Association. The rotating stairs on both sides of the entrance were added in this repair. In 2004 and 2013, the Harbin government successively repaired and transformed the buildings to satisfy the functions of chamber music appreciation, and meet the functional requirements of modern people for historical buildings. The building is a two-story brick-concrete structure. The doors and windows are designed to have a pointed arch and a round arch. A tower dome is set at the top of the building, and a hexagonal star sign is set at the spire. The interior area of the building is small, and its concert hall can satisfy more than 100 people at most. The interior decoration is unique, exquisite and charming. The combination of warm light design and symmetrical colonnade highlights the solemn and elegant spatial feeling. It is reported that during the more than half a century since the building was built, it had always been the management center of religion, culture, economy and society of Jews in Harbin, occupying a special position in the study of Jewish culture in the world. The Harbin Jewish General Hall from the past has become a witness to the intersection of Jewish people and Chinese culture, while today's old hall concert hall will become a place for the communication of Chinese and foreign art.

Located at the intersection of Jingwei Street and Jingwei Wudao Street in Daoli District, the Jewish New Synagogue is an old building with red and white as its main colors, having a huge golden dome on the roof and a hexagonal star erecting in the center. According to Harbin Records, the new synagogue is the largest synagogue in Northeast China, able to accommodate 800 people to worship at the same time.

别尔科维奇大楼旧址

犹太人中学立面局部

老会堂室内装饰局部

犹太人中学室内局部

老会堂立面

老会堂室内

武汉金城银行（现市少年儿童图书馆）
Wuhan Jincheng Bank (Currently the Municipal Children's Library)

金城银行位于武汉市汉口区南京路、黄石路与中山大道交会处。金城银行成立于1917年，总部曾设于天津，是当时中国重要的私人银行"北四行"之一（另三家为盐业银行、中南银行和大陆银行）。汉口金城银行大楼是一座四层楼的筋混凝土结构建筑，为中国第一代建筑师庄俊所设计，于1930年动工，1931年落成。

汉口金城银行大楼之正立面采用西洋古典廊柱样式，基座至门廊设21级台阶，廊柱高通达三层，在二层壁面位置开圆形拱窗，整体雄伟而富于变化。其柱头为简化的科林斯式与中国回字纹的组合，莨苕叶（Acanthus）之间又雕刻有西式兽头，似为美国新古典主义建筑的随性改造之举。

1938年日军攻陷武汉后，这里曾被强占为侵华日军的总司令部，1945年抗战胜利后收回。1957年，武汉市少年宫在此开设。

Jincheng Bank is located at the intersection of Nanjing Road, Huangshi Road and Zhongshan Avenue in Hankou District, Wuhan. Founded in 1917 and once headquartered in Tianjin, Jincheng Bank was one of the four important private banks in China then (the other three were Salt Industry Bank, China and South Sea Bank and Mainland Bank). Hankou Jincheng Bank Building, a four-story reinforced concrete structure, was designed by Zhuang Jun, the first Chinese architect. Construction was started in 1930 and completed in 1931.

The facade of Jincheng Bank Building in Hankou is of the western classical peristyle style, with 21 steps from the base to the porch. The peristyle is as high as three floors, and a circular arch window is opened on the second floor wall. It is magnificent and varied as a whole. The stigma is a combination of simplified Corinthian style and Chinese palindrome, and western-style animal heads are carved between Acanthus, which seems to be a casual transformation of American neoclassical architecture. In 1957, the Wuhan Children's Palace was opened here.

After the Japanese army captured Wuhan in 1938, this building was occupied as the headquarters of the Japanese invaders, which was recovered after the victory of the War of Resistance against Japanese Aggression in 1945.

建筑立面

立面构造局部

建筑立面旧影

柱头装饰细部

夜景灯光

建筑立面旧影

民国中央陆军军官学校（南京）

Central Army Academy of the Republic of China, Nanjing

中华民国陆军军官学校，前身为黄埔军校，1924年由中国国民党成立，1927年改制为"中央陆军军官学校"，1946年后改称"中华民国陆军军官学校"，位于南京市黄浦路，后为南京军区所在地。

1927年，南京国民政府决定将广州黄埔军校本部迁至南京，成立中央陆军军官学校。迁居南京办校之初，学校沿用原清陆军学校（老教育团）旧址为校舍，1928年至1933年，先后建造了大批新建筑，具有代表性者如1号楼、大礼堂、憩庐、122号楼等，总造价54928元。由张谨农设计，杨仁记营造厂承建。

大礼堂坐北朝南，长方形平面，大厅能容纳一两千人，有休息廊、舞台、准备用房，堂内北面设讲台，后有休息室等。

The Republic of China Army Academy, formerly known as the Whampoa Military Academy, was founded by the Chinese Nationalist Party in 1924, changed into the Central Army Academy in 1927, and renamed the Republic of China Army Academy after 1946. It is located in Huangpu Road, Nanjing, which has later become the seat of the Nanjing Military Zone.

In 1927, the Nanjing National Government decided to move the headquarters of Guangzhou Whampoa Military Academy to Nanjing to set up the Central Army Academy. At the beginning after the relocation, the academy used the former site of the Qing Army School (Old Education Regiment) as the school building. From 1928 to 1933, a large number of new buildings were built successively, such as Building No.1, Grand Hall, Qilu, Building No.122, etc., at a total cost of 54,928 yuan. These buildings were designed by Zhang Jinnong and built by Yang Renji Construction Factory.

The Grand Hall faces south and has a rectangular plane. The hall can accommodate one or two thousand people, having a lounge, a stage and a preparation room. There is a podium in the north of the hall and a lounge in

大礼堂采用钢筋混凝土结构。立面中央高三层、两侧二层，呈三段式形式，有文艺复兴时期法国府邸式建筑的影子，坡屋顶，上覆灰色波纹金属瓦。中央主要入口处门廊前矗立着八根爱奥尼克式巨柱，门廊顶部建有钟楼；东西两侧入口处各有一拱门，门侧墙壁上装饰四根爱奥尼克式立柱，其上各建一座高高的塔楼。中央主要入口处有拱门三，东西两侧入口处各有一拱门，占地1530平方米。大礼堂与1928年以后学校陆续建设的其他建筑一样，以西洋风格为主，形成西式风格的建筑组群。

抗战胜利后，直属陆军总司令部，改称陆军军官学校，其复校之后的重大历史事件，是这里成为受降仪式地点。

1945年9月9日上午9时整，象征中国人民抗日战争最终胜利的受降仪式在中央陆军军官学校大礼堂举行。此日，军校校门悬横匾，蓝底之上楷书"中国陆军总司令部"八个白色大字，礼堂大门口是金碧辉煌的"和平永驻"四个大字；厅内大屏风上缀一个硕大字母"V"，象征中国抗日战争彻底胜利。中央军校大礼堂从此名闻中外。

the back. The Grand Hall adopts reinforced concrete structure. The facade is three stories high in the center and two stories high on both sides. It is in the form of three sections, reminiscent of Renaissance French residence-style architecture, with a sloping roof and a grey corrugated metal tile. Eight Ionic-style columns stand in front of the porch at the main entrance of the center, and a bell tower is built on the top of the porch. There is an arch at the entrance of the east and west sides, and four Ionic-style columns are decorated on the side wall of the gate, on which a tall tower is built. There are 3 arches at the main entrance of the center, and there is an arch at the entrance of east and west sides. The hall covers an area of 1,530 square meters. The Grand Hall, like other buildings built by the academy after 1928, is mainly of the Western style. They form a complex of western-style buildings.

After the victory of the War of Resistance against Japanese Aggression, the academy was directly under the Army General Command and renamed as the Army Academy. The major historical event after the resumption of the school was that it became the venue of the surrender ceremony.

At 9: 00 am on September 9, 1945, the surrender ceremony symbolizing the final victory of the Chinese People's War of Resistance Against Japanese Aggression was held in the Grand Hall of the Central Army Academy. On this day, on the gate of the Academy, a horizontal plate was hung, on which eight white characters "Zhong Guo Lu Jun Zong Si Ling Bu" (Chinese Army General Headquarters) were written on the blue background. On the entrance of the Grand Hall is the resplendent four characters "He Ping Yong Zhu" (Peace Forever). A huge letter "V" was affixed to the large screen in the hall, symbolizing the complete victory of China's War of Resistance against Japanese Aggression. The Grand Hall of the Central Army Academy has become famous both at home and abroad since then.

大礼堂立面

受降仪式旧影

广场现状（组图）

沈阳中山广场建筑群
Buildings of Zhongshan Square in Shenyang

沈阳中山广场建筑群包括中山广场及周边历史建筑遗存。广场前身称中央广场，始建于1913年，到1919年改名"浪速广场"。原广场中央有日本人建的日俄奉天大会战纪念碑，抗战胜利后，国民政府将原纪念碑拆除，改名为中山广场。1949年后，仍沿用中山广场称谓。广场中心树立毛主席雕像，于1970年落成。中山广场建筑群名列第七批全国重点文物保护单位。

目前，除了原沈阳铁路局公安处被拆和改建外，中山广场周边尚存六大建筑，均保持了建筑原貌。

大和旅馆旧址（现辽宁宾馆）：始建于1927年至1929年，由日本南满铁路株式会社建设和经营管理。总体建筑地上六层，地下一层，是当时附属地一带最高建筑，为典型的欧式风格，也是沈阳地区最早的大型豪华宾馆之一。

大和警务署旧址（现沈阳市公安局所在建筑）：位于中山广场西北端，建于1906年，占地面积为30000平方米，是日伪统治时期对付沈阳人民的专政场所。

横滨正金银行奉天支店旧址（现中国工商银行沈阳市分行中山支行所在建筑）：位于中山广场西侧。1905年日本横滨

The architectural complex of Zhongshan Square in Shenyang includes the remains of Zhongshan Square and its surrounding historical buildings. Formerly known as Central Square, the square was built in 1913 and renamed as "Langsu Square" in 1919. In the center of the original square, there was a Japanese-Russian Fengtian War Memorial. After the victory of the War of Resistance against Japanese Aggression, the National Government demolished the original monument and renamed it Zhongshan Square. After 1949, the title of Zhongshan Square was still used. A statue of Chairman Mao was erected in the center of the square in 1970. Zhongshan Square complex ranks among the seventh batch of national key cultural relics protection units.

At present, apart from the demolition and reconstruction of the former Public Security Department of Shenyang Railway Bureau, there are still six buildings around Zhongshan Square, all of which keep their original appearance.

The former site of Yamato Hotel (now Liaoning Hotel): It was built from 1927 to 1929, and was built and operated by Nanman Railway Corporation of Japan. The whole building has six floors above ground and one floor below ground. It is the tallest building in the affiliated area at that time, and it is a typical European style. It is also one of the earliest large luxury hotels in Shenyang.

The former site of Daiwa Police Department (now the building Shenyang Public Security Bureau): Located at the northwest end of Zhongshan Square, it was built in 1906 and covers an area of 30,000 square meters. It was a dictatorship place against Shenyang people during the Japanese puppet rule.

The former site of Fengtian Branch of Yokohama Zhengjin Bank (now

广场卫星图

广场全景鸟瞰

广场中央雕塑旧影

广场中央雕塑局部

正金银行在沈阳（奉天）城内中街设立奉天出张所（支店），1921年该支店从城内迁入日本租界（南满附属地），1925年10月又迁至中央广场西侧。

东洋拓殖株式会社奉天支店旧址（现沈阳市总工会办公楼）：位于中山广场东侧。该会社创建于1908年，是日本官商合资的特殊金融机构，1922年在沈阳成立了东洋拓殖株式会社奉天支店，1926年迁至于此，由日本人设计建成。

三井物产会社旧址（现招商银行中山支行）：位于中山广场北端南京北街196号，总建筑面积6142平方米。该社创建于1918年，总会社设在日本东京，它是日本三大财团之一，是日本掠夺和剥削东北人民的重要工具。

日本朝鲜银行旧址（现华夏银行中山路支行）：位于中山广场东北角的两栋钢混办公楼是伪满洲兴业银行旧址，建于1920年，建筑面积为2881平方米。

the building of Zhongshan Branch of Shenyang Branch of Industrial and Commercial Bank of China): located on the west side of Zhongshan Square. In 1905, Masayoshi Bank of Yokohama, Japan set up Fengtian Chuzhangsuo (Fengtion branch) in the middle street of Shenyang (Fengtian City). In 1921, the branch moved from the city to the Japanese concession (attached to Nanman), and in October 1925, it moved to the west side of Central Plaza.

Former site of Fengtian Branch of Toyo Takushoku Co., Ltd. (now office building of Shenyang Federation of Trade Unions): located on the east side of Zhongshan Square. Founded in 1908, the company is a special financial institution jointly established by Japanese government and business. In 1922, Toyo Takushoku Co., Ltd. Fengtian Branch was established in Shenyang, and moved here in 1926. The building was designed and built by Japanese.

The former site of Mitsui & Co., Ltd. (now Zhongshan Sub-branch of China Merchants Bank): located at No.196, Nanjing North Street, north of Zhongshan Square, with a total construction area of 6,142 square meters. Founded in 1918, the Society is headquartered in Tokyo, Japan. It is one of the three major consortia in Japan and an important tool for Japan to plunder and exploit people in Northeast China.

The former site of Korea Bank of Japan (now Zhongshan Road Sub-branch of Huaxia Bank): The two steel-concrete office buildings located in the northeast corner of Zhongshan Square are the former site of Manchuria Industrial Bank, which was built in 1920 and covers an area of 2,881 square meters.

原横滨正金银行

原东洋拓殖株式会社

原大和警务署

抗日胜利芷江洽降旧址
The Former Site of Japanese Surrender Negotiation in Zhijiang

抗日胜利芷江洽降旧址，位于湖南省芷江侗族自治县城东3.5千米的七里桥，为第六批全国重点文物保护单位。

旧址含恰降会场、资料展览室、洽降纪念坊、中国陆军总司令部、何应钦办公室（三栋鱼鳞板瓦木结构平房）以及芷江军用机场旧址（原停机坪、指挥塔、美国空军俱乐部）等。芷江机场始建于1936年10月，在抗战时期为中美空军重要军事基地，有"远东第二大机场"之称。建造过程中，5000名中国农民付出了生命。至今，芷江机场旧址犹存十几个1.80米直径的巨大石磙，它们散落在荒野田间，是点缀在当年的机场指挥塔、美军俱乐部等建筑物前的纪念性雕塑。

1946年2月建造的洽降纪念坊，由四柱三门组成。柱粗1.16米见方，通高8米。中拱门高4.5米。坊宽10.64米，进深1.5

The Former Site of Japanese Surrender Negotiation in Zhijiang, located at Qili Bridge, 3.5 kilometers east of Zhijiang Dong Autonomous County, Hunan Province, is one of the sixth group of key cultural heritage sites under state protection.

The former site includes the surrender negotiation meeting place, the material exhibition room, the Surrender Negotiation Memorial Arch, the Chinese Army General Command, He Yingqin's Office (three bungalows of a wood structure with fish-scales tiles), and the former site of the Zhijiang Military Airport (former apron, control tower and American Air Force Club), etc.

The Zhijiang Airport, built in October 1936, was an important military base of the Sino-US Air Force during the War of Resistance against Japanese Aggression, and was known as "the second largest airport in the Far East". During the construction process, 5000 Chinese farmers paid their lives. Up to now, more than a dozen 1.80-meter-diameter huge stone rollers are still extant in the former site of Zhijiang Airport, which are scattered in the wilderness, and are commemorative sculptures in front of the airport control tower, American military clubs and other structures.

受降纪念坊

米。石料、青砖砌成，外涂水泥砂浆。中拱门上有横额"恰降纪念坊"。坊正、背面刻有当时军政要员题词，铭文略述日本侵略军投降之梗概。在立面构图上，此纪念坊以汉字之"血"字形构成三开间牌坊的正立面形象，故后人又称其为"血字纪念坊"。

中国陆军总司令部、何应钦办公室、会议室等，系按原貌复

The Surrender Memorial Arch, built in February 1946, consists of four columns and three gates. The column is 1.16 square meters thick and 8 meters high. The middle arch gate is 4.5 meters high. The arch is 10.64 meters wide and 1.5 meters deep, made of stone and blue bricks, coated with cement mortar. There is a tablet on which it is inscribed "Qiajiang Memorial Arch" on the middle arch. Inscriptions of military and political dignitaries at that time were engraved on the front and back of the arch, which outlined the surrender of the Japanese invading army. On the facade

受降堂洽谈席（组图）

芷江机场指挥塔外景

建,并陈设会谈的历史资料。

相关历史资料:1945年8月15日,日本政府接受《菠茨坦公告》,日本天皇裕仁宣布无条件投降。同年8月21日,代表百万侵华日军的今井武夫副总参谋长一行,奉侵华派遣军司令冈村宁次之命令,由南京飞抵芷江,向中、美军事当局何应钦、柏德诺(中印缅战区美军作战司令部参谋长)等洽降签字。中国战区受降全权代表何应钦在芷江部署全国16个受降区100处缴械点受降事宜,20天签发24份备忘录,日方交出了在华兵力部署图,接受了令其陆、海、空三军缴械投降命令备忘录。"芷江洽降"宣告了侵华日军的彻底失败,写下了我国近代史上抵御外敌入侵第一次取得完全胜利的光辉一页。为纪念这一重大历史史实,1947年2月,国民政府在洽降地建"受降纪念坊"以志纪念。

composition, this Memorial Arch forms the facade image of the three-jian archway in the shape of the Chinese character "血" ("blood"). Thus it is also later known as the "Blood Memorial Arch".

The General Command of the Chinese Army, He Yingqin's Office, and the conference room, etc. have been rebuilt as they were, with the historical data of the negotiations displayed.

Relevant historical materials: On August 15, 1945, the Japanese government accepted the Potsdam Proclamation, and Japanese Emperor Hirohito announced his unconditional surrender. On August 21st of the same year, Takeo Imai, deputy chief of staff representing millions of Japanese invaders, and his party flew from Nanjing to Zhijiang under the command of Okamura Ningji, commander of the Japanese troops invading China, and signed agreements with Chinese and American military authorities, including He Yingqin and Bert Nordberg (chief of staff of the US Combat Command in the China-India-Myanmar Theater). He Yingqin, the plenipotentiary of surrender acceptance in the Chinese War Zone, deployed in Zhijiang 100 surrender points in 16 surrender acceptance areas across China, issuing 24 memos in 20 days. The Japanese side handed over the deployment map of its troops in China and accepted the memorandum ordering its army, navy and air force to surrender. "Zhi Jiang Surrender Negotiations" declared the complete defeat of the Japanese invaders, writing a glorious page of the first complete victory against foreign invasion in the modern history of China. To commemorate this important historical event, in February, 1947, the National Government built the " Surrender Memorial Arch".

受降堂外景

机场官俱乐部外景

山西大学堂旧址

The Former Site of Shanxi University

山西大学堂旧址位于侯家巷9号,建于清光绪二十八年(1902年)。山西大学堂为全国最早的三所新式大学(山西大学堂、京师大学堂、北洋大学堂)之一。由中学专斋、西学专斋两部分组成。中学专斋为时任山西巡抚岑春煊在晋阳书院基础上创办,初名山西大学堂;西学专斋为英国人李提摩太利用"山西教案赔款"兴建,初名中西大学堂,后经双方协商,两校合并,1904年迁入新建的侯家巷校区。现存主要文物建筑为原工科教学大楼,以及同属旧址的围墙、东房、西房和西门。工学大楼坐北朝南,中部主楼五层,两侧翼楼二层,对称分布,其上建有女儿墙。楼体东西长79.92米,南北宽22.92米,中部高23.28米,东西高14.1米,建筑面积4672平方米。中部主楼面宽三间,设拱券式门洞,顶部建方形钟楼。两侧楼宽各十间,下层辟拱券式门窗,上层为方形窗,窗口装饰采用西式柱式,前后两坡面屋顶,由时任山西大学堂校长高时臻主持修建,属折中主义建筑风格。2013年山西大学堂旧址被公布为第七批全国重点文物保护单位。

The former site of Shanxi University, located at No.9 Houjia Lane, was built in the 28th year of the Guangxu Reign in the Qing Dynasty (1902), known as one of the three earliest modern universities in China (Shanxi University, Jingshi University and Beiyang University).The university consists of two parts, namely, the College of Chinese Learning and the College of Western Learning. The Chinese School was founded by Cen Chunxuan, then Governor of Shanxi Province, on the basis of Jinyang Academy, and was first named Shanxi University. The College of Western Learning was built by Timothy Richard, an Englishman, with the fund from the "Shanxi Massacre Indemnity", first named as Chinese and Western University. After consultation between the two sides, the two universities merged and moved to the newly-built Houjia Lane Campus in 1904. The extant main cultural heritage buildings are the original engineering teaching building, as well as the walls, east room, west room and west gate of the same former site. The engineering building faces south, with five floors in the central main park and two floors on both flanks, which are symmetrically distributed with parapets on them. The building is 79.92 meters long from east to west, 22.92 meters wide from north to south, 23.28 meters high in the middle and 14.1 meters high on the east and west sides, with a floor area of 4,672 square meters. The central main building is three-jian wide, with arched doors and a square clock tower at the top. The parts on both sides are of ten jians, with arch doors and windows on the lower floor and square windows on the upper floor. The windows are decorated with Western-style columns. And the roof have front and back slopes. Its construction was presided over by Gao Shizhen, then president of Shanxi University, with the eclectic architectural style adopted. In 2013, the former site of Shanxi University was announced as one of the seventh group of key cultural heritage sites under state protection.

室内布置(组图)

建筑全景俯瞰

建筑立面

铭牌

主入口

建筑立面

墙面窗饰

立面局部

北京大学地质学馆旧址
The Former Site of the School of Geology, Peking University

北京大学地质学馆位于北京市东城区沙滩北街 15 号，曾是嵩公府祠堂（清朝乾隆年间大学士傅恒的家庙）故地，后在此建造原国立北京大学地质学馆。20 世纪 30 年代初，北京大学从傅恒裔孙松椿手中购得此处房产，经过多方筹集资金，在此建成地质学馆。此建筑系建筑学家梁思成、林徽因于 1934 年设计，1935 年 8 月建成。

彼时正值建筑界西方现代主义与复古折中主义纷争之际，被后人视为"民族形式－复古主义"领军人物的梁思成先生却出人意料在设计地质学馆时采用了现代主义风格。

建筑平面呈曲尺形，为三层砖混结构，建筑外形完全服从内部功能。体形仅微量曲折，打破了立面的平直感，大玻璃窗洞使该建筑的外形显得清新、轻巧。建筑的主入口设在东南

The Former Site of the School of Geology, Peking University located at No.15, Shatan North Street, Dongcheng District, Beijing used to be the ancestral temple of Songgong Mansion (the home temple of Fu Heng, Grand Secretary in the Qianlong Reign of the Qing Dynasty). In the early 1930s, Peking University purchased this property from Sun Songchun, a grandson of Fu Heng, and built the School of Geology here after raising funds from various sources. This building was designed by architects Liang Sicheng and Lin Huiyin in 1934 and completed in August 1935.

At that time, during the dispute between Western modernism and retro eclecticism in architecture, Mr. Liang Sicheng, who was regarded as the leader of "national form-revivalism" by later generations, unexpectedly adopted the modernist style in designing the School of Geology.

With a curved plane, the building is a three-story brick-concrete structure. The shape of the building completely follows the internal functions. The shape is only slightly curved, breaking the straightness of the facade. The large glass window opening makes the building appear fresh and light. The main entrance of the building in the southeast corner, the door is wide and

入口立面

立面局部（组图）　　　　　　　　　立面旧影（组图）

角，门洞宽大内凹，混凝土挑檐简洁；门洞两侧的线脚、灯箱，以及台阶花池，都将入口突出为这座建筑的重点。主入口立面左上方的女儿墙局部高起处，设有旗杆，强调了入口的位置。建筑整体设计精细，窗间的墙上以砖块砌成简洁的凸凹横线，门窗比例、楼梯扶手、墙角的弧线等设计，都形成了明快和谐的现代造型艺术风格。

梁思成的这个设计，有人认为是囿于建设资金有限，实际上却是真正体现了梁思成对中国传统建筑设计理念的一贯认识：建筑之美应以适度、适用为基准。

concave, with a concrete overhang. The architraves, light boxes and the framed flower beds along the steps on both sides of the doorway highlight the entrance as the key point of the building. The parapet at the upper left of the facade of the main entrance is raised locally with a flagpole, which emphasizes the position of the entrance. The overall design of the building is exquisite. The walls between the windows are made of bricks with simple convex and concave horizontal lines. The design of the ratio of doors and windows, stair handrails, and arc lines in the corners all constitute a bright and harmonious style of modern plastic arts.

Some people think that Liang Sicheng's design is constrained by the limited construction funds, but as a matter of fact, this design reflects the architect's consistent understanding of traditional Chinese architectural design concept: the beauty of architecture should be based on moderation and applicability.

庭院局部 1

杭州西湖国宾馆
Hangzhou West Lake State Guesthouse

西湖国宾馆坐落在西湖的西面，三面临湖、一面靠山，庭院面积 36 万平方米，因环境优美、建筑精巧、陈设典雅而冠居西湖第一名园。这座 19 世纪末期建成的水竹居，在 20 世纪 50 年代重新设计改建，设计者是我国著名建筑设计师戴念慈先生。20 世纪 90 年代末期，西湖国宾馆又曾作部分翻建整修。

此宾馆整体为园林式布局，客房、办公区、游泳馆、餐厅等房屋建筑均为清水墙面的砖混结构，而檐部采用传统的反曲线挑檐大屋顶，但多数建筑采用灰瓦而不用琉璃瓦，形成淡雅的建筑主色调。其亭台楼阁、小桥水榭、曲廊修竹、古木奇石等江南古典园林常见的布置应有尽有，但集中在距离西湖水面稍远的坡地，形成园中之园。择此而居，可享春访桃花夏观荷，秋来赏桂冬瞻松之趣，更有竹风一窗，荷风半床

The West Lake State Guesthouse is located in the west of West Lake, facing the lake on three sides with the hills on the back, with a courtyard area of 360,000 square meters. It is celebrated as No. 1 Garden in West Lake because of its beautiful environment, exquisite architecture and elegant furnishings. This Water and Bamboo Residence (Shuizhuju), which was built in the late 19th century, was redesigned and rebuilt in the 1950s. The designer was Mr. Dai Nianci, a famous Chinese architect. In the late 1990s, the West Lake State Guesthouse was partially rebuilt and renovated.

The Guesthouse has a garden-style layout as a whole, and its buildings such as guest rooms, office areas, the swimming pool and restaurants are all brick-concrete structures with clear water walls, while the eaves adopt the traditional large roof with inverted curve overhangs, but in most buildings grey tiles instead of glazed tiles are used, forming a light and elegant main color of buildings. Architectural structures, small bridges, waterside pavilions, winding corridors, tall bamboos, ancient trees and strange stones and other common layouts in Jiangnan classical gardens are all found here, but they are concentrated on slopes slightly away from the water surface of West Lake, forming gardens within a garden. If you choose to live here, you can enjoy peach blossoms in spring, lotus in summer, Osmanthus in autumn

的清恬之境。抬眼东望，湖上十里尽收眼底。

今西湖国宾馆园区内的"西湖第一名园"，即是至今仍经常接待国家元首的区域，经 2005 年重新装修后，开始向公众开放。

and pine trees in winter. What's more, this is a tranquil place with bamboos at windows and lotus breeze over bedrooms. Looking up to the east, you can have a panoramic view of the lake stretching for miles.

"No.1 Garden of West Lake" in the West Lake State Guesthouse Park today, which still receives heads of state frequently, was renovated in 2005 and opened to the public from then on.

庭院局部 2

入口

庭院局部 3

建筑庭院环境（组图）

建筑夜景照明

自西湖水面观看国宾馆（组图）

杭州黄龙饭店
Hangzhou Huanglong Hotel

工程地点	杭州市
建筑面积	40000平方米
竣工时间	1986年
设计机构	杭州市建筑设计院
主要设计人	程泰宁

黄龙饭店选址于西湖风景区与杭州老城区之间，如何处理好建筑与环境以及功能相互间的复杂关系，是设计须要解决的主要矛盾，也为方案构思提供了契机。

设计从大环境出发，在总平面布置上借鉴中国绘画中的"留白"手法，使场地南侧的西湖和宝石山自然环境与北侧的城市空间渗透融合，追求整体气韵的连贯。同时，设计注重空间氛围与意境的营造：当宾客在华灯初上时进入大堂，透过通长的落地窗看到庭院后灯火辉煌的餐厅，仿若欣赏一幅立体而有现代气息的"夜宴图"长卷；宝石山色随移步换景在塔楼间时隐时现，传递着传统水墨山水的朦胧韵致。这些无形形态的营造，强化了建筑空间的艺术魅力。

Huanglong Hotel is located between the West Lake Scenic Area and the Hangzhou Old Town. How to deal with the complex relationship among architecture, the environment and functions was the main problem to be solved in the design, which also provided an opportunity for the scheme conception.

Starting from the general environment, the design has drawn on the "blank space" technique in traditional Chinese painting in the general layout, so that the natural environment of West Lake and the Baoshi Mountain on the south side of the site and the urban space on the north side are infiltrated and integrated, with attention focused on the coherence of the overall charm. Meanwhile, the design has focused on the creation of space atmosphere and artistic conception: when guests enter the lobby at the beginning of an evening when all lights are just turned on, they can see the brightly lit restaurant behind the courtyard through the long French windows, as if they are enjoying a three-dimensional and modern "night banquet painting" scroll. The scenery of the Baoshi Mountain changes among towers and buildings with each step one takes, conveying the hazy charm of traditional ink landscape paintings. The creation of such intangible forms has enhanced the artistic charm of architectural space.

远眺建筑全景

建筑全景鸟瞰

入口处铭牌

回廊及入口空间（组图）

穿过建筑组群看宝石山

庭院景观（组图）

建筑与远山融为一体

淮海战役烈士纪念塔
Huaihai Campaign Martyrs Memorial Tower

淮海战役烈士纪念塔位于江苏省徐州市，1965年落成。塔高38.15米。塔体正面九个镏金大字是毛泽东主席题写的塔名"淮海战役烈士纪念塔"。上端雕刻着由五角星照耀下相交的两支步枪和松籽绸带组成的塔徽。塔座正面镌刻着碑文，两侧为大型浮雕，右侧是中国人民解放军一往无前的英雄形象，左侧是人民奋勇支前的壮丽情景。塔下角亭环抱的围廊里，镌刻着党和国家领导人题词及三万多名烈士名录。

塔前有宽敞的平台，瞻仰纪念碑的人们需登上129级台阶，台阶分为10组，长250米，宽31米。台阶的最下面是直径150米的园林中心广场。当夏日暴雨时，大水沿台阶流下，当时的设计角度特殊，从台阶最底端看到大水形成千米远的水帘，台阶与平台搭配构成的斜面角度是特别计算的。

纪念塔周边还建有粟裕将军骨灰撒放处、纪念馆、淮海战役碑林、总前委群雕、国防园，以及青年湖、青年广场、中心花坛等景点。

淮海战役烈士纪念塔被国务院批准为全国重点烈士纪念建筑物保护单位。

The Huaihai Campaign Martyrs Memorial Tower, located in Xuzhou City, Jiangsu Province, was completed in 1965. The tower stands 38.15 meters high. On the front of the tower are nine gold-plated Chinese characters "Huai Hai Zhan Yi Lie Shi Ji Nian Ta" ("Huaihai Campaign Martyrs Memorial Tower") inscribed by Chairman Mao Zedong. The top is engraved with a tower emblem composed of two crossed rifles and pine-seed silk ribbon under a shining five-pointed star. The front of the tower is engraved with monumental writing, with large reliefs on both sides, the heroic images of the Chinese People's Liberation Army on the right, and the magnificent scene of the people bravely supporting the front on the left. In the gallery surrounded by angle pavilions at the foot of the tower, the inscriptions of the leaders of the Party and the state and the list of more than 30,000 martyrs are engraved.

There is a spacious platform in front of the tower, and people visiting the monument need to climb 129 steps, which are divided into 10 groups, with a length of 250 meters and a width of 31 meters. At the bottom of the steps is a garden center square with a diameter of 150 meters. When heavy rain falls in summer, a large amount of water flows down the steps. The design angle at that time was special. From the bottom of the steps, visitors can see the spate of water forms a water curtain spreading a thousand meters. So the inclined plane angle between the steps and the platform has been specially measured. Around the memorial tower, there are also other scenic spots such as the place where General Su Yu's ashes were scattered, the memorial hall, the Huaihai Campaign Forest of Steles, Group Statues of the Members of the Front Enemy Committee, National Defense Park, Youth Lake, Youth Square, and the Central Flower Bed, among others.

The Huaihai Campaign Martyrs Memorial Tower has been approved by the State Council as a key martyrs memorial building under state protection.

纪念塔及环境

纪念塔塔身

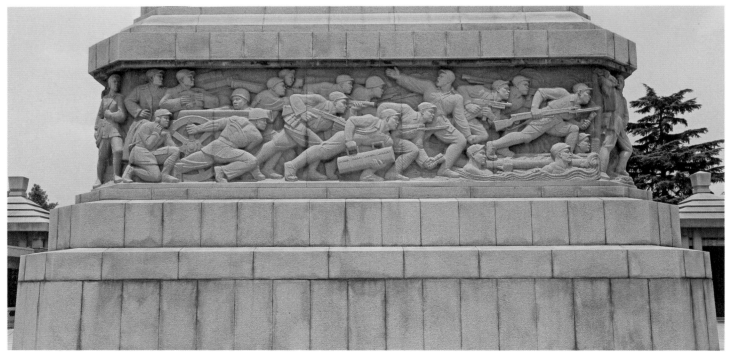

阶梯旁绿植及碑体浮雕

围廊局部

《淮海战役烈士英名录》

铭 刻 记

1995年4月,江苏省人民政府批准,在淮海战役烈士纪念塔围廊内铭刻淮海战役烈士名录,永志纪念。

淮海战役烈士纪念塔管理处根据参战部队和民政部门提供的烈士名单,进行了整理校核,共31006名。经电脑编排,同姓名者省略,实刻28391名。

徐州市人民政府

1995年11月6日

铭刻记

围廊内景

罗斯福图书馆暨中央图书馆旧址
The Former Site of the Roosevelt Library and the Central Library

工程地点：重庆市
建筑面积：2405 平方米
竣工时间：1941 年

罗斯福图书馆坐落在两路口长江一路 11 号（暨抗战时期的中央图书馆）始建于 1938 年，1941 年 1 月落成，建筑面积 2405 平方米，为一栋三层建筑，共有房舍 40 间。抗战初期，国立中央图书馆筹备处勘定在此建立"中央图书馆分馆"。1940 年 8 月，国民政府在此正式成立"国立中央图书馆"。抗战胜利后，原国立中央图书馆在江津白沙镇的各部门集中至此办公，1946 年，分批返回南京，此处改为办事处并继续开放。1946 年 7 月，为纪念罗斯福总统反法西斯作战的功绩，在该处设立国立罗斯福图书馆（筹备委员会），1947 年 5 月国立罗斯福图书馆对外开放。罗斯福图书馆是中国历史上第一个，也是唯一一个以外国总统名字命名的图书馆，并被指定为联合国资料寄存馆，是我国保存联合国资料最早的图书馆。

The Roosevelt Library (the Central Library during the War of Resistance against Japanese Aggression) located at No.11 Changjiang First Road, Liangjiangkou, was commenced in1938 and completed in January 1941. It is a three-story building with a floor area of 2,405 square meters and a total of 40 rooms. At the beginning of the War of Resistance against Japanese Aggression, the Preparatory Office of the National Central Library decided to establish a "Central Library Branch" here. In August 1940, the National Government formally established the National Central Library here. After the victory of the War of Resistance against Japanese Aggression, all departments of the former National Central Library in Baisha Town, Jiangjin were moved here, and in 1946, they were relocated to Nanjing successively. This place was changed into offices, continuing to be open. In July 1946, the National Roosevelt Library (Preparatory Committee) was set up to commemorate President Roosevelt's achievements in anti-fascist operations. The Library was opened to the public in May 1947. The Roosevelt Library, the first and only library named after a foreign president in Chinese history, has been designated as the depository of United Nations materials and was the earliest library in China to store United Nations materials.

入口及周边环境

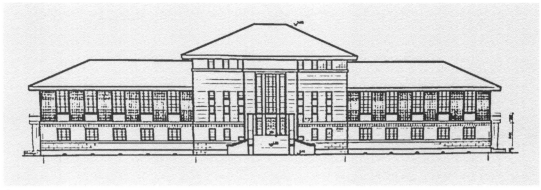

立面图

整座建筑依山而建，呈"丁"字形平面布局，主楼三层，门厅前设石台阶，两翼二层，半地下室一层，条石基础，砖木结构，灰砖黛瓦，古朴无华；馆内空间十分开阔，共有五层书库，连接各功能室的黑色楼梯大气沉稳，一排排可移动的棕色木质书架散发着浓浓的书卷气味。建筑周边绿树环抱，显得幽静、肃穆。此建筑外立面经过近年重新装修，外观改变较大，近期有提议作复原性修复。

The whole building has been constructed with the hills at the back, with a "T"-shaped plane layout. The main building has three floors, with stone steps in front of the entrance hall, two two-storey wings and one semi-basement floor. It is a brick-wood structure with a stone foundation, gray bricks and tiles, looking simple and unpretentious. The library is very spacious inside, with five levels of stack rooms. The black stairs connecting the functional rooms are elegantly balanced, and the rows of movable brown wooden bookshelves exude a strong atmosphere of learning and scholarship. Surrounded by green trees, the building is quiet and solemn. The facade of this building has been renovated in recent years, with significant changes in appearance. Recently, it is proposed to make restorative repair.

俯瞰建筑及环境

入口旧影

立面局部（组图）

入口前广场及周边建筑

洛阳拖拉机厂早期建筑
The Early Buildings of Luoyang Tractor Plant

工程地点 洛阳市 | 占地面积 645万平方米 | 竣工时间 1959年 | 主要设计人 中苏建筑师合作设计

中国第一拖拉机制造厂（简称"中国一拖"，习称"洛阳一拖"）位于河南省洛阳市涧西区。中国一拖是由毛泽东主席亲自敲定的厂址。自1954年起，开始在今址由苏联援建厂房，并在周边建造一批苏式建筑群（工厂配套的住宅、医院、学校、商店等）。按边建设边生产的原则，"中国一拖"的第一批正式产品"东方号"拖拉机于1958年投产并出厂（出厂日期为1958年7月13日），而厂区正式落成则迟至一年后的1959年11月1日。

"中国一拖"全厂占地645万平方米，厂房均为钢筋混凝土结构的单层双跨或多跨大型厂房，全部采用钢屋架，预制钢筋混凝土柱、梁、檩条及小型密肋屋面板。每幢厂房大小构件在5000个以上，重量在3000吨以上，大的构件一般在3吨至10吨。这种规模的大型厂房，在20世纪50年

China's No.1 Tractor Plant (referred to as "China 1st Tractor" or commonly known as "Luoyang 1st Tractor") is located in Jianxi District, Luoyang City, Henan Province. The site China 1st Tractor was personally finalized by Chairman Mao Zedong. Since 1954, plant buildings had been constructed with the support of the Soviet Union at this site, and a number of Soviet-style buildings (residential buildings, hospitals, schools and shops, etc.) were built around it. According to the principle of production during construction, the first batch of "Dongfang" tractors, the official products of "China 1st Tractor", was put into production and delivered in 1958 (the delivery date was July 13, 1958), while the Plant was officially completed as late as November 1, 1959, more than one year later.

The whole plant of "China 1st Tractor" covers an area of 6.45 million square meters, and all the buildings are single-story double-span or multi-span large-scale buildings with a reinforced concrete structure, all of which use steel roof truss, prefabricated reinforced concrete columns, beams, purlins and small multi-ribbed roof panels. There are more than 5,000 large and small components in each plant building, with a weight of more than 3,000 tons. Large components are generally 3 tons to 10 tons each. Large plant buildings of this scale were the first-class industrial buildings in China in the

厂区车间现状

入口立面局部（组图）

初期实属我国第一流的工业建筑。其中，第一期工程共完成建厂投资2.89亿元，完成厂房建筑面积30.31万平方米、宿舍建筑面积27.05万平方米。经国家验收委员会验收，工程质量被评为优等。

其标志性的建筑，当属厂区前部建筑。其厂区大门与两侧办公楼呈"凹"字形平面布置，均为苏式的红砖红瓦建筑，"凹"字的底部为两层高的楼房所形成的正门，两侧办公楼为五层瓦顶建筑，故立面也形成"凹"字形空间，与厂区前方的广场、广场上矗立的毛泽东纪念像等，形成有韵律的起伏、舒展、延伸。

正门上方有"第一拖拉机制造厂"金字，两侧为"生存勿忘质量""发展必须创新"白底红字标语，房檐正中上方是由镰刀、锤子、五星、旗帜、履带、麦穗、齿轮等元素组成的浮雕厂徽。据厂史记载："这是一拖的厂徽，与苏联哈尔科夫拖拉机厂的厂徽是一套图纸，是苏维埃文化和社会主义工业文化的标志。"

第一拖拉机制造厂工厂车间、厂房、职工住宅以及各种配套设施扎根在涧西区，历经数十年风雨，仍然为一拖的生产和生活发挥着重要作用。

early 1950s. Among them, the first phase of the project was completed with a total investment of 289 million yuan, featuring a plant floor area of 303,100 square meters and a dormitory floor area of 270,500 square meters. Upon acceptance by the National Acceptance Committee, the project quality was rated as excellent.

The landmark building is no other than the front building of the Plant. The gate of the plant area and the office buildings on both sides are arranged in a cancave shape, all being Soviet-style red brick and red tile structures. The bottom of the canave shape is the main entrance formed by two-story buildings, and the office buildings on both sides are five-story tile-roofed buildings, so the facade also forms a cancave space, constituting rhythmic ups and downs, stretching and extending with the square in front of the plant area and the Mao Zedong memorial statue standing on the square.

Above the main entrance, there are golden Chinese characters, "Di Yi Tuo La Ji Zhi Zao Chang" ("No.1 Tractor Plant"), and on both sides, there are red slogans on white background, meaning "don't forget quality for survival" and "we must innovate to develop". At the center of the eaves, there is an embossed plant emblem composed of a sickle, a hammer, five stars, a flag, a crawler belt, ears of wheat and gears. According to the plant history records: "This is the plant emblem, which shares the same set of drawings with the plant emblem of Kharkov Tractor Plant of the Soviet Union, and is a symbol of Soviet culture and socialist industrial culture."

The workshops, plant buildings, staff residential buildings and various supporting facilities of No.1 Tractor Plant are rooted in Jianxi District, and after weathering decades of trials and hardships, still play an important role in production and life of the plant.

厂区道路现状

入口局部

厂区全景鸟瞰

厂区配套苏式住宅区（组图）

中国 20 世纪建筑遗产保护进程及展望

The Evolution and Prospect of Preserving the 20th-Century Architectural Heritage in China

张松
Zhang Song

一、"20 世纪遗产"的由来

进入 21 世纪，世界范围内开展的文化遗产保护正成为一场全民运动和一项正当其时的创造性活动。然而，在 20 世纪 60 年代以前，从考古发现的远古遗址，到 19 世纪中期工业革命之前的文物古迹和古建筑一直是受到保护的。但工业革命以来形成的、特别是 20 世纪建筑遗产在联合国教科文组织（UNESCO）公布的《世界遗产名录》上鲜有出现。自从 20 世纪 70 年代以来，保护对象发生了变化，人们开始对历史性住宅、乡土建筑、工业建筑和历史环境（如传统街区、历史地段等）进行全面保护。

1981 年 10 月，第五届世界遗产大会审议了澳大利亚悉尼歌剧院及悉尼港申报世界文化遗产的议案，引起国际机构对战后建筑和晚近遗产（recent heritage）保护问题的关注，并在相关国际会议上针对晚近遗产的鉴定、评估、登录和保护等课题开展讨论。1985 年在巴黎召开的国际古迹遗址理事会（ICOMOS）专家会议，研究了有关现代遗产的保护问题。1988 年在荷兰的艾恩德霍文（Eindhoven）成立了现代主义运动记录与保护国际组织（International working-party for document and conservation of buildings, sites and neighborhoods of the modern movement，简称 DOCOMOMO）。

1989 年，欧洲委员会在维也纳召开了"20 世纪建筑遗产：保护与振兴战略"国际研讨会；1991 年欧洲委员会发表"保护 20 世纪遗产的建议"，呼吁以"遗产即历史记忆"的思想为指导，尽可能多地将 20 世纪遗产列入保护名录，并以遗产价值为基础确定其保护策略。1995 年，在美国芝加哥召开了主题为"保护晚近过去的历史"（preserving the recent past）的国际会议，主张对"晚近过去"（recent past）的文化遗产进行保护，将晚近建筑遗产作为文化资本、景观资源，进行积极保护与合理再利用。

1995 年和 1996 年，ICOMOS 分别在赫尔辛基和墨西哥城就 20 世纪遗产的保护课题召开国际会议（ICOMOS Seminar on 20th-Century Heritage）。1999 年，在墨西哥召开的 ICOMOS 大会上，会议收到不少有关保护现代遗产的提案（主要来自东欧和以色列），一些国家的报告中也反映出对 19 世纪后期和 20 世纪遗产保存状况的担忧。鉴于此，2001 年 9 月，ICOMOS 在加拿大蒙特利尔召开的工作会议制订了"20 世纪遗产蒙特利尔计划"（Montreal Plan for 20th C. Heritage），并将 2002 年 4 月 18 日国际古迹遗址日的主题确定为"20 世纪遗产"。

2011 年 6 月，ICOMOS 下属 20 世纪遗产国际科学委员会，简称 ISC 20C 在马德里召开了主题为"20 世纪建筑遗产干预方法"的国际会议，大会通过的《关于 20 世纪建筑遗产保护方法的马德里文件》（简称《马德里文件》）得到广泛传播和讨论。2014 年，ISC 20C 根据反馈意见已出版第二版，

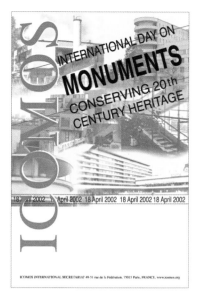

图1　2002年国际古迹遗址日主题招贴画

图2　20世纪遗产保护无锡论坛（2008年）

该文件是支撑20世纪遗产地保护和变化管理的技术指南。

二、20世纪建筑遗产理念的内涵

2008年4月，响应国际遗产保护潮流，国家文物局在无锡召开了以"20世纪遗产保护"为主题的中国文化遗产保护论坛，会上讨论通过了《20世纪遗产保护无锡建议》。同年5月，国家文物局发出《关于加强20世纪遗产保护工作的通知》，要求各地充分重视20世纪遗产的抢救和保护工作。此后，"20世纪遗产"开始成为国内传媒关注的热点话题，但不少媒体望文生义，认为"20世纪遗产"就是1900—2000年间不同类型的所有遗产。这样的看法其实并没有真正把握"20世纪遗产"这一新遗产类别的实质内涵。

我们正处在一个"遗产"概念扩展的年代，从考古遗址、纪念物、文物古迹、建筑遗产、城市遗产，到文化遗产、文化景观、工业遗产、无形文化遗产等，遗产的数量、类型都在不断扩展之中，而20世纪遗产概念的提出则与晚近过去的历史保护关系密切。1995年美国芝加哥会议所提出的"晚近遗产"还只涉及1920—1960年间的遗产，如今相关国际组织和机构已意识到晚近遗产，特别是与现代主义风格有关的建筑遗产，是人类共同遗产的重要组成部分，它们直观地反映了建筑和社会发展的主要过程。而"recent past"一词已成为通常用来描述那些不足50年历史的遗产资源和建筑遗产的一个专业术语。20世纪建筑遗产，包括所有样式和功能的建筑（新建筑、乡土建筑、再利用建筑实例），城市集合体（邻里小区、新城），城市公园、庭院和景观，艺术作品，家具，室内设计或大型工业设计，土木工程（道路、桥梁、水利设施、港口、工业综合体），纪念性场所，以及建筑档案、文献资料等。

现代建筑的开拓性技术和材料的应用往往存在无法预见的风险和问题，像轻混凝土拱顶、合成材料和大型玻璃板材等都在迅速老化。从具有象征意义的经典作品到普通建筑，都正在陷入衰败的境地或面临着被拆除的危机。正是因为如此，DOCOMOMO国际委员会的玛丽斯代拉·卡夏托等人在谈到这些现象时，将现代建筑称之为"一个处于高度危险中的遗产"（玛丽斯代拉·卡夏托，挨米莉·多尔热，2007）。

20世纪遗产概念的提出正是针对这类建筑遗产常常缺乏认可和保护，尤其是在与更加"古老"或更加传统的遗产相比照的情形下。晚近以来的创造似乎无法与历史悠久、声名显赫的古代遗产进行竞争，也许是晚近遗产数量众多，在如此广泛的范围内进行选择和保护也有相当的难度；也许还有某些现代建设曾破坏了自然环境和历史遗址，并没有给人们留下良好印象。

进入21世纪后，人们认识到20世纪的建筑也是人类遗产的组成部分，应该受到应有的保护。现代运动创造出的大量建筑已具有历史和文化价值，作为20世纪留给后世的文化遗产同样值得保护和关爱。现代运动中的建筑遗产，在今天面临的危机比任何时候都要大。这主要是由于建筑物的年限、经常性的技术革新、原有功能不适应新需求、文化观念上的误导等原因所引发。因此，20世纪遗产鉴定和评估过程必须考虑在可持续发展框架下遗产与当前和未来生活的关系，考虑以社区的普遍期望为基础进行遗产项目的评估，并特别关注环境、经济活动和文化生活。

三、中国历史建筑保护的简要历程

1988年11月，城乡建设环境保护部和文化部联合发布《关于重点调查、保护优秀近代建筑物的通知》之后，各地展开了不同程度的近代建筑调查工作。1989年，上海市根据该通知精神开展了近代建筑的调查、鉴定工作，于1991年在全国率先制定了保护优秀近代建筑的地方政府规章《上海市优秀近代建筑保护管理办法》，同时公布了第一批61处优秀近代建筑保护名录。

1991年7月，建设部、国家文物局在《关于印发近代优秀建筑评议会纪要的通知》中，要求各级有关部门继续做好近代优秀建筑的调查工作，对确有保护价值的建筑报请当地政府公布为文物保护单位。

1996年11月，国家在评选和公布第四批全国重点文物保护单位时，对文物保护单位的分类进行了调整，将"革命遗址及革命纪念建筑物"变更为"近现代重要史迹及代表性建筑"，新的分类方式的涵盖范围包括近现代重要史迹和代表性建筑。国务院公布的第四批全国重点文保单位共250处，其中古建筑110处，近现代重要史迹及代表性建筑共50处，这其中有10余处近现代代表性建筑，包括上海外滩历史建筑群、广州沙面建筑群、大连俄国建筑、青岛德国建筑等。此为第一次从建筑艺术价值或建筑学意义的角度针对近现代建筑进行评估并评定全国重点文物保护单位。《中华人民共和国文物保护法》（1982年）中规定受国家保护的文物类别中包含"与重大历史事件、革命运动或者著名人物有关的以及具有重要纪念意义、教育意义或者史料价值的建筑物、遗址、纪念物"，显然，在文物保护法律中考虑更多的是建筑的纪念意义、教育意义和政治意义。

2003年，建设部制定了部门规章《城市紫线管理办法》，要求城市人民政府在组织编制历史文化名城保护规划时划定城市紫线，即历史文化街区和历史建筑的保护范围界线。

2004年，建设部又发布了《关于加强对城市优秀近现代建筑规划保护工作的指导意见》（以下简称《指导意见》），要求加强城市近现代建筑的保护工作，其保护对象包括19世纪中期建造的具有较高历史文化价值的建筑。《指导意见》要求各地积极推进专项立法进程，做好近现代建筑的保护规划，并严格执行规划制定的保护措施。2004年8月，以马国馨院士为代表的中国建筑学会建筑师分会，向国际建筑师协会（UIA）提交了一份《20世纪中国建筑遗产的清单》，对那些存在损毁危险或需要抢救保护的现当代建筑提出关注（单霁翔，2008）。

2008年7月1日起施行的国家《历史文化名城名镇名村保护条例》，明确要求地方政府保护各地的历史建筑，即具有一定保护价值，能够反映历史风貌和地方特色的建筑物、构筑物（第四十七条）；要求各城市政府保护文物保护单位和登记不可移动文物以外的历史建筑，从保护资金、拆除及建设活动控制等方面做出了保护管理要求，为地方政府保护建筑遗产提供了法律依据。过去那种将文物保护单位以外的历史建筑全部当作"非法定保护对象"，不予保护的错误认识和大规模简单拆除的野蛮行为，开始得到一定程度改正。

2017年9月，住房和城乡建设部再次发布"关于加强历史建筑保护与利用工作的通知"，要求各地加强历史建筑保护与利用，做好历史建筑的确定、挂牌保护和建档工作，最大限度地发挥历史建筑的使用价值。

四、部分城市历史建筑的评定标准

2000年以来，在国家历史文化名城保护相关条例出台之前，上海、哈尔滨、厦门、苏州等城市考虑本地历史建筑的类型、地方社会经济特点，参照国外先进的历史建筑登录保护经验，制定了专门的历史建筑和历史文化街区的保护条例。北京、南京等城市通过历史文化名城保护条例，将历史建筑保护内容纳入其中。

文化遗产与历史有关，是前人留给子孙后代传承下去的东西，既包括文化传统，也包括人工制品。因而，在所有文化遗产的评估标准中，年代似乎（至少隐在地）是首要的衡量因子。下面，对国内地方保护法规中部分城市历史建筑登录的年代标准进行一些分析比较。

北京市2005年5月1日施行的《北京历史文化名城保护条例》明文规定：对尚未列为不可移动文物、反映一定时代特征、具有保护价值、承载真实和相对完整历史信息的四合院和其他建筑，应当认定为具有保护价值的建筑（第十四条），具有保护价值的建筑不得违法拆除、改建、扩建（第三十一条）。在保护条例中并未明确建设年代或建成时间的具体规定要求。其他更多的城市保护条例中所确定认定标准，其年限要求大致可分为：① 1949年以前；② 建成50年以上；③ 建成30年以上；④ 不做要求。（参见表1）

特别有趣的现象是，2013年武汉和杭州两座历史文化名城，同时对各自原有的保护办法进行了修订，并将政府规章上升为地方性保护条例。这其中，对历史建筑的年限标准，杭州市由原定的"建成50年以上"改为了不做具体的年限规定；

表1 部分城市保护条例中建筑年代标准的比较

城市	法规名称	施行时间	保护对象名称	认定标准中的年限规定*	备注
厦门	《厦门经济特区鼓浪屿历史风貌建筑保护条例》	2000年4月1日	历史风貌建筑	1949年以前	2009年3月20日修订
哈尔滨	《哈尔滨市保护建筑和保护街区条例》	2001年12月1日	保护建筑 保护街区	市人民政府制定保护建筑和保护街区认定标准	已废止
苏州	《苏州市古建筑保护条例》	2003年1月1日	控制保护建筑	（1）1911年以前（2）1949年以前	
上海	《上海市历史文化风貌区和优秀历史建筑保护条例》	2003年1月1日	历史文化风貌区 优秀历史建筑	建成30年以上	
武汉	《武汉市旧城风貌区和优秀历史建筑保护管理办法》	2003年4月1日	旧城风貌区 优秀历史建筑	建成30年以上	已废止
杭州	《杭州市历史文化街区和历史建筑保护办法》	2005年1月1日	历史文化街区 历史建筑	建成50年以上	已废止
北京	《北京历史文化名城保护条例》	2005年5月1日	旧城整体 历史文化街区 具有保护价值的建筑	认定标准另定	
天津	《天津市历史风貌建筑保护条例》	2005年9月1日	历史风貌建筑	建成50年以上	
南京	《南京市重要近现代建筑和近现代建筑风貌区保护条例》	2006年12月1日	重要近现代建筑 近现代建筑风貌区	19世纪中期至20世纪50年代	
哈尔滨	《哈尔滨市历史文化名城保护条例》	2010年1月1日	历史城区、历史文化街区、历史院落和历史建筑	未作规定	
南京	《南京市历史文化名城保护条例》	2010年12月1日	老城格局和城市风貌、历史建筑、历史文化街区、历史风貌区、历史街巷等	未作规定	

续表1

城市	法规名称	施行时间	保护对象名称	认定标准中的年限规定*	备注
武汉	《武汉市历史文化风貌街区和优秀历史建筑保护条例》	2013年3月1日	历史文化风貌街区 优秀历史建筑	建成50年以上	
杭州	《杭州市历史文化街区和历史建筑保护条例》	2013年10月1日	历史文化街区 历史建筑	不做要求	

（*认定标准中的年限规定只针对建筑，不涉及历史街区或风貌区）

而武汉市则将原定的"建成30年以上"标准，改为了"建成50年以上"，两名城的立法变化形成了鲜明的对比。过去相当长的时期内，一栋建筑如果没有超过50年历史就不会得到保护，在这一法则性规定约束下，大量有价值的建筑已被拆除或将要被拆除。如今虽说还没有统一的20世纪遗产的评估标准，但在1995年的赫尔辛基会议上各方已就相关问题基本达成共识。赫尔辛基会议试图总结出鉴定和甄别具有突出的普遍价值的20世纪遗产的方法，评判这些项目是否有潜力被列入《世界遗产名录》。会议总结强调，应将20世纪遗产的各种表现方式结合在一起来考虑。这些方式同时反映了20世纪的变化和革新，以及传统的形式和表达方式。会议一致同意"只有在特殊的情况下，才会提交讨论不足25年的项目列入《世界遗产名录》的议案，以保障历史透视和科学分析有充足的时间。"（ICOMOS，1995）。也就是说，建成25年以上的建筑或人工环境就可以参与申报世界文化遗产。

五、中国20世纪建筑遗产名录状况

正如"关于保护20世纪遗产无锡建议"中所指出的：20世纪遗产保护已逐渐进入工作视野，但仍未得到充分重视。在中国，年代较短的晚近建筑的保存状况不容乐观，认定标准的局限，法律保障的缺失，保护经验的匮乏，以及一些不合理的利用方式，导致大量具有重要价值的20世纪遗产正在加速消亡，抢救性保护工作日趋紧迫。

与此同时，一些历史文化名城也在不遗余力地试图保存一些过去的建筑和景观，虽然这些建筑和景观并不属于文化瑰宝或历史遗产，但积极有效地保护它们可以增添城市的魅力，提升市民的认同感。这其中，上海、苏州、北京等城市都做过一些积极的探索。2002年7月，上海市人大通过了《上海市历史文化风貌区和优秀历史建筑保护条例》，将此前的"优秀近代建筑"扩展为"优秀历史建筑"，保护建筑的年限规定由原定的1949年以前，扩展至建成使用30年以上。自1989年起至今，上海分五批公布了1058处，合计3075幢优秀历史建筑。

苏州市在保护数量众多的各级文物保护单位的同时，还通过地方立法和古城控制性详细规划对与古城风貌密切相关的古建筑物、古构筑物实行了切实有效的保护，古城区范围内的登录保护建筑总数达1347处（表2）。

2007年12月，北京市公布了《北京优秀近现代建筑保护

表2 苏州古城区内控制性保护建筑的类别及数量

类别		数量
古建筑物		557处
古构筑物		790处
其中	古井	639口
	古驳岸	22处
	古牌坊	22座
	古桥梁	70座
	砖雕门楼	37座

名录（第一批）》，在第一批71处保护建筑中新中国成立后建设的建筑51处，占72%，包括20世纪50年代建成的原苏联展览馆、798工厂，20世纪70年代建成的北京长途电话大楼、北京国际俱乐部、建国门外大街外交公寓等现代建筑均列入保护名录。这是一个非常好的开端，但在全国各地，新中国成立后的建筑、工人新村、"文革"十年中形成的公共空间和景观、乡土遗产、工业遗产、工程技术遗产、80年代经典建筑，等等，数量众多，亟待开展调查、登录和抢救性保护工作。

事实上，目前国内并没有一份针对20世纪建筑遗产保护的法定名录。2016年9月，2017年12月，中国文物学会和中国建筑学会联合发布了第一批、第二批《中国20世纪建筑遗产名录》。该名录是由中国文物学会20世纪建筑遗产委员会委员和建筑师等专业人士推荐，并通过投票选出。这份由学术团体和专业人士主导并公布的《中国20世纪建筑遗产名录》，虽然并没有相对应的遗产保护地位，也没有配套的保护法规政策，但在社会上依然引起了广泛的关注，名录的公布应当有利于提升民众对建筑遗产的认识，促进地方政府对其加强保护与管理工作。

名录的认定条件和程序，执行由中国文物学会20世纪建筑遗产委员会（CSCR-C20C）制定的《中国20世纪建筑遗产认定标准》（2014，以下简称《认定标准》）。该标准参照世界文化遗产登录标准，认为列入名录的"建筑应是创造性的天才杰作；具有突出的影响力；文明或文化传统的特殊见证；人类历史阶段的标志性作品；具有历史文化特征的居住建筑；与传统或信仰相关联的建筑"，《认定标准》中

图3　上海市第四批优秀历史建筑：1952年建成的曹杨新村（一村）

图4　北京市保护建筑798工厂

图5　丽江市中心红太阳广场（红色景观）

图 6　温州市江心岛上文革时期的摩崖石刻

图 7　湖北省洪湖市瞿家湾古镇的乡土建筑遗产

还详细列出了"在近现代中国城市建设史上有重要地位，是重大历史事件的见证，体现中国城市精神的代表性作品"，"能反映近现代中国历史且与重要事件相对应的建筑遗迹与纪念建筑，是城市空间历史性文化景观的记忆载体，同时，也要兼顾不太重要时期的历史见证作品，以体现建筑遗产的完整性"等 9 条具体标准规定。

从保护机制方面看，两批名单与已经公布的各级文物保护单位身份重叠较多，其中全国重点文物保护单位占 59.1%。在入选的第一批名录的 98 项建筑中，有 45 项是全国重点文物保护单位，占 45.3%；在入选的第二批名录的 100 项建筑中，有 72 项是全国重点文物保护单位，占 72%。另一方面，自评选第 4 批全国重点文物保护单位时强调对优秀近代建筑保护以来，特别是在公布的第六批和第七批全国重点文物保护单位中近现代建筑数量有明显的增加。而这份保护名录从一个侧面也反映了在评选全国重点文物保护单位的过程中，对近现代建筑保护逐步重视的实际情况。

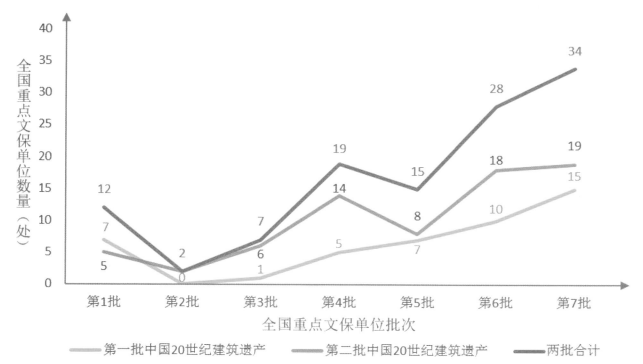

图8 中国20世纪建筑遗产中全国重点文保单位统计

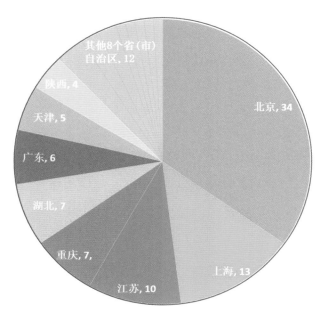

图9 第一批中国20世纪建筑遗产地域分布

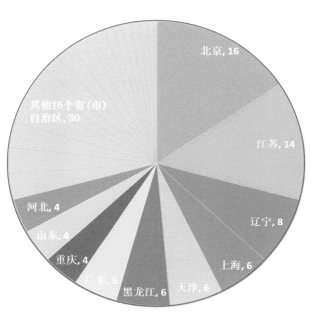

图10 第二批中国20世纪建筑遗产地域分布

从名录中项目的地域分布情况看，选录的项目多数集中分布于北京、上海、南京等大城市，在区域分布上不够均衡，在两批名录和计198项建筑遗产中，北京市共有50个项目入选，占总数的25.3%；江苏省24项，占12%；上海市19项，约占10%。这一实际状况既反映出这些大城市在近代以来的政治经济地位，也说明了近30年来对近现代建筑保护工作的重视，同时也反映了中小城市在历史建筑保护方面还需要得到更多的关注。主办方显然已经意识到名录地域分布上的不平衡问题，2017年12月专门将第二批名录公布仪式安排在安徽省池州市举办，对促进"文化池州"建设起到了触媒作用。

今后，中国20世纪建筑遗产名录遴选工作除了需要注意对地域分布上有所关注之外，对建筑物的年代标准也应当注意调整。从时间分期上看，按照《认定标准》的说法，"中国20世纪建筑遗产是按时间段予以划分的遗产集合，它包括着20世纪历史进程中产生的不同种类建筑的遗产"，"主

图 11 已拆除的 50 年代建成的教学楼（冯纪忠设计）

图 12 已拆除的 80 年代建成的高层公寓（戴复东设计）

要分成两段：中国近代建筑（1840—1949 年）；中国现当代建筑（1949—21 世纪初中叶）"。这一时间标准，远远突破了 20 世纪的时间范畴，向前推进了 60 年，向后延伸到当下（近 20 年），将建筑遗产登录保护外延到建筑文化传播弘扬，并不一定有利于强化地方政府对"晚近遗产"的保护工作。

因此，有必要更多地将建成 30~50 年的建、构筑物列入《中国 20 世纪建筑遗产名录》。从国家层面看，也应当加强对 1949 年之后与相关重大历史事件、经济建设相关的建筑遗产保护，以 30 年左右的时间来限定的话，时间下限可以考虑延伸至改革开放前 10 年至 20 年期间。在此，根据历史事件的形式和意义、建筑运动状况等因素，对中国 20 世纪建筑遗产项目的分期提出以下建议（见表 3）。

六、从世界遗产到世间遗产保护

人类共同记忆的链条，在 20 世纪这一环节上可能比更久远年代的建成遗产更加脆弱。近年来，中国对文化遗产保护越来越重视，但"年代久远"固定思维的影响依然很大。著名学者杨东平先生认为：我国的文物保护工作，一直存在"详远而略近"，"识大而不识小"，"因人害物，求全责备"，"崇假而贬真"的观念偏差（杨东平，1998）。由于认识上的局限性和行动上的功利性，一些地方虽说还存在数量巨大的 20 世纪建筑遗产，但一不小心往往就会成为了城市开发建设的牺牲品。

事实上，早在 1968 年联合国教科文组织（UNESCO）通

表 3　中国 20 世纪建筑遗产分期建议

时期划分	时代特征	代表性建筑遗产
"新中国"十七年（1949—1965 年）	人民当家作主、苏联援助、国家大建设	北京、上海等城市工人新村，苏联援华 156 项工程（武汉长江大桥），"三线"建设项目等
"文革"十年（1966—1976 年）	自力更生、艰苦奋斗、重点项目建设	南京长江大桥，河南林县红旗渠，"三线"建设项目等
改革开放前十年（1977—1988 年）	中国特色、百花齐放、现代化建设	深圳特区规划，京派建筑、海派建筑、岭南派等地域当代建筑

图 13 世界文化遗产：悉尼歌剧院

图 14 世界文化遗产：柯布西耶建筑作品

过的《关于保护受到公共工程或私人工程危害的文化财产的建议》中，就已明确指出："文化财产不仅包括已指定和已登录的考古学遗址、历史古迹和建、构筑物，而且也包括未登录或未分类的古代遗迹，以及具有艺术或历史价值的近代遗址和建筑。"1987 年 ICOMOS 通过的《华盛顿宪章》也认为："文化财产无论其等级多低，均构成人类的记忆。"人类应积极促进历史城市或历史地区的私人生活和社会生活的协调，并鼓励对这些文化财产的保护。

文化遗产范畴的拓展在时间轴上是由远及近的，在关注物质对象上是由经典之作到普通事物的，从这一层意义上讲，保护运动可以说正从关注世界遗产走向关注世间遗产，这种转变正如联合国教科文组织亚太地区文化顾问理查德·恩格尔哈特（Richard Engelhardt）先生总结的概要，的确耐人寻味和深思（表 4）。

近年来，在《世界遗产名录》上有越来越多的 20 世纪建筑遗产被列入。2007 年 6 月，在新西兰基督城召开的第 31 届世界遗产大会上，1973 年落成的悉尼歌剧院作为 20 世纪的建筑杰作被列入《世界遗产名录》；2008 年 7 月在第 32 届世界遗产大会上，德国柏林现代住宅区被列入《世界遗产名录》。2016 年 7 月 17 日，在土耳其伊斯坦布尔召开的世界遗产大会上，"勒·柯布西耶现代建筑系列作品"被列入《世界遗产名录》，该项目包括分布在德国、阿根廷、比利时、法国、印度、日本、瑞士等 7 个国家的 17 个作品。国际遗产保护组织的这些行动，正是对 20 世纪建筑遗产保护这一理念的最好诠释。

针对量多面广的 20 世纪遗产，无论是延续功能发挥作用，还是适当改造再利用，都必须依靠社会、依托社区居民的广泛参与。一方面，就地方城市而言，社区需要设立一些熟悉的标志，以便能够在快速变化的世界里与集体记忆继续保持联系，在城市建设中，要在保护自然环境的同时保护好历史环境，其中最为重要的就是城市建筑、乡土建筑和历史文化街区，它们已成为地方的标识性景观或城镇特色的重要组成

表 4 文化遗产保护的范畴转变

时间	过去	现在
保护对象	王室、宗教和政治的纪念物	普通人的场所与空间
管理部门	中央行政机构管理	社区、社团管理
利用情况	精英的使用方式	普通用途

（出处：Richard Engelhardt 的演讲稿 ppt）

图15 东京文化会馆（前川国男设计，1961年）

图16 东京国立代代木体育馆（丹下健三设计，1964年）

部分；另一方面，由于20世纪遗产形成年代较晚，未经历过多的自然侵蚀，因此许多20世纪遗产至今仍然保持着生命力。因而需要社会各界积极参与到建筑遗产保护的相关行动中来。在日本，日本建筑学会设立开展现代建筑的调查、评估和保护工作的专门委员会。著名建筑师前川国男、丹下健三等人的代表作品已被列入建筑遗产保护名录或国家登录有形文化财。

七、结语：今后方向的展望

人类社会发展的历史一直充满着变化与挑战，而且各种变化的周期越来越短、变化的节奏越来越快。在快速变化的过程中，人类失去了赖以生存的稳定时空感、失去了人与人、人与物之间的恒常关系。正如让·鲍德里亚（Jean Baudrillard）在《消费社会》中所描述的，"我们生活在物的时代：我是说，我们根据它们的节奏和不断替代的现实而生活着。在以往的所有文明中，能够在一代一代人之后存在下来的是物，是经久不衰的工具和建筑物，而今天，看到物的产生、完善与消亡的却是我们自己"（让·鲍德里亚，2000）。

国际上，针对20世纪早期遗产的保护方法已日渐完善，但是保护和保存第二次世界大战后形成的建成遗产的方法还在摸索之中。因此，我们在探索20世纪建筑遗产的保护方法上，既要借鉴国外先进的保护经验与技术方法，又要针对我国20世纪遗产的特殊情况，从学术研究、价值认定和保护技术等方面积极开展多学科合作，尽快建立起20世纪建筑遗产整体性保护的理论体系。我们在制度建设、普查研究、保护修缮技术、公众参与、经济政策以及规划管理等方面上亟待加强和改进，应当鼓励成立更多的非盈利机构参与保护运动，鼓励公众参与建筑遗产保护，尽快改变民众在保护20世纪建筑遗产方面普遍缺乏认知的状况。

在各地申报世界遗产热潮不减的势头下，我们更应该学习、借鉴世界遗产登录评定的标准、分类方式等，建筑遗产的价值评估应当充分考虑到对文化的多样性和文化的可持续发展的尊重。在评定各级文物保护单位和历史建筑名录时，更多地考虑到遗产资源的综合性、整体性，将建筑遗产、文化景观、乡土建成遗产等类别的20世纪遗产更多地列入各级、各类保护名录中。与文物古迹重视历史事件，重视名人故居、纪念地等红色记忆的纪念意义、教育意义相对应，20世纪建筑遗产保护管理，更多地从建筑学、建筑人类学、建筑文化传播学等方面进行分析、研究、鉴定和评估，应当更多地将建成不足50年的历史建筑、建成环境和文化景观列入保护名录。

20世纪建筑遗产的保护管理与活化利用，应当从维护场所精神，提升环境品质，规划建设宜人的生活环境的角度制定战略性保护管理规划，切实通过建筑遗产的保护与利用（或再利用），来提升城市和社区的吸引力，增强市民的认同感。将历史文化资源作为可持续旅游和城市创新发展的重要资源，在实现高质量发展和高品质生活的城市发展目标中发挥重要的积极作用。

（除图1外其他图片均由作者拍摄或绘制。）

参考文献

[1] ICOMOS.Montreal Action Plan on 20th C. Heritage，2001.

[2] ICOMOS.The Conclusions of ICOMOS Seminar on 20th Heritage，Helsinki，1995.

[3] The ICOMOS International Scientific Committee for Twentieth Century Heritage，Approaches for the Conservation of Twentieth Century Architectural Heritage, Madrid Document，2014.

[4] 玛丽斯代拉·卡夏托，挨米莉·多尔热，孙凌波. DOCOMOMO国际委员会：分享现代建筑的专业知识，世界建筑，2007（6）:114~115.

[5] 中国古迹遗址保护协会秘书处.中国古迹遗址保护协会通讯：国际古迹遗址理事会关于二十世纪遗产文献特刊[J].2008（1）.

[6] 单霁翔.20世纪遗产保护的实践与探索[J].城市规划，2008（6）：11~32+43.

[7] 张松.历史城市保护学导论——文化遗产和历史环境保护的一种整体性方法（第二版）[M].上海：同济大学出版社，2008.

[8] 张松，周瑾.论近现代建筑遗产保护的制度建设[J].建筑学报，2005（7），5~7.

[9] 马国馨.百年经典亦辉煌[J].建筑创作，2006（4）：132~133.

[10] 中国建筑学会建筑师分会.20世纪中国建筑遗产，[J].建筑创作，2006（4）134~147.

[11] 中国文物学会20世纪建筑遗产委员会.中国20世纪建筑遗产名录（第1卷）[M].天津：天津大学出版社，2016.

[12] 金磊.20世纪遗产是当代城市的标志 首批98项"中国20世纪建筑遗产"浅析[J].建筑设计管理，2017,34(2)：17~21.

[13] 杨东平.未来生存空间[M].上海：上海三联书店，1998：162~163.

[14] 让·波德里亚.消费社会[M].刘成富，全志钢译，南京:南京大学出版社，2000：2.

从建筑学视角
看第二批中国 20 世纪建筑遗产项目中的现代建筑

Obsening the Modern Architecture in the Second Batch of Chinese 20th Century Architectural Heritage Projects from the Perspective of Architecture

戴路 李怡

Dai Lu, Li Yi

20 世纪的中国历经风云变幻，中西方文化激烈碰撞，日益交流。在白云苍狗之后回望，这段历史并非无迹可寻，建筑凝固文化，定格时间。中国现代建筑在此阶段内弱势起步，历经被动输入转向主动发展的过程。但由于其建成时间不过几十年，似乎面孔更为亲和，其价值往往更易被忽略。这并非中国一国之难，国际社会同样面临相似的问题。随着越来越多的 20 世纪建筑遗产进入世界遗产名录，现代建筑的重要价值逐渐为世人瞩目。中国文物学会 20 世纪建筑遗产委员会于 2014 年组建，并于 2017 年公布第二批中国 20 世纪建筑遗产名录，100 处遗产中大量现代建筑入选，对第一批进行补充，这无疑是对入选建筑的遗产价值所给予的极大肯定。本文从时间、地域、功能、风格等方面对这些现代建筑进行统计和分析，并从教学和实践两方面探讨其价值启示。以建筑学视角深入观察，探究其入选意义。

现代建筑融合时代特色，彰显时代个性，研究现代建筑的意义不仅在于知晓历史，更在于指导现在和未来。以建筑学视角对第二卷中入选的现代建筑进行观察，对其基本信息进行统计和分析，能够以最直观的方式呈现遗产价值，更进一步为实践和教学提供新思路。

一、统计分析

1. 时间推进，记录现代建筑发展

19 世纪和 20 世纪之交，中国大门被迫开启，资本主义列强疯狂进行资本输出。正在发展的现代建筑作为西方文化的一部分，其传入成为了必然。新材料新技术新风格的舶来，冲击了中国固有的建筑体系，本意在于文化侵略的外国建筑，却也让中国建筑乘上了通向现代的快车。历经百年求索，中国现代建筑逐渐实现了吸收和创新。在第二批中国 20 世纪建筑遗产名录中，除少数中国传统建筑外，有 93 项是或含有中国现代建筑[1]。它们具有极精湛的设计水平，代表当时先进的设计理念，记录现代建筑发展的过程。

历经力量积蓄，中国现代建筑于 20 世纪 20 年代正式登台亮相，这一阶段入选的现代建筑也成为遗产名录中数量最多者（图 1）。大量的殖民建筑入侵华夏之后，中国社会逐渐对其接受，于是之后接受现代建筑也就成为自然。西方社会的影响、建筑师的主动引进，使得中国发展出带有自身特质的现代建筑。20 世纪 40 年代，遭受过战争洗礼的中国现代建筑并未就此止步，他抖落身上的炮火灰尘，缓缓站起，以蹒跚的步伐开始了新时代的探索。新中国成立后的"一五计

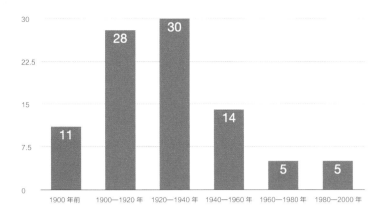

图 1　各时间段内现代建筑数量[1]

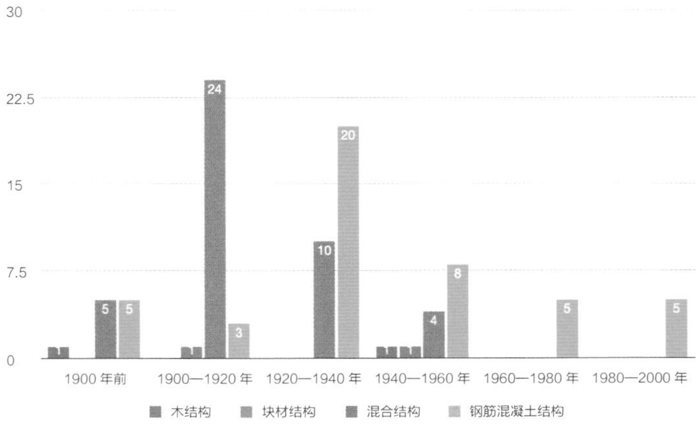

图 2　各时间段内现代建筑结构

划"等加速了中国的工业建设,在数量上虽有减少,但在苏联等国的援助下,产生了技术上的更多进步。进入 80 年代后,现代建筑入选数量与前 20 年相同,这些建筑从时间上看,显得过于"年轻",与当今建筑基本无异,多数还不够成为"遗产",但它们都是经济发展的重要体现,也是改革开放的见证者。

建筑结构的革新是现代建筑的重要体现。对入选名录中的现代建筑进行结构统计[3]可发现(图 2),自 20 世纪 20 年代现代建筑正式出现后,钢筋混凝土结构的运用成为了必然和趋势。木结构、块材结构和混合结构逐步在数量上走向消亡。正是技术上的革新引发形式的变革,"铁筋洋灰"的框架结构为建筑灵活的空间划分提供实现的可能性。剪力墙、空间屋顶,这些新结构促使高层和大跨建筑的出现。技术成就形式的革新,使得这些建筑成为所在年代最"摩登"的体现。

2. 地域体现,聚焦各地建筑特点

不同的自然环境、人文环境以及社会环境使地域文化各具特色。建筑作为文化载体,更能反映出仅属于该地区的独特风貌与民族特质。随着时代发展,地域文化不可避免地面临着"全球化"的强烈冲击,"全球化"浪潮强势地席卷各地[4],现代建筑的"方盒子"遍布世界。纵然实现了科学技术的实践,但城市特色和建筑个性却在逐渐消失,"千城一面"令人叹惋。

通过统计名录中入选的现代建筑所在地(图 3),可发现其真迹遍布我国 24 个省(市、自治区)。从地处南国凝结华侨智慧的开平碉楼,到冰封北疆充满俄罗斯风情的马迭尔宾馆;从位于东南体现海航文化的马尾船政,到坐落西北反映旧唐

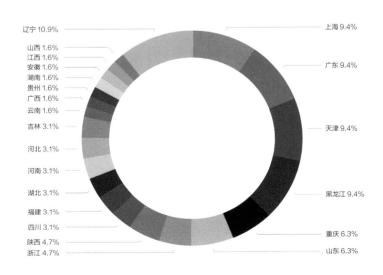

图 3　各地区现代建筑数量占比

风情的大雁塔风景区三唐工程,无一不是建筑地域性的体现。名单中现代建筑地域的多样性,反映出现代人对当今快速建设模式的反思,以及对文化认同和城市记忆的追寻。

从数量上看,入选名录的现代建筑多分布于沿海或边境,这些地区交通发达,经济快速发展,并且多数都曾有过殖民统治的历史。殖民文化与传统风格相遇,在建筑中予以清晰体现。如福建省鼓浪屿近现代建筑群中,建于1895年的日本领事馆旧址已部分使用钢筋混凝土结构。其外墙用英式砌法砌筑,四周环绕新文艺复兴风格的连续半圆拱券外廊,而屋架部分却采用西洋双柱架同中国传统抬梁屋架相结合的做法,融合西方与中国当地技术[5]。同样,建筑样式在融合发展的过程中,也会重新组合,形成新的建筑地域性。如天津入选的南开中学,其中范孙楼主要体现古典复兴,同时也包含世界多种建筑形式,形成天津近代独有的地域性表现[6]。将建筑收入遗产名录,一定程度上将促进建筑宣传,甚至成为当地的符号与标志。进而成为建筑相关从业者研究学习的好范例,成为游子心中故乡的明月光,成为游客打卡观赏的必到处。凡此种种,都旨在寻求到属于地域和城市中独有的建筑特质与魅力。

3. 功能多样,丰富现代建筑体系

20世纪的中国由手工业社会转变成工业社会,对建筑的需求也从简单走向复杂。工业发展和城市扩张促使新功能新类型的建筑层出不穷,公共建筑随之兴起,建筑设计更看重实用性和生产性建筑设计。入选的现代建筑中,除去功能复杂的建筑群体,对其中的86项进行功能统计分析[7](图4),可发现文教建筑、工业建筑和居住建筑所占比例最高,三者总和几乎占据半壁江山,体现出20世纪的时代主题。而一些新兴建筑类型的产生,如行政办公建筑、旅馆建筑、交通建筑等,更是成为了现代建筑的功能宣言。

通过时间上的横向对比(图5),可发现每类建筑遗产都有

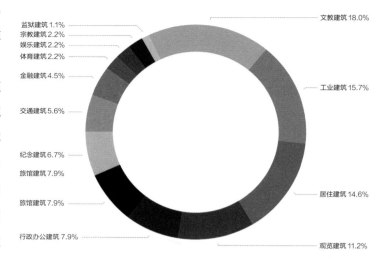

图4 各地区现代建筑功能占比

较为集中的发展期,这需要与当时的社会背景结合来看。如20世纪20年代前后出现文教类建筑的热潮,从最早的传教建筑,逐步因高等教育之需而发展成为现代教育建筑,风格上也逐渐在西式折中里加入中国样式,以建筑群体的方式共同呈现在校园中。而相对更为安定的高校环境,为建筑的完整留存提供了可能。

建成在20世纪各个年代的工业建筑,能够提供一扇回望工业社会中现代建筑发展的窗口。厂房的大空间、生产劳动的工作环境,对建筑本身提出了更高的要求。简洁明了的建筑结构,合理明晰的建筑形式,从一开始就是在践行现代建筑的理念。其中,最为人们熟悉的要数798近现代建筑群。其原为"一五"期间的北京重点工程,由民主德国一家建筑机构负责设计,是典型的包豪斯建筑。其规模之大、风格之典型,使得德国前总理施罗德在参观798艺术区时深深感叹:"几十年前的包豪斯建筑在德国都很少发现了,今天居然在北京存在,真是太难得了!"[8]机械加工车间中的大跨度拱梁,防止太阳炫光的北向天窗,连同厚重墙体和大方立柱[9],使功能技术与风格艺术实现完美结合。虽然厂房今

	文教建筑	工业建筑	居住建筑	观览建筑	行政办公建筑	旅馆建筑	纪念建筑	交通建筑	金融建筑	体育建筑	娱乐建筑	宗教建筑	监狱建筑
1900年前		3	2					2	1				1
1900—1920年	8	2	4	4	2	1		2				2	
1920—1940年	6	3	5	3	4	1	2	3	1	1			
1940—1960年	1	4	2	2	1	2	2						
1960—1980年		2		1		2				1			
1980—2000年	1					2		1			1		

图5 各阶段现代建筑功能数量

日已成为艺术社区,但工业功能早已被烙进建筑之中,自内而外都成为现代建筑理念的物化,成为中国现代建筑的代表。

在建筑功能类型上,交通建筑、体育建筑等也有入选,这类建筑一定程度上更能反映出现代技术的进步。大量人流的聚集不仅要求功能流线合理组织,更对建筑大空间的实现提出要求。钢筋混凝土框架结合各类屋顶空间结构形式,满足建筑功能需求,丰富现代建筑体系。

4. 风格流变,反映时代趋势影响

入选名录中的现代建筑风格多样,随着时间的推移,逐渐脱去繁复外衣——现代建筑风格愈发明显。最先登录中国的西方建筑,用新古典主义和折中主义炫耀威力,而后又经新艺术运动和装饰艺术的装饰。在形式上并不只有简化,建筑师在实践中,也将西式风格同中国传统建筑结合,"大屋顶"建筑便是一种主动探索。

环境对建筑师的影响在此时体现得淋漓尽致——留学归国的建筑师出于文化本位进行设计,以此来表达对中国传统的追寻;而西洋建筑师的探索甚至更早,四川大学建筑群体即为一例,英国建筑师弗雷德·荣杜易在外观上着力突出中国特色,呈现出中西结合、包容有序的格局,成为中式折中主义的代表。

入选的 20 世纪 50 年代建筑多体现苏式风格,最具代表性的十大建筑中又有三项入选,成为新中国成立初期"民族传统"的风格表现。中轴对称的平面与造型设计,彰显苏式建筑的严谨性,体现出庄严稳定的形象。当时间继续推进到改革开放前后,中国现代建筑已不再是集仿,而形成了既符合现代建筑原则,又有中国特色的新风格。干净洗练的线条,简明真实的表达,是对现代建筑理念的努力践行。

丰富的旅馆建筑创作尤其体现出技术和理念的结合。广州白云宾馆是从高层技术方面对现代建筑予以诠释,而杭州黄龙宾馆则通过多层单元分散布置,使体量空间尺度与自然景区相协调。

名录中的现代建筑成为每个时代的缩影,在风格上具有极高的辨识度,不仅留下了建筑风格变革的痕迹,也成为了时代背景的见证者。这些作品使得各种建筑风格并不只是抽象的概念,而是以真实的反映和宝贵的遗存而清晰立体。

二、价值启示

1. 指导实践,激发建筑设计灵感

中国现代建筑从百年前走来,不同于传统建筑的沧桑,其功

图 6 各阶段现代建筑风格代表[10]

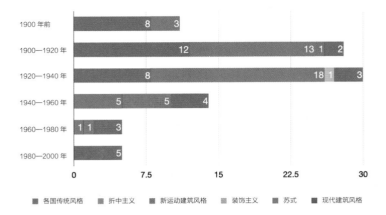

图 7 各阶段现代建筑风格

能明确,生命鲜活,始终在体现进步;也不同于传统建筑的繁复,其线条简约风格明畅,始终在反映时代。它们是活着的遗产,其产生背景、建造过程、修缮状况等均有据可查,基础资料相对较为完备[11],这就能够为当代建筑师的方案设计提供指导。所谓"以史为鉴",其意义即在于取其精华为我所用。

进入新世纪,中国曾经在发展的过程中走得太快,"欲与天公试比高"的建筑物拔地而起;大量的建设需求使这片土地容纳了过多,它们的质量参差不齐,风格不明,引发城市特色失落,建筑成为了仅供消费的产品。在失去方向之时,为什么我们不停下来回头去寻找?入选20世纪建筑遗产的现代建筑,历经时间检验而成为经典。它们依旧矗立,成为遗产不应只停留在保护层面,更应深入到学习的程度。以1995年建成的甲午海战纪念馆为例,距离现在不过二十余年,却因象征主义在建筑形象中的成功运用而收获赞誉无数。建筑师将空间体验与建筑叙事紧密结合,使游客从远望建筑到进入再到离开的过程中,实现浸入式游览。

建筑船体与带有动感的雕塑人物将爱国情绪点燃,渲染出无限悲壮的气氛。反观当下,有多少所谓"象形"的建筑形象,色彩艳丽,直白粗俗,这种表达方式当真令人啼笑皆非。

英国20世纪学会前主席G.斯坦普曾说:"命运的车轮正在不断地加快转动,创新和复兴之间的距离正在消失"[12]。当代建筑师的责任,并非争于朝夕,"打造"出新的风格,而在于接合记忆链条,延续文化血脉。在观形探义间,提升建筑审美,进而激发设计灵感,成就这一个百年不朽的建筑诗篇。

2. 促进教学,拓宽建筑教育思路

大量现代建筑入选这一现象,必将引起社会关注,推动建筑遗产保护进程。从建筑学角度来看,这一进程也将会蔓延到建筑教学环节,为建筑教育提供新思路。这个时代瞬息万变,信息爆炸,新事物更易引发学生兴趣,但这些历史的、经典的建筑,却能够为其补上文化的缺失。眼界的开拓必将引起更多思索,遥想20世纪20年代,中国第一代留学宾夕法尼亚大学的建筑师群体,在校接受古典建筑教育,再通过欧洲游学学习现代建筑样式。这成就了他们归国后的事业,使他们能够得心应手地设计各类建筑,从而推进现代建筑发展进程。杨廷宝、梁思成等建筑师或所在事务所设计的作品有多个入选遗产名录,这是对建筑师的极大肯定,而这些作品无疑这也都得益于当年的所学所闻。

现代建筑遗产保护或许能够催生出建筑学科的更多发展,如:现代建筑遗产的研究与保护、现代建筑遗产的规划与修缮、现代建筑遗产的结构与技术研究等方面,从而培养更多专业人才,提升对现代建筑的关注,保护身边的建筑遗产。

具体到教学环节,20世纪建筑遗产中的现代建筑同样也能够为课程学习提供更多内容,如:围绕现代建筑遗产背景所进行的建筑设计、同现代建筑相关的建筑历史研究或对于现代建筑遗产保护相关的技术途径。通过教育的力量,促进建筑遗产的保护进程。

三、结语

第二批中国20世纪建筑遗产项目中的现代建筑,反映了百年来中国建筑行业的发展,反映了从被迫接受到主动创新的历程。将其列入遗产保护名单中,并不是在单纯怀旧,也不是为过去的成就而沾沾自得。以建筑学的角度对这些作品进行分析,重新审视它们在现代建筑传播进程中所体现出的创新性和探索性,寻找到这一群体的入选意义,才是回顾与分析的最终目的。在这其中,不同阶段、丰富功能的建筑范例成为时间与历史的见证。它们遍布全国,带有强烈的地域性格,不仅在承袭过去,也是在引领未来。

这些建筑曾经被认为无关轻重,但这些建筑早已在不断践行现代建筑的思想,它们详尽地为今人提供了建筑设计的样本,提供了美的典范。对其进行保护，给予价值认同，一方面能够避免中国建筑史上这一段历史付之阙如，另一方面也能够鞭策建筑师，促进引导好设计的横空出世。同样的，对于建筑学教育方面，这些建筑实体将为学科课程发展带来新思路。珍视，研究进而再创作，入选的现代建筑精品值得被如此对待。面对曾经被毁坏、被改造的现代建筑，邹德侬心痛地写下："……改造过的建筑，已经把最有意义的信息源给清除掉了，因而有关这座建筑的原有内含都已经荡然无存，那些内含是重要的时代见证。假若有人发问：中国有现代建筑文化吗？它的实例在哪儿？我们该指哪儿说起呢？"[13]那么这些列入20世纪建筑遗产的作品，就是珍贵的答案了。

注释：

[1] 此93项中不包括保定陆军军官学校、井冈山革命遗址、西柏坡中共中央旧址、国殇墓园、抗日胜利芷江洽降旧址、于田艾提卡清真寺、三坊七巷和朱紫坊建筑群。

[2] 以建筑群体中大部分建筑建造完成的时间代表该项的完成时间。

[3] 以建筑群体中数量最多的结构形式代表该项的结构形式。

[4] 引自袁秋平 徐强. 基于地域文化的现代建筑设计初探——以青岛市火车站为例[J]. 青岛理工大学学报，2013,34(6)：39-43.

[5] 引自钱毅. 鼓浪屿百年建筑风格流变及其背后的文化意义[J]. 中国文化遗产，2017(04)：16-31.

[6] 引自孙博怡. 地域性的设计方法在天津当代建筑实践中的应用研究[D]. 天津大学，2009.

[7] 86项中的庐山会议旧址及庐山别墅建筑群因功能不同而再分为2项、大雁塔风景区三唐工程再分为3项，总数共计为89。

[8] 引自洪志坚，郭弟强. 感受"798"的现代艺术[J]. 艺术与设计，2009,2(09)：340-342.

[9] 引自苏金河. 中国包豪斯建筑的典范之作——798工厂[J]. 建筑，2011,14：76-77+4.

[10] 以建筑群体中数量最多的风格代表该项的风格。

[11] 引自单霁翔. 活着的20世纪建筑遗产亟待保护[J]. 建筑，2012（22）：14-16.

[12] 引自金磊. 认同与保护20世纪建筑遗产. 人民日报，2018,10

[13] 引自邹德侬. 文化底蕴，流传久远——再读"文远楼"[J]. 时代建筑，1999（01）：59-61.

参考文献：

[1] 刘东旭，张勃，孙立伟. 浅谈地域性历史建筑遗产的保护[J]. 江苏建筑，2011（1）：1-3.

[2] 尹志鹏，姜殿成. 从建筑学角度看高校教学建筑的发展[J]. 山东建材学院学报，2000(04)：300-302.

中国 20 世纪建筑遗产的保护价值评价体系建构
The Construction of the Evaluation System of the Protection Value of Chinese Architectural Heritage in the 20th Century

林娜 张向炜 刘军

Lin Na, Zhang Xiangwei, Liu Jun

中国的 20 世纪是一个政治变革、经济创新和文化变迁的时代，在这个带有强烈转折性的世纪中，建筑承载了社会思想的智识史和文化史。托尼·朱特在其出版的《思虑 20 世纪》中提到，"20 世纪被贴标签、阐释、援引和抨击要多于任何一个世纪"，围绕着"20 世纪的教训、记忆和成就是什么？哪些东西留存下来了，又有哪些需要挽回"展开探讨，无疑具有承前启后的意义[1]。的确，作为新近的过往，20 世纪建筑中有哪些是我们已经失去的，又有哪些是被我们遗忘的，还有哪些才是至关重要应该完好留给未来的，是我们未来应该思虑的问题。本文通过对 20 世纪建筑遗产的定义、特点、价值认定标准等因素进行分析，旨在通过对遴选指标架构、评价方法选择、评定实施策略等方面的研究，建构保护工作模型，从而逐步建构成熟的中国 20 世纪建筑遗产价值评价体系。

一、建筑遗产保护价值评价体系共识的缺失

20 世纪的中国，经历了社会、经济、文化前所未有的大发展，建筑理论、建筑艺术、建筑技术以及近代城市规划理论均得到了空前的发展。1999 年国际建协第 20 届世界建筑师大会上，由吴良镛院士起草的《北京宪章》中提到，"20 世纪是一个大发展和大破坏的世纪，大规模的技术和艺术创新无疑极大地丰富了建筑史，但同时许多建筑又难尽人意。"[2]恰恰就是在这种激荡之中，20 世纪的中国建筑取得了前所未有的成绩，留下了许多优秀的作品。

对于建筑遗产而言，不同国家地区或不同学者都有不同的价值评价体系和评价方法，其中存在共识，但也有很多分歧。英国学者费尔顿在《世界文化遗产地管理指南》[3]中指出历史建筑的价值包括情感价值、文化价值和使用价值，并作出了相应的描述。他认为遗产保护与价值评价工作中应充分引入多学科的交叉互动，在确立保护目标后再确认遗产的价值，并以价值的优先顺序作为确定干预等级的决定因素，从而保存遗产对象的关键价值要素和信息。费尔顿价值体系最重要的特征在于将使用价值置于更为重要的地位。在尊重建筑遗产文化价值的基础上，通过合理使用或是功能更新与再利用来延长建筑的使用寿命，实现动态的更新保护。俄罗斯学者普鲁金在其论著《建筑与历史环境》[4]中将古建筑的价值体系分为历史价值、城市环境价值、建筑美学价值、艺术情感价值、科学修复价值、功能价值六个方面。他的特别之处在于提出功能价值，还将城市环境价值置于同等重要的地位，这就拓展了建筑遗产保护的空间和维度，将其置于更大范围的区域，思考其影响贡献与价值意义。

目前，对 20 世纪建筑遗产而言，这些承载了离我们最近的历史的建筑作品，受到来自社会各界的关注，但在关于中国 20 世纪建筑遗产保护价值的评价体系建构上，专业支撑、政策引导、大众认知三个层面上并未形成统一的共识。因此

始终无法建立一个完善的评价制度。在专业支撑方面，尚未形成指标遴选、权重方案、评价方法、评定实施一系列过程上的共识，使得保护的专业性、准确性、标准化明显不足。同时，由于社会大众对于20世纪建筑遗产保护自觉性的缺失，导致保护与发展工作缺乏群众的土壤和文化自觉意识的根基，最终使得目前的保护与发展工作步履维艰。邹德侬教授在2012年发表的《需要紧急保护的20世纪建筑遗产：1949至1979》中特别提到，"20世纪下半叶建筑的遗产保护尚没有真正起步，其中，1949至1979年间优秀建筑，由于它们天生脆弱，正在当前的建设洪流中飞快消失。认识和保护共和国头30年的建筑遗产，已是迫在眉睫了。"[5]因此，正是由于社会各界对于建筑遗产保护价值认知和评价体系缺乏共识，使得我们无力减缓那些优秀的20世纪建筑在眼前消失的速度。

由于20世纪建筑遗产其历史年代的特殊性，主要具备以下几个特点。第一，建筑类型丰富，保存完整。作为刚刚发生的历史，大部分的20世纪建筑保存相当完整，充分的重视和积极的维护可以让这些建筑继续服务于社会。第二，它们是城市文脉的情感寄托。20世纪建筑遗产是20世纪中国城市历史特征的主要组成部分，是城市传统风貌文脉价值的延续，置身于历史的脉络之中，才能真正理解建筑遗产的意义和价值。第三，它们具有时代特殊性下的纪念意义。20世纪的中国经历了从短缺经济到迅猛发展的转折时代，在这些历史特殊时期下的建筑作品，不管今天看起来是否适用，都具备承载时代的纪念意义。第四，建筑形式多元化。20世纪适逢西方现代建筑思想进入中国，建筑师在传统与现代这个话题上的尝试与探索，集结成了许多里程碑式的建筑作品。通过对20世纪建筑遗产这部分资源特点的总结，充分认识到其保护价值的深层次内涵，对于历史研究而言，具有不可或缺的史料价值，对于建筑本身发展也有积极的促进作用，同时对于未来城市化进程中建筑与周边自然、人文环境之间的关系处理，也是具有重要借鉴意义的。

二、20世纪建筑遗产保护价值的认知

西方建筑史学家尼古拉斯·佩夫斯纳在《欧洲建筑纲要》中写道："时代精神渗透了（一个时期）社会生活、它的宗教、它的学术成就和它的艺术。"[6]而建筑则是"变化着的时代的变化的精神"的产物。20世纪的中国，恰逢西方文化强势进入，对中国本土文化带来的冲击与震荡让20世纪的建筑呈现出转折性的走向，从技术的创新、艺术的多元和中西方文化融合等多元化角度审视20世纪中国现代建筑，都让我们越来越清晰地认识到，20世纪建筑遗产所具有的保护价值与研究价值。

2000年，联合国教科文组织世界遗产委员会在濒危遗产报告中，针对一些优秀当代建筑处于被废弃或被改造的境地，

表达出对20世纪建筑遗产命运的忧虑。在全球化的语境下，虽然各国对于建筑遗产的保护价值的评判标准略有不同（表1），但通过对各项认定标准的对比研究不难发现，这些指标都包含了历史原真性、建筑类型样式、保存完整性、创新材料技术等因素，主要涉及建筑遗产的历史价值、建筑价值和环境价值方面的内容。建筑所传达的历史信息是其首要价值，其中不仅包括重要历史事件、历史人物的信息与思想，同时也承载了城市发展的脉络。其中，历史原真性，即历史信息的真实性、完整性、准确性，也成为建筑遗产保护工作的基础出发点。除了历史价值，建筑本身的文化、艺术、技术价值，也是建筑遗产认定的重要方面。例如，英国对于战后登录的建筑，与建筑风格是传统还是现代相比，更加注重建筑的品质，而美国也特别强调要具备历史属性并体现较高的艺术价值。因此，对于建筑遗产保护而言，对建筑本身价值的挖掘，有利于总结并促进之后的建筑发展。建筑并不是孤立的存在，而是处于某种环境之中，城市环境、自然环境都是建筑存在发展的重要影响因素，而对于建筑遗产的保护，不应仅仅局限于单体建筑，而是应该将目光放在建筑群体组合价值上。

《中国文物古迹保护准侧》中强调："保护的目的是保存并延续其价值。"建筑遗产的保护价值应主要包括纪念性价值和现实价值，其中，纪念性价值是指岁月、历史、目的性纪念所带来的价值，而现实价值，即是指延续使用价值以及关联性艺术价值。《中国20世纪建筑遗产认定标准2014》中指出，"中国20世纪建筑遗产项目不同于传统文物建筑，其可持续利用是对文化遗产'活'态的尊重及最本质的传承，保护并兼顾发展与利用是中国20世纪建筑遗产的主要特点。研究并认定20世纪建筑遗产重在传承中国建筑设计思想、梳理中国建筑作品，不仅是敬畏历史，更是为繁荣当今建筑创作服务。"

由现有评定标准可见，中国20世纪建筑遗产与以往其他时期的建筑遗产相比，保护价值中所彰显的"兼顾发展与利用"是我们应该关注的核心问题，这与费尔顿的评价体系中的"将使用价值置于更为重要的地位"不谋而合。现在，我们需要

表1 各国现有建筑遗产价值认定标准

	中国[7]	美国[8]	英国[9]	法国[10]
历史价值	见证重大历史事件，体现城市精神的代表作品 反映近现代中国历史且基于重要事件相对性的建筑遗迹与纪念建筑，体现遗产的完整性	与那些形成我国历史广阔特色的历史事件相关联 与杰出的历史人物有关联 从中业已发现或可能会发现史前或历史上的重要信息	建筑展示国家的社会、经济或军事历史或与重要的历史人物密切相关	不仅包括设计师作品，还应包括一半建筑师、一半历史时期的见证
建筑价值	具有建筑类型、建筑样式、建筑材料、建筑环境，乃至建筑施工工艺等方面的特色及研究价值的建筑物或构筑物 具有时代特征、地域文化综合价值的创新型作品	体现某一类型、某一时期或某种建造方法，具有鲜明特色的作品，或某个大师的代表作、或具有较高艺术价值的作品 重建的建筑，但与现存环境相契合，并作为一个整体修复规划的一部分，且并无其他类似建筑存世	建筑必须在建筑设计、装饰或工匠技艺上有重要意义 特殊的建筑类型、技术（比如显示了技术变革）和重要的平面形式	考虑建筑在美学、技术、政治、文化、经济以及社会演变过程中所具有的价值
环境价值	对城市规划与景观设计诸方面产生过重大影响	历史资产、历史环境的再生与再利用 具有重要群体组合价值		保护范围上进一步扩大，不仅包括建筑单体，还包括建成环境组成群体

在现有中国 20 世纪建筑遗产标准认定的基础上，细化指标遴选，制定权重方案，选取评价方法，最终进行评定实施，使中国 20 世纪建筑遗产保护工作体系化地展开。总而言之，中国 20 世纪建筑遗产的认定标准应该是在全局意识逻辑下的整合，不仅尊重 20 世纪建筑遗产的历史价值、建筑价值和环境价值，同时还要结合现存建筑的年代、类型、数量、结构完整性、样式特征等，对建筑遗产进行分层保护，针对大量现存的 20 世纪下半叶的建筑遗产，通过合理使用或是功能更新与再利用的方式来延长建筑的使用寿命，实现建筑遗产的价值最大化发挥的目标。

三、中国 20 世纪建筑遗产价值评价体系建构

可持续的价值评价体系建构，需要以社会各界的共同关注为前提，并经历一个漫长的过程逐步进行完善，并可持续地发展下去。评价体系的内容应该包括指标遴选、评价方法和评定实施。而要研究建筑遗产价值的评价方法或者体系，必须首先要厘清建筑遗产评价应遵循的几个原则：首先是科学性原则，从评价指标的确定，到评价方法的选择，再到评定后工作的实施，都应该本着科学的、严谨的、有理有据的原则；其次，完整性原则，即对建筑遗产多元化角度进行评估，做到全面且有细节；第三，层次性原则，分清轻重缓急是建筑遗产保护的重要环节，这一环节直接决定了建筑遗产的保护是否及时、有效、深层次地达到预期目标；第四，可操作性原则，这一原则是确保建筑遗产评价可以可持续性地发展下去的关键，整个评价体系的建立才有意义，不然就变成水中花、镜中月，再美好也是空谈。

1. 指标遴选

建筑遗产的评价首先需要通过指标遴选，对建筑进行初步鉴定评估。就指标遴选而言，首先确定遴选指标的各项要素，然后再进行指标权重分析，最终建立精准的指标遴选体系。建筑的价值主要体现在历史性、科学性、艺术性、社会性和文化性等方面，因此，遴选指标的确定应从这几个方面入手，逐步拓展深化。建筑历史性，主要包括建筑的年代，结构完整程度，相关历史事件和人物；建筑科学性，主要包括建筑材料的合理使用，施工与结构技术的突破与特点；建筑艺术性，主要涉及建筑的美学价值、建筑形式上的突破，建筑细部装饰以及与周边环境之间的呼应；建筑社会性，是否对社会发展产生积极的影响，公众参与度有多高，达到多大的社会效益；建筑文化性，主要包括传统文化的传承，对外来文化的融合与再生，本土文化多样化的体现。

基于中国 20 世纪建筑发展特点，结合《关于 20 世纪建筑遗产保护办法的马德里文件 2011》中对于 20 世纪建筑遗产价值的鉴定与评估标准，以及《中国 20 世纪建筑遗产认定标准 2014》中涉及申报资格的详细描述，归纳出适合

表 2　建筑遗产评价体系的遴选指标

综合指标	要素指标	指标解释	遴选等级
建筑本体	基本物理信息	建筑物的地理区位、设计与建成时间、建筑类型、建筑层数、建筑面积	辅助性指标
建筑本体	建筑形式特色	结合当时建筑发展背景，建筑形式所具备的与众不同的特点，并对之后建筑发展起到积极的意义	关键性指标
	结构与施工特点	结构的代表意义以及施工工艺上的突破性创新	关键性指标
	材料与技术创新	新材料、新技术的创新性尝试	关键性指标

续表 2

综合指标	要素指标	指标解释	遴选等级
建筑与周围环境	与自然环境的关系	建筑设计与自然环境之间的内在联系与相互影响，自然环境要素（如风、光照、污染等）对建筑物的损耗情况	辅助性指标
	城市发展进程中的重要性	建筑在城市发展中扮演的角色，社会大众对于建筑的关注度以及参与度	关键性指标
建筑价值	历史价值	建筑历史的重要性，与历史事件、历史人物之间的关系，是否能反映当地文化和历史背景的特性，包括建筑的岁月价值和纪念性价值	关键性指标
	科学价值	科技发展过程中的里程碑式产物，承载着科技的前沿与创新，主要涉及建筑的技术经济问题，比如结构特色、施工质量、经济把控等	辅助性指标
	艺术价值	主要体现在艺术个性与建筑风格	辅助性指标
	社会文化价值	作为文化的载体，在传播现代生活方式以及承载社会效益与价值方面所彰显的文化精神，以及所衍生的建筑文化认同，社会参与对于这部分价值具有格外重要的意义	关键性指标
建筑保护	整体结构基础现状	整体结构完整性、建筑地基基础、变形裂缝的情况以及非承重构件的损毁程度	关键性指标
	主体材料损伤程度	包括天然损坏和人为损坏，天然损坏主要指腐蚀性损坏、微生物损坏、风化等，人为损坏则包括使用不当造成的损坏以及二次修葺不合格带来的损坏。	关键性指标
	功能可持续使用情况	建筑功能布局对建筑发展的突破性贡献，是否符合现阶段使用需求，侧重持续使用情况	辅助性指标
	保护等级评定	基于以上指标的评价，初步给予保护等级评定，是为了给今后的分层次的保护工作打下基础	关键性指标

中国 20 世纪建筑遗产评价的遴选指标方案（表 2）。指标遴选对于建筑遗产的维护与发展具有非常重要的意义，在建筑本体方面，对于所在实践、地点与功能的物理记录，以及建筑形式、结构、技术特点的剖析，是后期建筑维护、修缮、利用的基础；在与周边环境之间的关系方面，更侧重于对建筑的社会性、文化性层面进行评估，对相关景观与环境的贡献，还有对人为、自然环境下建筑遗产的损伤程度与再利用方式，都具有指导意义；在建筑价值评估方面，主要依据历史、科学、艺术、社会、文化等要素将建筑遗产进行归类，在宏观层面上对建筑遗产进行评价；最后，就是建筑遗产的保护措施，基于以上指标的界定，保护工作应该从几方面进行展开，最终形成保护等级评定机制。

2. 评价方法

在指标遴选的基础上，进行指标权重分析的方法很多，但对于建筑遗产的评价而言，很难规避主观因素，只从客观科学角度精确地进行权重分析。因此，需要一方面通过统计计数的方式进行评价，一方面还应该结合解释性比较好的专家经验打分方式进行评价。层次分析法（简称 AHP），是目前为止在各个领域权重分析中最频繁应用的方法，即结合定性与定量分析，层次分明地针对全系统的指标决定进行分析的策略。顾名思义，就是将问题阶梯层次化，通过建立判断矩阵、决定优劣而将层次排序，进行决策性评价，这种方法适用于多准则、多目标复杂问题的决策分析。[11]由于建筑遗产的指标权重分析评价具有特殊性，不应单纯地依靠一种评价方法，而应运用层次分析法与专家经验打分法相结合的方式。同时对于判断矩阵以及优劣层次的分析，专家经验（知识结构、知识水平和认知程度）是一种很好的补充，使指标评价更具客观性与合理性。

采用层次分析法与专家经验打分法相结合的方法，其评价过程应由以下几个步骤组成：确定评价对象；分析研究对象时间、空间等基本信息，选取遴选指标；运用层次分析法对遴选指标进行分层；建构价值评估指标矩阵；针对矩阵中的分层指标选择适当的无量纲化公式或评价方法，确立每个矩阵中的每个指标权重值，这部分建构是结合指标重要性以及专家打分意见进行的，判断矩阵表示出同一层次上各个元素相互的影响力大小的对比；整合各指标的权重值，最终将其转化为评价指标值，建立建筑综合评价模型；利用综合评价模型，专家根据评分基准与参评对象的性能情况对实际项目对象进行打分评估。

3. 评定实施

20世纪建筑遗产价值评价的难点主要包括三方面的内容。首先，建筑遗产本身具有多方面的价值，在保护案例甄选过程中，遴选指标和评价方法方面需要初步形成共识，才有可能继续下面的工作；其次，保护价值评价方法众多，涉及多个学科交叉，选取符合现阶段20世纪建筑遗产保护实际情况的评价方法是问题的关键；第三，建筑遗产的实际情况随着时间推移而发展变化，建筑遗产本身、所在环境以及科学技术的发展都是动态变化的，并且还牵扯到保护价值与经济之间的平衡，尤其是大批量的20世纪60、70、80年代的建筑，需要进行合理有效的保护价值的经济可行性评估，进而建构保护工作模型。因此，基于20世纪建筑遗产目前的实际情况，以及其目前存在的问题与矛盾，评定实施的可操作性是评价体系初步建立的基础。

目前，对于20世纪下半叶建筑遗产评定后的实施，并未形成保护策略上的共识。大量的建筑建设于中国社会、经济大发展和文化转型期，在目前城市建筑环境中所占的比例也相当高，且面临的不仅是单纯遗产保护的问题，更多的是能否可持续使用的问题。因此，如何进行有效的价值评价并进行评定实施，是目前20世纪建筑遗产保护的重要任务。经济方面，需要考虑如何在不动原有结构的前提下，适应现代生活的基本需求。由于主体结构对整体经济投入影响巨大，所以建筑功能调整都需要尽量基于原有结构之上。此外，在传承文化使命方面，如何赋予旧建筑以新的文化、空间生命力，且比新建建筑更具历史感，于城市环境融合得更为紧密，哪里需要改，哪里需要动，尺度的把握非常重要。

因此，对于建筑遗产的本体信息数据的采集是一项非常庞大的工程，同时价值甄别对专家的数量和水平都有很高的要求，面对如此庞杂且数量巨大的20世纪建筑，首先应通过层次分析法对20世纪建筑进行初步遴选，并通过专家评审的方式，进行分批次的评选，逐渐形成中国20世纪建筑遗产名录，并在名录的基础上进行分级评选，结合专家意见，进行价值等级评定。价值等级评定是为了形成有层次的中国20世纪建筑遗产名录，进而结合法制层面推进中国20世纪建筑遗产保护工作，这是一个符合现阶段实际情况的优化评定实施方案。基于这一评定实施方案，可以进行下一步的保护策略和措施分析，对建筑遗产价值的经济评估和文化传承使命进行剖析，从而逐步建构成熟而又完善的中国20世纪建筑遗产评价体系。

四、结束语

在社会经济迅猛发展的今天，中国20世纪建筑遗产评价体系的建构，无疑能让站在21世纪的我们更加清晰地审视这段刚刚发生的历史。建筑遗产作为遗产的一种类型，其特殊

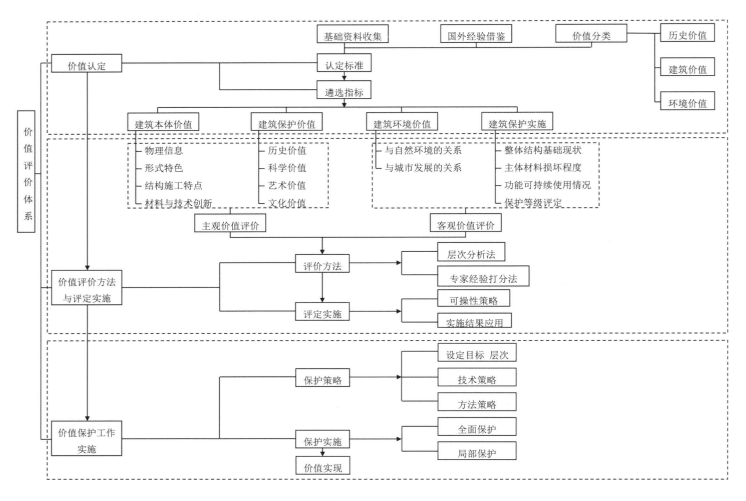

中国20世纪建筑遗产价值评价体系建构图（作者绘制）

性更在于体现着文化自我认同、科技发展的轨迹和艺术水平的提升。中国20世纪建筑遗产的保护价值也不仅仅体现在其厚重的历史价值，同时还涉及不容忽视的建筑价值、艺术价值、社会价值、文化价值等。对于这部分遗产的保护需要社会各界的关注，同时更需要有完善的、成熟的建筑遗产评价体系，使保护工作更加扎实地可持续进行下去。

任何一个领域的评价体系的建立都需要经历一个漫长的研究、实践、调整、再研究的过程，值得庆幸的是，在社会各界的关注与努力下，20世纪建筑遗产评价体系的建构，正在有序地飞速发展。依据国际标准，借鉴各个国家对于20世纪建筑遗产保护工作的经验，结合自身特点与发展现状，初步制定一套遴选指标并探索适合的权重方法，进而建构中国20世纪建筑遗产的遴选方式、评价方法、评定实施策略以及保护措施的模型，为未来中国20世纪建筑遗产评价体系形成共识、逐步成熟打下基础。在这个庞杂的评价体系建构系统中，理清事情发展的核心价值观并积极地进行推进，是现阶段刻不容缓的工作。

参考文献：

[1] 托尼·朱特. 蒂莫西·斯奈德. 思虑 20 世纪 [M]. 北京：中信出版社，2016.

[2] 支文军. 世纪的回眸与展望——国际建协第 20 届世界建筑师大会（北京,1999）综述 [J]. 时代建筑,1999,(03):92-98.

[3] FEILDEN. B.M, JOKILEHTO. J. 世界文化遗产地管理指南 [M]. 刘永孜，刘迪，译. 上海：同济大学出版社，2008.

[4] 普鲁金. 建筑与历史环境 [M]. 韩林飞，译. 北京：社会科学文献出版社，1997.

[5] 邹德侬. 需要紧急保护的 20 世纪建筑遗产（1949 至 1979）[N]. 中国建设报,2012-08-31(003).

[6] NIKOLAUS PEVSNER. An Outline of European Architecture [M]. London, Penguin Press, 1996.

[7] 中国文物学会 20 世纪建筑遗产委员会. 中国 20 世纪建筑遗产名录 [第一卷][M]. 天津大学出版社，2016.

[8] 王红军. 美国建筑遗产保护历程研究 对四个主题事件及其相关性的剖析 [D]. 上海：同济大学，2006.

[9] 宋雪. 英国建筑遗产记录及其规范化研究 [D]. 天津：天津大学,2008.

[10] 费小坤. 建国后（1949—1978）历史建筑保护价值及其评价体系研究 [D]. 武汉：华中科技大学，2016.

[11] 张军，王室程. 建筑遗产价值评估方法 [J]. 哈尔滨工程大学学报,2017,38(10):1661-1668.

苏联建筑风格及其影响下的中国 20 世纪建筑遗产

Soviet Architectural Style and 20th Century Architectural Heritage in China Under its Influence

陈雳　张翰文

Chen Li, Zhang Hanwen

新中国成立之后在城市建筑发展上深受苏联的影响。20 世纪 50 年代，在当时特殊的时代背景下，大量的苏联专家来华指导中国经济建设，其中包括一些建筑专家。在他们的帮助下，我国很多城市建造了许多高质量的苏联风格建筑作品，它们被保留至今，成为重要的 20 世纪建筑遗产，人们对其研究和保护的同时，也在回顾这段不平凡的历史。

一、苏联建筑风格

在 20 世纪初的建筑发展历程中，俄罗斯建筑不仅传承了俄罗斯民族风格、古典主义风格，还接受了体现现代主义精神的新艺术运动建筑风格，在世界建筑历史中独树一帜。

15 世纪，拜占庭式建筑是俄罗斯建筑的主要风格，集中式布局和穹顶是其典型标志，此后受文艺复兴影响，教堂的尖顶成为另一标志性形象。16—17 世纪，莫斯科在俄罗斯的地位被确定，呈现出各民族融合的折中建筑风格，但在技术上已经远远落后于西方。18 世纪俄罗斯开始采取激进的改革措施，向西欧学习先进技术。19 世纪初莫斯科发生大火，城市重建统一采用了古典风格，形成了城市建筑新的特色。1917 年俄国的十月革命，完成了人类历史上一次巨大的社会转变，苏联在经济上、政治上和思想上都实现了极大的发展，同时在建筑上也发生了重大的变革，在欧洲兴起的现代建筑运动影响到了苏联，撼动了古典主义和新艺术运动的基础，形成了著名的苏俄构成主义风格。十月革命到 20 世纪 30 年代，是苏联建筑蓬勃发展的新时代，出现了像塔特林、维斯宁兄弟等一批优秀的苏俄建筑师。

二、斯大林建筑风格

1930 年苏联开始策划建设苏维埃宫，虽然最后没有完成，但苏维埃宫的设计方案成为了斯大林式建筑风格的起点。斯大林式建筑风格是指在斯大林领导时期形成的一种建筑风格，始于 1933 年，结束于 1955 年苏维埃建筑学会。虽然盛行时间并不长，也不像古典主义那样在漫长的时间里受到世界的瞩目，但它对于全苏境内建筑风格的影响是显而易见的。20 世纪 40—50 年代，苏联建成了大批斯大林风格建筑的高楼大厦，还影响了同时期的中国、朝鲜和东欧社会主义国家，如东柏林的卡尔·马克思林荫大道建筑群、罗马尼亚布加勒斯特的出版大厦、华沙科学文化宫等一大批建筑都曾受过斯大林建筑风格的影响，许多建筑已经成为文化遗产保留至今。

1945 年之后，作为社会主义国家的苏联倡议"为共产主义的理想社会秩序做出贡献"，斯大林式风格的建筑在渐渐兴起，其建筑形式气势磅礴，高耸雄伟，布局对称，大多设有钟楼，装饰富丽堂皇，充满了"多余装饰"，艺术风格更多地呈现了当时强烈的意识形态特征，彰显共产主义的革命激

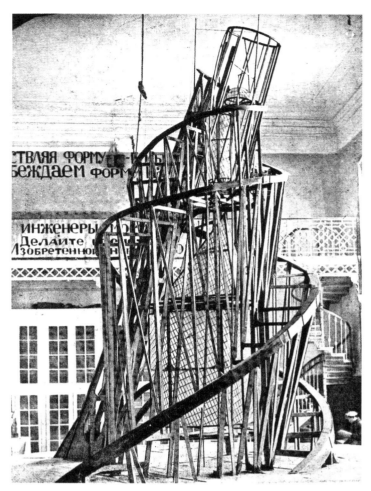
第三国际纪念塔

情与荣耀。斯大林式风格的代表建筑有苏维埃展览馆（莫斯科地铁的起始站）、华沙科学文化宫、"莫斯科七姐妹"等。"莫斯科七姐妹"是 1950 年为纪念莫斯科建市 800 周年所修建的高楼，包括希尔顿莫斯科列宁格勒酒店、皇家拉迪森酒店、俄罗斯外交部大楼、国立大学、艺术家公寓、红门大楼、库德林广场大楼等重要建筑。建筑外形呈阶梯状，衬托中央塔楼的上升之势，外立面也效仿哥特教堂和莫斯科巴洛克式城堡浮华的外饰面，它的尖塔设计起初是为了区分斯大林式建筑与美国 30 年代的摩天大楼的区别，后来也成为了斯大林式建筑风格的重要特点。

三、著名建筑师

苏俄建筑的代表建筑师有：塔特林、梅利尼科夫、里兹斯基、车尔尼雪夫、维斯宁兄弟、列奥尼多夫等。弗拉基米尔·塔特林（Vladimir Tatlin，1885—1953 年）是构成主义运动的重要人物，他对建筑材料和结构有着深入的研究，1920 年塔特林完成了第三国际纪念塔的设计，这也是是构成派的代表作品。这座钢铁、玻璃和革命元素构成的纪念塔从未被世人忘记，莫斯科、巴黎、伦敦等各个国家都先后重现塔特林塔的复刻品，其对欧洲新建筑运动的产生也形成了深远的影响。

20 世纪，我国有一批留学苏联归来的建筑师，他们在我国的城市建设中发挥了重要的作用，其中蔡镇钰（1936—2019 年）就是留苏建筑师的重要代表。1959 年 10 月，蔡镇钰进入苏联莫斯科建筑学院学习，1963 年获得博士学位，归国后进入了华东建筑设计研究院。改革开放后，蔡镇钰向同事们介绍了他自 20 世纪 30 年代以来的建筑历程，以及自己在苏学到的建筑理论和总结出的正反面的经验。其代表作品有上海新世界商城、上海电信大楼、毛里塔尼亚伊斯兰共和国的国家体育场、上海元代水闸遗址博物馆等。他也在 1994 年被评为全国工程勘察设计大师。作为一名中国建筑师，他更着手于结合先进的科学技术，学习和发扬我国传统哲学理念、朴素的生态观和优秀的建筑文化进行创作。对于

俄罗斯国民成就展览馆

北京展览馆

中国传统建筑中哲学理念的研究，包括人与自然界的和谐、风水等他都进行了深入的研究，生态观设计理念在其住宅设计中尤为突出。

四、苏联建筑风格传入中国

中国自鸦片战争以来进入近代发展阶段，在政治、经济、文化等各个方面渐渐西化，建筑领域也开始逐渐接受西方现代建筑的影响。20世纪50年代，中国受到苏联现实主义建筑理论的影响，中国现代建筑发展也转向了其影响下的模式。

新中国成立之初，毛主席在《论人民民主专政》中提出了"一边倒"的外交政策，这也明确了新中国将加入到社会主义阵营，成为社会主义阵营中的最主要的国家。苏联对中国的经济建设提供了大量的援助。20世纪50年代，苏联派来大批各个领域的技术人员和专家，对中国提供直接帮助，在提供帮助和技术的同时，也将其建筑文化输入到了中国。新中国成立伊始，莫斯科城市规划和建设的专家组来到北京，对北京市的城市建设工作进行指导工作。如专家阿比拉莫夫在与梁思成交流的过程中重点提及了民族形式的重要性。此后几年，数位专家来北京参加城市建设的工作，他们把苏联建筑

莫斯科大学

南开大学主教学楼

发展的理论介绍给了中国的同事。阿谢普可夫在清华大学系统地教授工业建筑和苏维埃建筑史课程。他还在课堂上详细地讲述了苏联建筑由20世纪初到"二战"之后的发展历程，主要的思想内容大致有"社会主义现实主义"，"民族形式，社会主义内容"，"反对结构主义，反对形式主义"，"对人的关怀，以人为本"等。随后又有一批苏联专家来到中国，向中国建筑师介绍他们的建筑理论，如：巴拉金、克拉夫秋克等。

五、对中国的影响

20世纪50年代，梁思成一直积极地推广和诠释苏联建筑理论，特别强调民族形式的建筑设计理论。苏联专家批判北京新修建起来的和平宾馆和西郊办公楼都是方盒子、大玻璃的结构主义作品，他认为中国建筑师中了"资本主义的毒"。梁思成在1952年底的《人民日报》上发表文章，将苏联社会主义建筑的理论思想进行了概括总结，其中苏联关于建筑民族性的观点和梁倡导的建筑思想非常契合。在当时特殊的

圣彼得堡列宾美术学院

改造前后的圣彼得堡大学（组图）

历史背景下，我国对他们的建筑理论全面学习和接受，并且积极付诸实践。

1959年，为了庆祝新中国成立十周年，在北京建设了"国庆十大工程"。早期的北京展览馆1953年开始修建，1954年9月竣工，是具有明显苏式风格的公共建筑。建筑坐北朝南，平面呈"山"字形，主体建筑为一层，内部局部设有二层。外部立面中心对称，钟楼高耸，箭塔高达87米，塔尖设有一颗巨大的红星，塔基平台四角各设有一个金顶凉亭。建筑前设有喷水池和广场，建筑内部设有中央大厅、文化馆、工业馆、农业馆、电影馆、露天剧场等场馆，各个展厅内部装饰十分华丽、精致，直观地体现了苏式建筑设计的思想与理论。北京展览馆不仅体现了"对人的关怀"，它的内部和外部也都体现了较高的艺术性，满足了最重要的建筑的民族性原则。上海展览馆也是同样的风格，堪称北京展览馆的姊妹篇。

当时我国许多高校建筑也采取了苏式建筑的风格，如：武汉理工大学余家头校区主教学楼、黑龙江大学主楼、哈尔滨工业大学主楼、西安电子科技大学主楼、南开大学主楼、四川大学理科楼、合肥工业大学的主楼……此类风格的教学建筑甚多，但至今尚缺少一个全面的调研和整理。

这些高校建筑模仿的原型是苏联莫斯科国立大学的主楼，它们在规划上多处于大学正入口的位置，强调中心轴线对称。建筑的立面采用古典建筑的比例，竖向三段式，横向五段式，强调韵律感，比例协调，庄严肃穆。有的中心部分突出，强调向上的气势，屋面楼板挑出压顶。在立面材质上，底层有花岗岩材质的基座和入口踏步，或有混凝土砂浆饰面，二层以上为花岗岩、抹灰或红砖砌筑。内部层高较大，一般为4.5米以上，高大宽敞。因为这是一个时代的标志，具有不可磨灭的记忆，很多建筑已经成为了国家、省市一级的重点文物保护单位，很多大学也以此作为悠久校史的见证。

六、评价

苏联式风格建筑对于中国20世纪50年代前后的建筑有着很大的影响，是我国建筑现代化的一次有意义的实践，也是现代建筑转型历史中不可忽视的一页。中国仿照其将建筑与

深圳北理莫斯科大学主楼

意识形态联系在了一起，建筑作为艺术有了立场问题，建筑不仅有质量的区别，更有了对和错之分，很大程度上被艺术框架理论所束缚。苏式风格更加注重建筑的民族性，资本主义国家的建筑从意识形态上被完全否定，使中国在现代建筑探索的路上受到了一定的阻碍。建筑装饰上虽然一再简化但还是小心翼翼地区别于西方的现代建筑。因此，中国建筑师的创作思维被建筑形式所禁锢，拷贝和模仿成为了那一时期中国建筑设计中一种非常普遍的现象，建筑的局限性显然易见。尽管如此，苏联的建筑理论在我国的实践对我国城市建设发展的推动作用是不可否认的。

20世纪80年代改革开放之后，西方国家的当代建筑理论进入中国，建筑思想迎来了全面的解放，建筑创作也从禁锢中走了出来，建筑界迎来了百花齐放的局面。苏联援助我国时所完成的建筑设计与之所体现的建筑思想也成为了珍贵的20世纪遗产被保留了下来，成为特殊岁月的历史见证。

[本文得到教育部留学基金委青年骨干教师出国研修项目（项目编号：201809960009）的支持。]

参考文献：

[1] 邹德侬. 中国现代建筑史[M]. 天津：天津科学技术出版社，2001.

[2] 利亚布申，谢什金娜，苏维埃建筑[M]. 北京：中国建筑工业出版社，1990.

[3] 朱涛. 新中国建筑运动与梁思成的思想改造：1952—1954 阅读梁思成之四[J]. 时代建筑，2012(6)：130-137.

第二篇

中国 20 世 纪 建 筑 师

由于各种原因,《中国 20 世纪建筑遗产名录 [第二卷]》一书是个迟到的贡献,但也为时未晚,因为有什么材料,写什么历史;有什么建筑作品,才能挖掘项目背后的设计师其人。记得令建筑界、文博界、城规界及公众瞩目的"致敬中国建筑经典"展览,先后举办过三届,一是 2017 年在山东威海举办的"致敬中国建筑经典:中国 20 世纪建筑遗产事件、作品、人物、思想展览",二是 2018 年在福建泉州举办的"中国 20 世纪建筑遗产名录展览";三是在 2020 年北京副中心举办的"致敬中国百年建筑经典——北京 20 世纪建筑遗产"。从各方反馈及媒体报道来看,大家关注的是如何通过建筑师去丈量历史并在其中找寻到深埋的设计痕迹,从而在其中思辨出设计者当年的创作历程乃至某些"暗伤"。岁月如歌,中国 20 世纪建筑史的百年史诗是由作品筑就的,但只有建筑师不拘囿于一隅的多风格创作,才有建筑百花园的大千气象。致敬中国建筑经典,最本真、最耀眼的是,围绕 20 世纪时代主线,呈现主创建筑师和工程师大家的设计理念及创作生涯。在《中国 20 世纪建筑遗产名录 [第一卷]》中共记载了 30 多位前辈,为了人物不重复,在本卷中共选取了 24 位建筑师与工程师,希望这一篇章筑起的是一个"人文化"的设计创意工程与历史,也筑起 20 世纪建筑遗产的中国丰碑。本篇记载的建筑人物没有传记艺术的书写,更没有虚拟的建筑评论,而是力求尽可能详实地解读出建筑师的榜样之力。如果说,文化的光芒温暖并照亮了这些建筑,那么沉淀下来的每个建筑故事,就必然延续了创作者的足迹与思考。何以守护住彰显城市文脉的建筑遗产之根与魂?保护、传承与更新要找寻初创者留下的那些标识。20 世纪建筑遗产对年轻建筑师意味着什么?希望 20 世纪建筑事业的繁荣、经典项目的认同,可以让大家在理性对话中找准当代受用的技法与审美,体会最珍贵的建筑师前辈的心路历程。

Unit Two

The 20th–Century Chinese Architects

For various reasons, *The List of 20th-Century Chinese Architectural Heritage(Vol 2)* is a belated contribution, but not too late, because a history can be complied only with materials, and designers behind projects can be explored only from architectural works. The Exhibition "Tribute to Chinese Architectural Classics", which draws the attention of the architectural circles, cultural and expo circles, urban planning circles and the public, has been held for three times. The first one was "Tribute to Chinese Architectural Classics: Exhibition of Events, Works, Figures and Thoughts of Chinese Architectural Heritage in the 20th Century " held in Weihai, Shandong Province in 2017, and the second one was "Exhibition of Chinese Architectural Heritages in the 20th Century" held in Quanzhou, Fujian Province in 2018; and the third one was the "Tribute to Chinese Centennial Architectural Classics–20th Century Architectural Heritage in Beijing" held in the sub-center of Beijing in 2020. The feedback from all parties and media reports lets us know that people are concerned about how architects measure history and search for deep traces of design in it, so as to identify the designers' creative process and even some "dark wounds" during the creative process. Bygone years are like songs, and the centennial epic of Chinese architectural history in the 20th century has been built by works. However, only the architect's multi-style creation, which seeks to innovate, can develop diversified architectural magnum opuses. While paying tribute to Chinese architectural classics, the most authentic and magnificent attempt is to present the design concepts and creative careers of the main architects and engineers in the 20th century. In addition to more than 30 predecessors recorded in *The List of 20th-Century Chinese Architectural Heritage(Vol 1)*, 24 architects and engineers have been selected in this volume in order to avoid repetition. I hope this chapter will build a creative design project and history of "humanistic culture" and erect a Chinese monument of architectural heritage in the 20th century. For the architects recorded in this chapter, no artistic writing of biography is followed, let alone virtual architectural criticism, but great efforts have been made to interpret the power of architects' role models as detailed as possible. If the light of culture has warmed and illuminated these buildings, every architectural story distilled will inevitably carry on the creators' footprints and thinking. How should we protect the root and soul of architectural heritage that highlights urban cultural heritage? We should protect, inherit, update, and look for those signs that the founders have left behind. What does the architectural heritage of the 20th century mean to young architects? It is hoped that the prosperity of architecture in the 20th century and the recognition of classic projects will enable us to identify the techniques and aesthetics applicable to contemporary times and the most precious mental odyssey of the predecessors in architectural design through rational dialogues.

库尔特·罗克格（Curt Rothkegel，1876—1945年）

罗克格，1876年生于德国，1903年在青岛建立事务所。其丰富的人生及建筑创作经历，使他成为20世纪初期为中国建筑做出贡献的代表人物之一。

第二批中国20世纪建筑遗产推介项目：北京国会旧址

近代在华的德国建筑师当中最著名的当属库尔特·罗克格。罗克格1876年5月24日出生于德国西里西亚的大斯托里茨市（Gross Strehlitz，现波兰境内），1903年来到青岛，1905年与建筑师伯赫曼（W. Borchmann）合作成立了建筑事务所"营造式画师罗克格"，1907年在天津成立了罗克格营造公司，1914年罗克格参加了保卫青岛的战斗，被俘囚禁在日本，1919年被释放回到中国，在沈阳继续从事建筑活动，1929年返回德国，先后居住在南提洛尔和波茨坦等地，1945年在列支敦士登去世。

他一生设计了大量的作品，遍布青岛、北京、天津、沈阳、厦门等多座中国城市，甚至远涉邻国朝鲜。他的专业水平不仅受到了德国教会和私人业主的赏识，还得到了政界人士的肯定，朝鲜高宗李熙、清朝醇亲王之子载洵、庄亲王载勋、袁世凯、张作霖等都与他有过交往。罗克格的作品，功能类型多样，风格千差万别，综观他在华的工作经历可谓是近代中国早期建筑发展状态的缩影。

在青岛的德籍建筑师在展示自己创作理念的同时，又要受当时的政治文化等因素影响。罗克格精通新罗马风建筑的设计，在天津德国会馆的设计中将传统元素运用得淋漓尽致。他的青岛德国海军俱乐部和医药商店的设计是纯粹的青年风格派式样，两座建筑墙面朴实，窗口和山花采取流畅的自然曲线，摆脱了新罗马风的定式。

在青岛基督教堂的设计中，罗克格采用德国青年风格派的设计理念，塔楼用优雅的曲线造型，形式简洁明快，但是在建造时，教会对原方案进行了修改，将塔楼形体拔高，造型改为纯粹的新罗马风式样，放弃了罗克格的初始设计。罗克格方案的入选和修改过程耐人寻味，体现了起源于德国的两种建筑思想的竞争，最终代表德意志民族主义的新罗马风式

样得以胜出,从建筑发展角度讲,这一结果不能不说是一个倒退。

罗克格还主持了清朝资政院大厦的设计,方案造型为纯古典的形式,建筑面积两倍于德国国会大厦,后因辛亥革命爆发,工程开工不久就告夭折。能够在中国的政治中心设计建造西方古典主义的大型公建,这说明罗克格设计水平被政府充分认可,也说明了当时中国社会某种程度的开放。

在中国民族风格走向现代化道路的过程中,罗克格无疑是被当今很多人忽视的先行者,虽然当时的设计还有些生硬,但他深入思考,大胆尝试,为后人的继续探索提供了借鉴。

罗克格主持了北京正阳门的改造工程,他的方案将原城门增设大踏步,窗口部位增加装饰线脚,完工之后城门形象更加完整,细部刻画也更丰富,但建成效果也引发了一些争议,许多中国人认为这是一次不伦不类的设计。罗克格还收到了在北京东华门的旧址上仿建一座勃兰登堡式城门的设计邀请,考虑到城市整体风貌,罗克格婉言回绝了这一项目。可见当时外籍建筑师已经具有了传统街区风貌的整体保护意识。

1911年万国医疗卫生博览会在德国德累斯顿举行,罗克格主持了中国展馆的设计。中国馆包括中国塔和展厅部分,采取了简化的中国传统风格。在建造过程中,他将现代材料与传统式样进行了大胆结合,并取得了成功,这是近代早期建筑师对中国传统建筑现代化转型的一次有益探索。

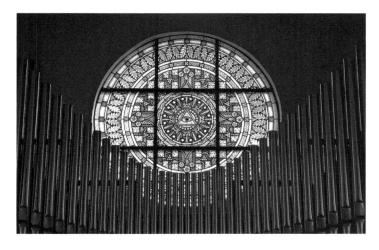

李守先（1878—1940年）

李守先，字继孟，汉族，云南泸西县人，建筑师，市政建设者，云南辛亥革命、护国运动的革命先驱。

第二批中国20世纪建筑遗产推介项目：云南陆军讲武堂旧址

1898年，20岁的李守先考取廪贡生；1903年在昆明考取云南高等学堂（五华书院改办）；1904年9月考取官费生，赴日本留学，先入经纬学堂学习日语，后入路矿学堂学基础课；1905年考入日本工业学校，攻读土木建筑工程专业，并在日本岩仓铁道学校兼修铁道勘测设计，还曾在日本东京早稻田大学进修过法律和经济学。1906年，李守先加入同盟会，1907年毕业后参加云南建设，负责云南陆军讲武堂的建筑设计和施工，同时兼任讲武堂的文职教官，讲授数学、测绘等课程。

1915年至1916年，护国战争胜利后，李守先在云南省政府实业司、民政司、政务厅等单位历任参事、技正及机械制造所所长等职。1919年至1922年，李守先被委任云南陆军讲武堂监修委员和护国战争胜利纪念建筑群护国门、护国桥、护国路的设计建筑专员。在此期间，李守先参与了多项建设项目，他被委派勘测修建铁路，勘测整修省城六河及多条护城河，修筑石桥、城墙、新式马路，改建城市设施，修建水闸，勘测抢修水患造成山崩泥石流的黑（盐）井河龙川大河堤坝，勘测修复地震灾区的一些受灾建筑，修建五华山旁有艺术特色的八角形大礼堂屋顶，兼任东陆大学会泽院施工顾问。1940年12月，李守先在昆明逝世，享年62岁。李守先的设计思想有着兼收并蓄的风格，设计实践中他既尊重中华传统，又吸收外域优点，把中国传统的建筑样式与近代西方和日本的建筑优点互补融合，设计出了风格独特、气势雄浑、庄重威武、美观实用的云南陆军讲武堂。

思九生（Robert Ernest Stewardson）

思九生，英国建筑师，1904年12月5日成为英国皇家建筑师学会准会员（ARIBA），1921年初被选为英国皇家建筑师协会荣誉会员（FRIBA）。

第二批中国20世纪建筑遗产推介项目：上海邮政总局

1894年至1898年，思九生在爱丁堡实习。学徒期结束后，他搬到伦敦，在那里担任威廉姆·亨利·怀特（William Henry White）的助理一年；而后先后在威斯敏斯特的英国工程处（H.M. Office of Works）、威廉·爱德华·莱利（William Edward Riley）领导的LCC建筑师部等地工作。1905年，他被阿斯顿·韦伯（Aston Webb）选中，被派往南非布隆方丹的公共工程部门，建造格雷学院（Grey College，南非自由州大学的前身）的校园建筑，随后，在1908年，他移居上海，担任沃尔特·司各特（Walter Scott）首席助理，并于1913年4月开始独立执业。

思九生于1911年创办思九生洋行，1919年将洋行与他人合作经营，曾改称"Stewardson & Spence""Stewardson, Spence & Watson"，不久，合伙人退出，仍恢复使用R. E. Stewardson英文名，但是中文名始终未变。

思九生洋行是上海著名的建筑事务所，承接建筑设计、测绘检验，经营上海的房地产业务。思九生洋行在上海的作品很多，如益丰大楼（设计于1911年）、怡和洋行大楼（Jardine, Matheson & Co.，外滩27号，1920—1922年）、卜内门洋行大楼（The Brunner, Mond & Co. Building，四川中路133号，1921—1922年）、嘉道理爵士住宅、大理石大厦（静安区延安西路64号，1924年）、汉口江海关（汉口沿江大道86号，1921）等都是思九生洋行的杰作。

合作建筑师：
史滨生（Herbert Marshall Spence，1883—1958年）

史滨生曾在英国纽卡斯尔进行职业训练，在加入英国工程办公室（HM Office of Works）之前曾与多家公司合作，1911年在上海英国工程处中国分公司工作。1919年，与思九生合伙，加入了思九生洋行，公司英文名改称 Stewardson & Spence。

沈理源（1890—1951年）

沈理源，浙江杭州人，近现代著名建筑师、建筑教育家。早年于意大利那不勒斯大学攻读数学，后转入建筑学科。1915年回国，初在黄河水利委员会工作，后创办华信工程公司，又先后任教于北平大学建筑系、天津工商学院建筑系和津沽大学建筑系（教授、系主任）。所译弗莱彻《比较世界建筑史》部分章节，以《西洋建筑史》之名出版于1944年。

第二批中国20世纪建筑遗产推介项目：天津市解放北路近代建筑群

沈理源，浙江杭州人，中学时代是在1896年创办的上海南洋中学度过的，这是一所享誉海内外素有"国人自办第一校"之称的中学。有文献记载，十多位两院院士和文学泰斗巴金、国际摄影大师朗静山、外交家顾维钧等，建筑界沈理源、庄俊、佘畯南等均毕业于此。早年于意大利拿波里（那不勒斯）大学攻读数学和建筑学科。2010年前后，中国建筑工业出版社原副总编杨永生编审召集了一个"中国建筑名师丛书"编委会，选出十几位中国第一、二代建筑先贤准备出版传记体著作，2012年5月共有《沈理源》《虞炳烈》《梁思成》《杨廷宝》《陈植》五册出版。从中国文物学会20世纪建筑遗产委员会推荐的20世纪建筑遗产项目看，虽英年早逝，但沈理源对中国建筑作品、建筑思想、设计流派、建筑教育等均贡献巨大。身为沈理源的学生，1945年毕业于北京大学工学院建筑系的清华大学建筑学院教授王炜钰在书的代序中评价"……他1915年学成回国，比第一代建筑家梁思成、杨廷宝等还早十多年。他设计的建筑在20世纪20年代初就已颇具影响，他测绘的胡雪岩故居平面图是目前发现的，由我国建筑师用现代测绘方法完成的最早的建筑测绘图。"沈理源的职业建筑师生涯于1920年创办的华信工程司开始，后又与多家事务所合作从事建筑创作。沈理源最晚于1920年主持华信工程司工作，到1932年9月，北平市工务局准予执行业务并发给执照的土木技师（建筑师）技副共14人，有沈理源、钟森、关颂声、杨廷宝等人，至此他的业务范围遍及京、津、沪、杭等地。研究发现北京市建筑设计研究院的姚丽生、欧阳骖等都是沈理源的门徒。需要说明的是，沈理源乃一介平民建筑师，所谓"平民"系他未将精力花在疏通关系上，也不攀权贵，靠信誉招徕生意，1937年七七事变后，他也未随富商与文人转移到重庆陪都，而是留下来继续工作。

沈理源先生一生历经清朝末年、北洋政府、国民政府、新中国几个历史阶段，除投身建筑创作外，还曾先后在国立北平大学艺术学院、北京大学工学院、天津工商学院任教授及建筑系主任。作为建筑教育家，为传播世界建筑理论与文化，他亲自编译出版英国人班尼斯特·弗莱彻（Banister Fletcher）著的《弗莱彻建筑史》。沈理源先生英年仙逝，业界与他蒙面的人可能不多，但他在短暂的35载创作生涯（1915—1950年），竟在国内完成了一个又一个经典作品。王炜钰教授将其归纳为："……从西洋古典建筑风格、折中主义风格到摩登建筑风格；从普通砖混结构住宅到大跨结构的剧院、银行、商场等类型的公共建筑，论质与量，在当时历史背景下，国内很少有建筑师能与他相比。"

沈振森、顾放所著《沈理源》一书虽仅135千字，但成为让后人认识这位走进建筑圣堂大建筑师的优秀读物。据书中介绍，沈理源先生首开中国20世纪建筑的先河，创造剧院辉煌，并对摩登建筑、清华园内现代风建筑及天津五大道附

近的洋楼等建筑做出尝试，如位于天津和平区河北路277号的周氏旧宅（周馥之子北方著名实业家周学熙的长子周明泰之宅），建于1933年的西洋风格小楼，是一座典型的英国外廊式庭院住宅，该建筑被戏曲史学界称为"几礼居"（因为周明泰对史学与戏剧文献学有深度研究）。《沈理源》一书提出了这样的建筑学思考：虽沈先生有丰富多彩的建筑风格与类型设计，如北京万安公墓等，但他罕有中国古典式样的建筑作品。在20世纪20—30年代，中国建筑界在追随西方建筑形式时，也出现了持续十多年时间的以国粹派著称的"中国固有式建筑"思潮，如吕彦直的中山陵，杨廷宝、董大酉等的中山陵、音乐台、市政府大厦等。1930年，政府公开向社会招标进行上海市中心区域规划设计，沈理源参加了，但他的方案并没有采用中国固有式建筑设计风格，据业界专家评价，他很少发表建筑评论，只是用设计实践验证自己坚守的建筑观。

沈理源先生所设计的极具20世纪建筑遗产价值的代表性作品包括：清华大学早期建筑（化学馆、机械馆、航空馆、电机馆、体育馆扩建、新林院住宅区等）、北京大学图书馆（现《求是》杂志社）、天津解放北路近代建筑群（有"东方华尔街"之美誉）。在天津这条著名的金融贸易街上，他设计的著名银行建筑有：盐业银行（1926年，天津赤峰道12号）、新华信托银行（1934年，解放北路10号）、浙江兴业银行（1922年，和平路237-1号）、金城银行（1937年，解放北路108号）、中华汇业银行（1926年，解放北路119号）等。其中以盐业银行设计最为精彩。其外立面无论柱头窗套，还是入口门头都运用精细的文艺复兴式古典雕刻，而室内的顶棚、墙面、门窗口、踢脚线等更采用西式手法，从而实现室内外的统一，就连室内彩色花窗也绘制成有津地域风情的盛产芦盐的盐业主题场景。

2020年恰逢沈理源先生诞辰130周年，辞世69周年，无论从中国20世纪建筑遗产上颇有作为的建筑师角度，还是中国早期建筑文化实践的先驱者看，都应该全面纪念并缅述他。

柳士英（1893—1973年）

柳士英，籍贯江苏省苏州市，20世纪初赴日本学习，参与创办我国第一所建筑专科学校——苏州工专。20世纪50年代筹建中南土木建筑学院，长期从事建筑教育事业，为我国的建筑教育和建筑设计发展，做出了极为重要的贡献。

第二批中国20世纪建筑遗产推介项目：湖南大学早期建筑群

柳士英幼年丧父，家境贫寒，童年就读于地方善堂附设小学，1907年考入江南陆军学堂。于1914年考入东京高等工业学校建筑科预科，1916年升入本科。五四运动期间也以满腔爱国热情积极响应，回国参加活动，后回到日本。1920年毕业后，他回国到上海，初在日本人所设的"东亚公司""冈野建筑师事务所"任技师，1922年与留日同学、挚友刘敦桢、王克生、朱士圭等人创办华海建筑公司建筑部，任设计部主任。华海公司设计工程布及上海、苏州、南京、芜湖等地。1923年他参与了苏州工业专门学校的创办，担任建筑科主任、教授，与刘敦桢、朱士圭先生等共同办学，开创我国现代建筑教育之先河。

1928年，柳士英任苏州市市政工程筹备处总工程师，并首任苏州市工务局长，为苏州现代城市建设奠定了发展基础。1930年他重返上海，继续华海建筑公司业务，并受聘执教于上海大夏大学等学校，同时完成上海中华学艺社等建筑设计。1934年，柳士英任湖南大学土木系教授，之后长期为湖南大学以及湖南地区的发展建设做出贡献，新中国成立后任湖南大学土木系主任。1952年他受命筹建中南土木建筑学院，任筹委会主委，一年后任中南土木建筑学院院长。1958年后，他先后任湖南工学院院长，湖南大学副校长，高教部教材编审委员会委员等职。柳士英完成了多项建筑工程的设计，包括：上海同兴纱厂，杭州武林造纸厂，安徽芜湖中国银行，上海中华学艺社，上海大夏大学校舍，上海大夏新村，上海王伯群故居，湖南电灯公司办公楼，湖南大学第二、三、四、七、九宿舍，至善村教工住宅区，胜利斋教工宿舍、大礼堂、图书馆、工程馆、科学馆（加顶层），主持武汉华中工学院（现华中理工大学）校园规划，华中工学院诸多教学楼、实验楼、教职工宿舍、动力实习工厂等项目，曾编有《西洋建筑史》《五柱规范》《建筑营造学》《建筑制图规范》等教材。

柳士英先生是我国建筑教育的开拓者、先行者，在几十年的工作中，他教书育人，无私奉献，勤奋开拓，默默耕耘。他始终坚持教学与生产实践相结合，不断创新提高，其作品融汇中西文化精髓，具有独特风格，为国家培养了许许多多建筑人才。

柳士英1952年加入中国国民党中央委员会，曾任民革中央委员和湖南省委员、湖南省人民代表、湖南省政协委员、第四届全国政协委员。1951年他筹建湖南省土木建筑学会，任理事长。

刁泰乾（1897—1969年）

第二批中国20世纪建筑遗产推介项目：重庆大学近代建筑群

刁泰乾为重庆市江津区白沙镇人，1920年留学法国，就读于法国巴黎大学土木工程系，毕业后回国。自1926年开始，刁泰乾历任万县市政府工程师丰都县市政公所工程师、重庆大学教授兼工程师、四川省建设厅工程师、西康省裕边公司总工程师、四川省水利局工程师和国立自贡工业专科学校教授；新中国成立后，曾任自贡市粮食公司工程师、川南建筑公司工程师、四川省建设厅生产技术处处长、成都工学院图书馆馆长和四川省人大代表（1954—1966年）。

刁泰乾和弟弟刁泰升合作设计的重庆大学工学院建于1935年，带着浓厚的欧式新古典主义风格又具本土色彩。墙体全部用条石砌筑，开创了重庆石建筑的先例，而楼内各层屋架均用杉木制做。建筑呈L形布局，入口处的六边形塔楼最为出彩，各层屋架均用杉木制作，是影响重庆的十大经典老建筑。

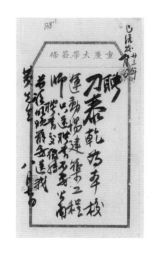

杨锡镠（1899—1978年）

杨锡镠，字右辛，江苏吴江人，祖籍为苏州吴江县（今吴江市）桃源镇，1922年6月毕业于南洋大学土木工科，获学士学位。杨锡镠长期在北京市建筑设计研究院工作，曾任北京市建筑设计院总建筑师兼三室主任，为20世纪50年代北京建院"八大总"之一。

第二批中国20世纪建筑遗产推介项目：北京工人体育馆

杨锡镠1922年自南洋大学毕业后，在上海东南建筑事务所任工程师，后于1924年与他人合办凯泰建筑事务所，参与建筑设计，并成为合伙人之一。1927年到广西任职。1929年，经范文照和李锦沛介绍加入中国建筑师学会，并于同年自设杨锡镠建筑事务所，成为独立开业的建筑师，同时兼任（上海）沪江大学商学院建筑科教师。1930年任中国工程师学会执行部总务；1934年，担任《中国建筑》杂志发行人和《申报·建筑专刊》主编；成为中国工程师学会会员。后因抗日战争爆发，杨锡镠关闭了自己的事务所，并于1948年复入凯泰事务所。杨先生于新中国成立前后来到北京，最初担任联合建筑师工程师事务所的合伙人之一，后因1953年"公私合营"，改制后进入北京市城市规划管理局设计院（北京市建筑设计研究院前身）任总建筑师兼三室主任。1964年杨锡镠随老院长沈勃调入北京市规划局，不久后又回到设计院工作，历经退休和返聘，直至1978年，杨先生于北京去世，终年79岁。

杨锡镠先生在上海东南建筑公司主要参与项目有：上海南洋大学体育馆（现上海交大老体育馆），南京东南大学科学馆（原国立中央大学科学馆，现东南大学吴健雄学院）等项目。在凯泰事务所里，杨先生设计了古典中式风格的窦乐安路鸿德堂（今多伦路鸿德堂），获得了南京中山陵设计竞赛三等奖。在广西柳州期间，他参加了广西梧州中山纪念堂竞赛，获二等奖。在自办事务所的时间里，设计完成主要作品的有：上海中法报台、上海特区法院、南京饭店和"远东第一乐府"百乐门歌舞厅、上海商学院、无锡经茹堂（方案）、湖北省政府大楼（未建成）等。复入凯泰事务所后设计的较为著名的项目有：大陆游泳池、大都会歌舞厅等。1953年后，杨锡镠先生作为总建筑师主要负责的项目有：北京太阳宫体育馆（今北京体育馆）、中国科学院物理所、北京陶然亭游泳池和网球馆、北京工人体育馆、北京红领巾湖室外游泳场、北京展览馆剧场加顶改造、北京市工人俱乐部、苏联大使馆、北京航空学院、北京医学院、北京半导体厂等。杨锡镠先生曾连任第二、三、四、五届北京市人民代表大会代表。

全国工程勘察设计大师熊明在接受采访时曾回忆：

我做的第一个项目就是工人体育馆。杨先生是总建筑师了，同时又是院里技术委员会的，所以对那个方案杨先生也是提了意见的。人民大会堂三层观众席的视线设计以及它的参数，都是杨先生指导设计的。杨先生个人一是钻研问题很深，二是待人很谦和。他平常生活交往很轻松，围棋下得也很好。杨先生学结构出身，但他自己喜欢建筑，在建筑的科学技术方面，比别的建筑师研究得多，因为他的理工学得比较好。那时在北京，提倡学习苏联，社会主义的内容，民族主义的形式，北京体育馆就是一个例子。杨先生对各种运动的体育

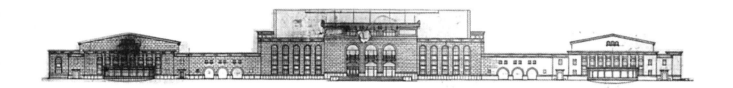

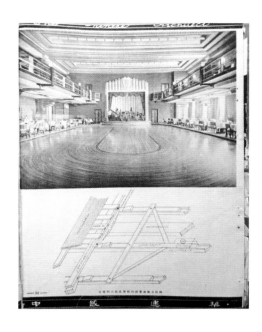

规范，各种空间、尺寸、高低，各种设备的规则标准，都了解得比较多。刚提到的视线包括两个，一个是视觉质量，一个是视线。视觉质量包括方向，比如长形的场地，在长边一侧视觉质量最好，但自从透明栏板出现了以后，有的人就喜欢在端头的位置，看运动员上篮、扣篮等等。视觉质量还包括一个视距，视觉的距离，距离的远近。另一方面视线主要强调两个，一个是无障碍，后面的人视线要能越过前面人的头顶；另一个是半无障碍，座位错位排列，每排座位与前后座位错开半个位置。这个适合电影院，但是对球类运动就不行了。运动员在场地上活动范围大，当跑到端头时，错开座位的脑袋仍然会挡住视线。座椅篮球、足球场这类的座位最好是全无障碍。所以杨先生对这类建筑的视线设计还是进行了很深入的研究的。他阐明了怎么选用视觉差，视觉差指的就是后一个人的视线与前一个人的头顶的高差。这些公式不一定是他创造的，但他运用得很熟练。怎么选择这个视线、高差、设计的视点，他还有一套简单的算法，甚至编辑成表册，一下就可以查到。怎么调整这个曲线，这曲线又和座位的高低有关系。身高太高的话，坐在后面的人会觉得不安全，太低又会遮挡视线。而且，在前面看台较缓和的时候，大台阶之间距离不大，到后面很陡的时候，每一个大台阶之间就要设置踏步，每个大台阶之间的进深和高差、放多少个踏步，杨先生都算得非常细致、全面、透彻。当时国内也没有规范，他可能凭借一些相关资料，自己搞了一套速算的办法、表格，

如何选择各个参数，这一套他都运用得特别熟练。一直到现在，北京所有的体育建筑，都是用的是他这一套办法。这个表格现在没有了，但是这套计算方法，简单的公式，很多书上都有。

疏散也是杨先生研究比较深的。疏散的时间、宽度，每一个人需要疏散的尺寸，包括楼梯的坡度，这些参数他都研究得很全面。那个时候没有现在的规范，但是很多参数计算很简单，计算的原则就是来去一样，即来了多少人，都能及时地走出去，由此来决定过道的宽度。从看台里每一排出来的人都到过道里，集中了三四排，人就多了，要做到来去一样。所有出口，包括安全门的出口，大门的出口，从各方面汇集来的不论多少条人流，都可以做到出去刚刚合适，否则就会浪费或者不够。

奚福泉（1902—1983年）

奚福泉，字世明，上海人，1922年赴德国留学，1926年毕业于德国德累斯顿工业大学，获特许工程师，1929年获得德国柏林工业大学博士学位，长期在上海从事民用建筑设计工作，后期致力于我国工业建筑设计与研究工作。

第二批中国20世纪建筑遗产推介项目：国民大会堂旧址

奚福泉回国之初在上海公和洋行从事建筑设计，1930年7月加入中国建筑师学会，1931年4月加入启明建筑公司，1935年创办公利工程司。

1935年9月，奚福泉设计的南京国民大会堂获第一奖。国民大会堂位于南京长江路，坐北朝南，左右对称，主体建筑地上4层，地下1层。建筑整体分为前厅、剧场、舞台3个部分，建筑面积5100平方米，前厅为砖混结构，剧场为钢筋混凝土柱网结构，型钢屋架。在设计中，奚福泉采纳了西方剧院的造型和简洁明快的近代建筑风格，并在建筑内外巧妙地运用了中国传统纹饰进行装饰，这使得南京国民大会堂突显中西合璧的"新民族主义"建筑风格。抗战胜利，国民政府还都后，对国民大会堂进行了修复，增加了观众席，并在每一个座椅的背后设有3个按钮，用于会议表决使用。1949年后，国民大会堂改名为人民大会堂。1985年后，大会堂内部进行了全面整修，后又进行了多次维修及保养。

1950年以前，奚福泉先生还设计了上海福煦路（今延安中路）福明村、上海福煦路浦东大厦、上海虹桥疗养院、浦东同乡会大厦、上海四明银行、上海都城饭店（现新城饭店）、河滨大厦（七层）、南京交通部新署、上海正始中学、第五师阵亡将士纪念塔、中国国华银行南京分行、中国国货银行、欧亚航空公司上海龙华站及南京故宫站、上海时报馆、上海欧亚航空公司龙华飞机棚厂、南京国民大会堂、国立戏剧音乐院（方案）、上海梅园别墅、上海杨树浦底复兴岛内行政院物资供应局仓库和南京、汉口、芜湖、沙市、宜昌等地的邮局大楼以及住宅、工业、文化等多项设计。

1950年，奚福泉成为中国建筑师学会登记会员、理事长，创办奚福泉建筑师事务所；1951—1952年任上海联合顾问建筑师工程师事务所建筑师，1953年入轻工业部设计公司华东公司（中国轻工业上海设计院），后任轻工业部上海轻工业设计院副总工程师；1957年任中国建筑学会第二届理事。他还是上海市第三、四、五、七届人大代表，中国建筑学会上海分会理事。新中国成立后，奚福泉先生主要从事工业建筑设计，如佳木斯造纸厂、南平造纸厂、芜湖造纸厂、甘谷油墨厂、南京钟表材料厂、西安风雪仪表厂、援助阿尔巴尼亚造纸厂、五金与塑料制品厂，援助几内亚火柴卷烟厂和菲律宾、尼泊尔等轻工业工程，为中国工业建筑的发展及对外援助工程作出了杰出贡献。

杨永生先生编《建筑百家书信集》刊登《陈植致阎子祥信》中所言：兹陈述者关于《中国人名词典》不知能否将奚福泉列入。此君埋头工作，设计才能高超，因为大家对他了解不多，但他与夏昌世（现已去德国定居）两人乃中国建筑师中仅有的得博士学位者，梁工的博士学位则属赠予的名誉学位。……奚老在轻工业部设计期间，各地奔波，从佳木斯到广州，从乌鲁木齐到越南边境，不辞辛苦。

陆谦受（1904—1991年）

陆谦受，广东新会人，1930年毕业于英国建筑学会建筑学院（Dip.A.A），成为英国皇家建筑学会会员(A.R.I.B.A)，长期在上海从事民用建筑设计工作，1949年后赴香港。

第二批中国20世纪建筑遗产推介项目：中国银行南京分行旧址

陆谦受幼年时跟随家人去英国，大学毕业后回国，任上海中国银行建筑科科长，1935年当选为中国建筑师学会副会长，20世纪50年代曾将自己的事务所与甘铭、周耀年及李礼元等建筑师事务所合作参与大型村屋——苏屋村的部分设计。

主要完成的建筑设计有：上海中国银行总行、上海同孚大楼（上海中国银行西区分行，与吴景奇合作）、上海中国银行虹口银行大厦新屋、汉口路华商证券交易所证券大楼、极司菲尔路（今研究院路）中国银行宿舍、上海中山路住宅、大嘉宝乡村医院、南京住宅区新建住宅、南京珠宝廊中国银行、青岛中国银行宿舍、苏州中国银行、南京金成银行、南京警厅后街中国银行住宅等。

陆谦受与吴景奇合著有《我们的主张》（《中国建筑》26期，1936年7月）

《我们的主张》（节选）　　陆谦受　吴景奇

人类有生以来就需要居住的地方。所以住的问题，上自天子，下至庶民，都要想出一个解决的办法。在上古时代，人类的生活很简单，住的条件也是很简单。只要找到一处能遮风避雨及防止野兽侵入的地方，一个巢或一个穴，住的问题，就算解决。到后来，人类的生活续渐复杂起来，住的条件也跟着发生变化了。大概起先是完全根据生活的需要来进展，其后便与美术发生关系。

因为人类的心理，是富于情感的，在各项生存的条件得到满足之时，一时的情感，便得要找一条出路。于是文学、音乐、美术，以及凡可以作为抒情工具的东西，都应运而生。整个人类的生活，因此更加丰富。在这种情况下之下，建筑当然不能是例外。

说到情感，大家都知道它是完全受环境所支配和影响的。环境不断地在变迁，所以感情也跟着不断地在变迁。然则一切发挥情感的东西，所谓抒情的工具者，决不能从古至今，丝毫不变，岂不是很明显的事实吗？

所以当我们看到在每一个时代和每一处地方的文学、绘画或雕刻，我们便可以推测当时当地社会的一切情形；至于建筑也有同样的作用。

因此，我们对于建筑艺术的主张，是一个很复杂的问题，得到一种答案了。我们以为派别是无关重要的。一件成功的建筑作品，第一，不能离开实用的需要；第二，不能离开时代的背景；第三，不能离开美术的原理；第四，不能离开文化的精神。

所谓实用的需要，就是说，建筑要能满足我们特别的需要。譬如一间戏院，就要能够使我们舒舒服服地看到演员的动作和听到歌唱的声音。

所谓时代的背景，就是说：建筑要能充分地显出我们这一个

时代的进化的特点,不要开倒车,使人家怀疑着现在是唐还是宋。

所谓美术的原理,就是说:建筑的结构、颜色、形式,都要合乎美术的原理,不要因为标新立异,就不顾一切的将奇形怪状的东西都弄出来。

所谓文化的精神,就是说:建筑要能代表我们自己文化的精神,不要把中国的城市,都变成了欧美的城市。

所以在这四种原则之下,我们就应该努力创造一个新的风格出来,作为我们对于这一个时代文化的贡献,我们自己应当争点气,下点苦功,做点事业,不要老是跟在别人的后面。必需这样,我们的建筑艺术,才有出头的日子。

(选自《建筑百家杂识录》)

合作建筑师:
吴景奇(1900—1943年)

广东南海人,1925—1931年在美国宾夕法尼亚大学建筑科学习,获硕士学位。在美国实习2年,期间在费城 Adin Benedict Lacy 建筑师事务所任绘图设计员,回国后在范文照建筑师事务所任助理建筑师。后任职于上海中国银行建筑科。

徐敬直（1906—1982年）

徐敬直，广东中山人，出生于上海，1924—1926年在上海沪江大学学习，1927年转入美国密歇根大学建筑系学习获学士学位，1931年毕业，获硕士学位，20世纪40年代末到香港。

第二批中国20世纪建筑遗产推介项目：国立中央博物院（旧址）

徐敬直大学毕业后在美国曾跟随芬兰建筑大师沙里宁（Eliel Saarinen）工作。1932年回国后，徐敬直加入范文照建筑师事务所，后经范文照、赵深介绍，加入中国建筑师学会。1933年他与李惠伯、杨润均合办兴业建筑师事务所，任总经理兼建筑师。

1933年，在蔡元培的倡议下，国民政府教育部设立了中央博物院筹备处，1934年成立中央博物院建筑委员会。1935年举办了中央博物院方案征选活动，徐敬直设计的方案为当选方案（与李惠伯合作）。

中央博物院的建筑设计思想，力图体现中国传统建筑风格，以弘扬中华民族传统文化精神，同时又区别于周边已有建筑。博物院建筑委员会经过研究，决定采用辽代的式样来建造博物院。徐敬直的原设计方案，是仿清代建筑式样的，在梁思成、刘敦桢两位顾问的指导下，徐敬直重新设计了方案。建筑整体布局强调深层次的对称轴线，主体建筑离中山东路主干道较远，留下宽敞的空间，用以设计广场、草坪及绿化带。大殿前设有三层平台，以衬托出主体建筑的雄伟高大。大殿仿辽代建筑蓟县独乐寺山门的形式，其结构多按《营造法式》设计，一些细部及装修采用唐宋遗风。屋顶上铺棕黄色琉璃瓦。陈列室仿自美国某博物馆，为平屋顶，外墙加中国古典式挑檐，使之与主体建筑风格相协调。整座建筑设计科学合理，比例严谨，是在满足使用功能的前提下，采用新结构、新材料建造的仿辽代建筑的优秀项目，受到了社会各界业内外人士的好评。

建设进行了一年多后，因抗战爆发，工程停工；1946年后仍按原设计方案进行施工。1948年4月，第一期工程及附属工程竣工。1949年后，中央博物院被正式命名为南京博物院。后人民政府拨款，对博物院建筑进行整修、增建，至此，博物院第一期工程才算最后完成。博物院占地9万多平方米，其中陈列室面积4800多平方米，文物库房及文物阅览室面积10000多平方米，图书馆库房及阅览室面积800多平方米。

徐敬直于1935年成为中国营造学社社员，1945年任中华营建研究会编辑委员会名誉编辑，1948年完成香港建筑师注册登记；1950年主持（香港）兴业建筑师事务所，1956—1957年任香港建筑师学会第一任会长。著有英文版中国建筑之古今（Chinese Architecture Past and Constemporary，1964年）。在兴业建筑师事务所期间，设计完成南京中央农业试验所、上海实业部鱼市场、南京中央博物院、重庆沙坪坝中央大学校园、上海徐家汇路上海铅笔厂厂房、南京中山东路中央博物院宿舍、南京丁家桥中央大学附属医院修建门诊部工程、昆明王振宇仓库、昆明王振宇住宅、贵阳环城路兴业新村汇利企业公司、贵阳环城路兴业新村林雅横洋房住宅、王兴周汇利烟草厂、内江液体燃料管理委员会加油站、中国银行昆明分行职员宿舍及住宅等医疗、银行、工业等方面的多项工程设计。

米哈伊尔·安德列维奇·巴吉赤（Mikhail Andreevich Bagich，1909—2002年）

1909年出生于黑山共和国，1926年考入哈尔滨华俄工业专科学校（现哈尔滨工业大学），毕业后从事建筑设计。

第二批中国20世纪建筑遗产推介项目：马迭尔宾馆

俄国著名设计师米哈伊尔·安德列维奇·巴吉赤1909年出生在南斯拉夫黑山共和国。其父曾任海参崴驻防军第八东西伯利亚步兵团中尉，母亲是贵族出身，是省长的女儿。1917年俄国十月革命爆发后，经历了漫长的动乱，其父1922年被苏军枪毙，其母带着年仅14岁的巴吉赤和年幼的弟弟、妹妹，逃亡到哈尔滨。巴吉赤深知生活艰辛，于是发奋学习，在中学就取得了金质奖章。1926年，他如愿考上了哈尔滨最高学府——哈尔滨华俄工业专科学校（现哈尔滨工业大学）攻读建筑专业，期间经常参加建筑设计竞赛，半工半读完成了学业。1933年，巴吉赤大学毕业后从事建筑设计。1936年，日本近藤林业公司投资100万日元在哈尔滨市中心兴建新哈尔滨旅馆（现国际饭店），巴吉赤受聘担任建筑师。1937年他主持马迭尔宾馆装修改造工程。1940年，巴吉赤承担秋林俱乐部的设计工作。1945年苏联红军出兵东北，让他率领建筑营的士兵为苏联红军设计纪念塔。其后长春、哈尔滨等地的苏联红军纪念塔全部由巴吉赤设计，由此奠定了他"碑塔之王"的地位。

1947年6月6日，东北行政委员会决定在哈尔滨现道外八区长青公园内筹建烈士纪念馆和纪念塔。巴吉赤用20天完成了纪念塔的设计，在诸多募集方案中斩获一等奖。1948年10月10日纪念塔落成。新中国成立后，巴吉赤出任哈尔滨市建筑工程设计院工程师，这一时期是巴吉赤事业的巅峰阶段，设计精品接连问世。1952年，由他设计的东北农学院教学楼落成。这是一栋具有分水岭式意义的建筑，与新中国成立前的哈尔滨欧式建筑有了明显的区别，被称为苏联社会主义民族建筑风格。建筑造型语言简练统一，方与圆对比强烈。而后，巴吉赤又设计了东北林业大学教学楼。1958年，他又为纪念哈尔滨市人民战胜1957年特大洪水设计了防洪纪念塔。22.5米高的抗洪英雄雕像，在20根7米高科林斯圆柱组成的半圆形罗马回廊映衬下，显得格外挺拔，成为哈尔滨市重要标志性景观。

1958年，巴吉赤一家移居澳大利亚。1962年，他被录取为澳大利亚皇家建筑大学成员。虽然有着丰富的从业经验，但在竞争激烈的国度里，巴吉赤再没有惊世之作问世。2002年，巴吉赤去世，享年93岁。

孙秉源（1911—2013年）

1911年生于镇江，1929年至1936年半工半读，20世纪30年代中期至60年代设计有多个项目，90年代从事设计管理，著有《消防设计119图例》。

第二批中国20世纪建筑遗产推介项目：北京工人体育馆

孙秉源，籍贯江苏镇江，家境贫寒，1929年经人介绍到上海杨锡镠建筑师事务所任职，工作期间到上海沪江大学建筑学科学习，1936年毕业。1938—1949年，孙秉源在上海凯泰建筑师事务所任职，1950—1951年在上海华东工业部华东建筑工程公司工作，1951—1952年在北京联合顾问建筑师事务所工作，1952年到北京市建筑设计院工作。1950年前，他曾参与上海百乐门舞厅、南京饭店、上海高等法院、国立上海商学院、大都会舞厅、金都电影院、大陆游泳池以及公寓、商场、住宅等工程项目的设计。1950年后参与、主持设计的项目有淮南煤矿医院、汉阳纺织厂住宅区、北京人民印刷厂礼堂、铁道部研究所办公楼、炮兵司令部办公楼、北京体育馆比赛馆、苏联驻华大使馆、北京工人体育馆、叙利亚体育馆等工程项目。他还曾主持北京市通用构配件的设计、定型工作，还参加斯里兰卡班达拉奈克国际会议大厦施工图设计和审核工作。

胡庆昌（1918—2010年）

胡庆昌，籍贯浙江金华，出生于北京，1937年于天津工商学院毕业，1952年进入北京市建筑设计院工作，在建筑结构抗震设计领域做出重要贡献。

第二批中国20世纪建筑遗产推介项目：民族饭店

胡庆昌家境殷实，出生于书香门第。他本人天资聪慧，受家庭环境影响，自幼酷爱读书。幼年时在家私塾补习，1931—1933年在天津工商学院附属中学学习，1933—1937年在天津工商学院工科土木工程系学习。毕业后他曾任家庭教师，1938—1946年在唐山开滦矿务局土木工程处任工程师，之后在天津平塘公路改善工程处、善后救济总署、安徽淮南煤矿、天津美孚公司工程部、北京新生建筑公司设计部等处任工程师，1952年进入北京市建筑设计院工作，为北京市建筑设计研究院建筑结构专业的发展进步贡献终生。他主要完成了北京同仁医院、北京友谊宾馆大礼堂、民族文化宫、民族饭店等工程的建筑结构设计，指导完成毛主席纪念堂结构设计方案、北京昆仑饭店、北京西苑饭店结构设计、北京亚运会英东游泳馆、首都速滑馆结构设计与研究等工作。在民族饭店、民族文化宫的建筑结构设计中，他在国内首次采用全装配式高层抗震框架剪力墙结构体系。

从20世纪50年代开始，胡庆昌就进行了地震灾害调查，进行建筑结构抗震设计理论研究，参加了山西太原地震、河北邢台地震、广东河源地震、辽宁海城地震、河北唐山地震等震害调查，编写震害分析报告，开展钢筋混凝土装配式框架节点抗震性能试验研究，20世纪60年代主编《多层框架结构抗震设计若干问题》等专著，主持编写京津地区抗震鉴定标准及京津地区抗震设计暂行规定，主持装配式框架节点的定型研究，主持编写《国家标准建筑抗震设计规范》部分内容，编写《高层钢筋混凝土抗震墙结构的抗震设计》，编制《建筑物抗震构造详图》图集CG—329，并被全国通用建筑标准设计协作委员会评为"全国优秀建筑标准设计"等多项称号。其研究设计成果多次获得国家及省部级奖励。他多次参加国际建筑结构专业会议并作学术报告，担任多个国家及省部级学术组织委员及领导，1994年被授予全国工程勘察设计大师称号。

人生总有一些刻骨铭心——追忆父亲（节选）/胡明

20世纪50年代的北京，以北京十大建筑为标志的建设项目，父亲亲自参加了友谊宾馆、民族文化宫、民族饭店等几个项目的结构设计。在当时，我们国家整体是在全盘借鉴和学习苏联的政治环境下，苏联技术专家的任何指令仿佛都是正确的。在北京民族饭店的设计过程中，因要考虑为中华人民共和国成立十周年献礼的政治需要，在结构设计和施工工期的技术问题上，父亲在借鉴和学习苏联的经验之外，还依靠他深厚的英文功底，学习参照欧美先进的抗震结构设计理论，并与当时国内施工能力和水平的实际情况相结合，开创性地研究和应用了加快施工进度、节省成本的预制装配式结构体系。通过研究、试验和设计施工等几个重要环节，将预制装配式抗震结构体系应用到民族饭店的实际工程中。父亲作为

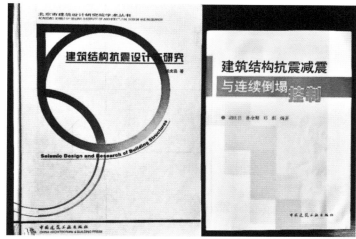

一位学者和工程技术人员,并不是一味按照政治需要去追风奉承,他本着一个爱国知识分子的良知和对现代科学技术的尊重,夜以继日地工作,通过研究和试验,反复论证、分析在工程应用中的可行性和可能会出现的问题。在民族饭店工程施工最紧张的时候,父亲几乎每晚都要12点以后才能回家,有时刚刚睡下,就又被叫到工地去解决问题。通过努力,大家按时完成了国庆十周年献礼的民族饭店工程。在1966年邢台地震和1976年唐山大地震波及下的北京,民族饭店经历了地震的考验。装配式抗震结构体系作为我国最早的高层建筑抗震设计,是极具特色并创新了全新装配式钢筋混凝土框架剪力墙结构的设计与施工技术。时至今日,采用这一先进技术施工的民族饭店,依然以其独特的时代特色,耸立在长安街上,同时也开创了我国建筑结构装配式抗震结构体系实际应用的先河。

选自胡明、金磊主编的《厚德载物的学者人生——纪念中国结构工程设计大师胡庆昌》(天津大学出版社,2018年7月)

王时煦（1920—2012 年）

王时煦，出生于黑龙江省齐齐哈尔市，1942 年毕业于哈尔滨工业大学电气专业，是北京市建筑设计研究院第一代建筑电气工程师，新中国民用建筑防雷设计研究先驱。

第二批中国 20 世纪建筑遗产推介项目：民族饭店

王时煦在大学期间曾经在齐齐哈尔发电厂实习，大学毕业后到本溪湖煤铁公司实习，后在此公司就职。1946—1950 年在北京造纸厂工作，设计 1025KVA 变电所，并带领工人施工将此变电所建成。1950 年，他进入永茂建筑公司（北京市建筑设计研究院前身）工作。在几十年的工作中，王时煦先生完成了北京小汤山疗养院、军委测绘总局、总后勤部办公楼、北京信托公司、前门邮局、六一幼儿园、中央民族学院、北京人民医院、北京精神病院、北大牙科医院、北京友谊宾馆、北京友谊医院、北京同仁医院、北京积水潭医院、北京地质学院、北京外语学院、北京工业学院、文化部办公楼、水产部办公楼、北京自然博物馆、北京前门饭店、北京新华印刷厂、北京铁匠营电影院、北京第二汽车修理厂、斯里兰卡大会堂、越南洗印厂等工程的照明、动力、变电等室内外电气设计。他是人民大会堂电气设计人兼电气设计专业负责人，设计完成了人民大会堂照明、动力、舞台、防雷、变电所等全部电气工程。

从 20 世纪 50 年代起，王时煦先生为天安门、颐和园、天坛、景山、北海、十三陵、长陵、潭柘寺、劳动人民文化宫、中山公园等北京市重要古建筑设计了防雷装置，并指导防雷装置的施工安装工程。

作为永茂建筑公司第一位电气专业设计人员，他还建立了永茂建筑公司直至北京市建筑设计研究院电气设计的设计、计算、绘图、概算、施工说明、审核、竣工检查等一整套管理方法和步骤，参与院内电气专业的图纸审查和指导，并进行防雷、照明、变电所自动控制的专业领域的深入研究，特别是深入进行建筑物防雷的研究，调查、积累了 190 多个雷害事故的案例，并与中国科学院、中国电力科学研究院等单位共同制作了几十个雷击模型进行试验，总结雷击规律，制定防雷措施，并编写出《建筑物防雷设计》专著，此专著后多次再版发行。他多次参加国家建委、建设部等各主管部门编制设计规范的编制和审查工作，其中包括我国第一本电气设计规范《电气装置暂行规则》。20 世纪 60 年代他就参加国家十年科研规划的编制工作，提出电气专业的发展方向及科研课题。王时煦曾任北京土木建筑学会理事、北京照明学会理事长、中国照明学会副理事长北京消防协会理事、北京市科学技术协会委员。

《中国建筑文化遗产》主编金磊曾撰文写到：中国工程院院士、北京市建筑设计研究院有限公司总建筑师马国馨所著《南礼士路 62 号》一书虽只有 25 篇文章，但它涉及的中外建筑作品数百项，从南礼士路 62 号走出的建筑师、工程师不下数百位。与马院士交流，书中的人物除他的学生英年早逝的王兵建筑师外，均属业界设计先贤，仅工程师中就有 50 年代"八大总"的结构总师杨宽麟（1891—1971 年）、结构总师胡庆昌（1918—2010 年）、建院结构前辈梅葆琛（1925—2008 年梅兰芳之子）、设备总师杨伟成（1927 年—）、结

构总师程懋堃（1930—2017年），还有2020年正值百年诞辰的电气高工王时煦（1920—2012年）。马院士将王时煦称为行业泰斗。事实上，我与王总很熟悉，从某种意义上也属"师徒"关系。其一，由于我20世纪80年代中后期利用设计之余研究建筑安全与城市防灾，所以不仅关注王总积毕生心血从事的建筑防雷（尤其是古建筑防雷），同时充当他的助手，先后陪同参与防雷设计检测的工程有：故宫博物院，天安门城楼，中央党校图书馆等。1995年前后为捍卫《国家建筑防雷规范》的科学性，并贯彻国际电工委员会（IEC）标准，在王总领导下，我们与全国三十位防雷专家共同抵制伪科学的"消雷器"产品。其二，在张镈大师《张镈：我的建筑创作道路》书中，有一个新中国以来的设计作品表，在50多个项目中，王时煦任电气总设计师的项目超十个，包括人民大会堂与民族饭店，此均为"国庆十大工程"，王总多次带我参观并讲述人民大会堂他亲自设计的大礼堂照明、大宴会厅照明。其三，他是北京建院老一辈最支持我从事防灾减灾设计研究的专家，我们曾共同为《电杂志》撰文，共同编译《最新防雷建筑设计资料集》。从有贡献的奋斗者之遗憾看，王时煦先生考系新中国50年代开启故宫防雷设计第一人，确应有专门讲述他事迹的著述，从此意义上讲，马总的文章为我们带了个头。在文章中，详细撰写了王老对人民大会堂26年如一日的奉献，从当年的电气设计到运营中不断关注其安全及扩容，他都默默地工作。1985年12月7日中办人民大会堂管理局专门给北京市建筑设计院发来了表彰函，表扬王时煦老26年来作为人民大会堂的主要设计者兢兢业业的付出与职业操守。马院士说这封信乃一部历史文件，证明着建院人的敬业精神。但据我了解，王时煦老知晓这一切，他极度谦逊，真不以为功，这是令人学习的可敬精神。此外，马院士的文章还特别用一定篇幅介绍了王时煦的父亲王宾章（原黑龙江省教育厅厅长）1936年7月26日遭日本宪兵队杀害的事迹。其实，在王时煦生前他也几次谈及他父亲曾遭受的不公正评价。从马院士的文章中看到，1986年7月26日在王宾章先生牺牲50周年之际，齐齐哈尔市有关方面对王宾章这位抗战志士予以纪念，还了英雄的本来面目，也了却了王时煦老先生的"心病"。

严星华（1921—？）

严星华，浙江鄞县人，汉族，1945年毕业于中央大学建筑系，曾任广播电视部设计院副院长、广播电视部设计院顾问总建筑师。1984年创办北京市中京建筑事务所，任董事长兼总建筑师，是改革开放后最早一批开办建筑师事务所的创业者之一。

第二批中国20世纪建筑遗产推介项目：全国农业展览馆

上大学期间，严星华曾参加学校组织的申新纱厂俱乐部及宿舍的设计竞赛，荣获第一名。后该工程按设计予以实施。大学毕业后曾到基泰工程司工作。1949年来到北京后，严星华长期从事民用建筑设计、创作和研究工作，曾任建工部北京工业建筑设计院副主任工程师、山西临汾地区设计室副主任、广播电视部设计院副院长、广播电视部设计院顾问总建筑师。1984年严星华创办北京市中京建筑事务所，任董事长兼总建筑师。

他的主要代表作品有中南海主要领导办公及住宅、北京医院门诊楼、北京国际饭店（原外交部招待所）改扩建、中国驻朝鲜大使馆、北京电影制片厂、莫斯科北京饭店、西安电影制片厂、新疆电影制片厂、郑州中州宾馆、中国驻柬埔寨大使馆、广州越秀山电视塔、唐山铁道学院、位于柬埔寨的第一届亚洲新兴力量运动会建筑、山西临汾电影院、山西临汾炼钢厂、内蒙古广播电视中心、江西省广播电视中心、山东省广播电视中心、中国红十字会总会培训中心、北京城乡贸易中心、北京港澳中心等建筑。

严星华先生曾任全国政协委员、中国建筑学会常务理事、中国建筑学会建筑师学会副会长、中国圆明园学会理事、北京土木建筑学会副理事长、北京勘察设计协会副理事长、建筑学报编委会主任委员等。

严星华先生设计的全国农业展览馆、中央彩色电视中心，分获"50年代北京十大建筑"和"80年代北京十大建筑"，他是唯一一位建筑作品两次获得"十大建筑"称号的建筑师。其中全国农业展览馆荣获中国20世纪建筑遗产项目。谈到自己的设计历程，严星华曾经对媒体有过如下回忆："1958

年，我全身心投入到全国农业展览馆的设计中，深入接触项目后发现，各路建筑大师在很短的时间内纷纷拿出设计方案，据不完全统计全国农业展览馆已经有了 97 份设计方案。但这些方案都存在一个显著的缺点，就是在设计原则上与当时国内规模最大的综合性展览馆——北京展览馆过于雷同，表现形式与方法也没有太大的区别，从根本上说就是没有表现出农业展览的专业特点。另外，各方案的平面布局不尽合理，或过于分散，或过于集中，或设计主旨与城市规划的要求不相符合。全国农业展览馆馆址与优美秀丽的水碓公园相邻接，除了满足展览馆的功能要求之外，还要考虑娱乐和休闲的因素，体现建筑和自然环境间的和谐统一。当时的东直门一带高层建筑较少，全国农业展览馆正处在东直门外大街的尽头，方案中应有一个醒目突出的主体建筑。"

严星华曾参观过全苏农业展览馆，全苏农业展览馆采用的分类法是按照地区原则进行的，每个加盟共和国各有一个馆。这样的设计给严星华留下了比较深刻的印象。在严星华的方案中，将全国农业展览馆的正立面设计为一座有 33 米高绿色琉璃瓦覆盖的三重檐八角形亭阁的综合馆（一号馆），气势十分雄伟。同时，设计突破了全苏农业展览馆的固有模式，采用了专业分类法。除设置一个综合馆为主馆外，还设计了特产馆（四号馆）、畜牧馆（六号馆）、气象馆（七号馆）和水产馆（八、九、十号馆）等专业展览馆。这些不同的展馆有着各自不同的造型和特点，建筑形式丰富多彩，建筑师将其合理地安置和摆放，构合成一组园林式的建筑，于统一中又体现其各自的个性，布局巧妙。如今，当人们站在东直门向东遥望，远远就可以看到一座墙体为米黄色、正中有绿色琉璃瓦覆盖着的三重檐八角形楼阁的建筑，这就是全国农业展览馆。

欧阳骖（1922—2003年）

欧阳骖，出生于河北怀来县（今北京延庆），广东中山县人，1946年于北洋大学（现天津大学）建筑工学系毕业，后长期在北京市建筑设计研究院工作，任北京市建筑设计研究院教授级高级工程师。

第二批中国20世纪建筑遗产推介项目：中国人民革命军事博物馆

欧阳骖大学毕业后在沈阳东北交通部、北平龙虎建筑师事务所、浙江大学总务处工程部等机构工作，任技术员、职员等。1949年新中国成立后，他进入北京永茂建筑设计公司、北京市建筑设计院工作，先后任工程师、高级工程师、教授级高级工程师，北京工业大学建筑系客座教授，1985年任北京建华建筑设计合资公司总建筑师、顾问总建筑师，国家一级注册建筑师，中国建筑师分会会员、北京市古建协会会员。欧阳骖主持设计了多项大型民用建筑工程，主要作品有：北京市六一幼儿园、北京市积水潭医院、中国人民革命军事博物馆、北京工人体育场、中央电视台（广播大厦东侧）、北京三里屯使馆区1#~10#使馆、北京昆仑饭店（方案）、北京市西山八大处佛牙舍利塔、北京鲁迅博物馆接待楼、山东蓬莱阁宾馆、中央广播电台、北京广播剧场、电视台洗印车间、咸阳发射塔、通县八号发射台、乐器厂消音室、郑州肉类联合加工厂、玉器厂、慕田峪长城缆车站房、东郊火葬场殡仪馆以及天安门广场隔离栏等。

20世纪80年代末90年代初，欧阳骖受聘于北京工业大学建筑系客座教授，负责教授毕业班设计，1997年以后，受聘于北京市市政工程设计研究总院，任顾问总建筑师，先后指导完成了多项市重点工程设计。其中平安大街建设工程获市优秀设计一等奖。他生前完成的最后一项设计是天安门广场栏杆制作工程。

1982年加入九三学社，历任第六、七届市委委员、工程技术委员会副主委，1992年8月任九三学社北京市第八届委员会联络部副部长。

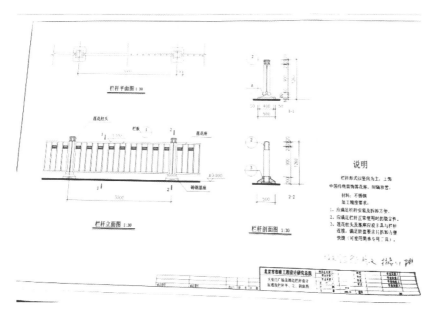

忆往（节选）/ 欧阳骖

北京工人体育场是北京市总工会的任务，于1957年委托设计。设计任务为：一，容50 000人的体育场；二，包括四个游泳池的游泳场（其中一个在室内）；三，容4000人的体育馆；四，总面积为共20 000平方米的体训楼（运动员宿舍）；五，其他附属建筑。总用地面积约为36 432公顷。

我们按甲方设想做了方案甲，按我的设想做了方案乙。

在当时的体制与工作方法下，设计人员只是单纯按甲方要求做具体设计，不参与决策、可行性研究、立项等有关设计前期工作。但作为一名"人民建筑师"（当时的一种称呼），我出于责任感，出于要对国家重点任务高度负责的心态，我不但在技术上要多考虑问题，而且也要在政治上多考虑问题，要高举爱国主义旗帜。

首先我认为新中国十周年十大建筑中要弘扬民族志气。我认为容50 000人座位太少。当时战败国日本帝国主义正要举办亚运会，正在建设一座能容70 000人的运动场。我们在庆祝新中国十周年时所建体育场只容50000人是不合适的。我就向当时北京市市委书记刘仁同志反映我的想法，建议改能容75 000人，他很同意。后来在画图过程中又改为80 000人，连同场地内看坐20 000人在内，故在政治集会时号称是十万人即由来于此。而5万与8万人并不是在造价上增加60%，而实际最多增加30%（因手头无此资料故具体数字不能详述）。但其政治影响，实际使用价值，扩大使用功能，延缓另建大体育场等意义重大。工人体育场经历几届全运会使用，军运会使用，还成功地举行了亚运会开闭幕式。这样节约了大量土地和资金。这是一项重大评估决策的结果。影响之大非同小可。

然后各方又将甲、乙两个总平面布置方案进行比较。方案甲单建体训楼20 000平方米的建筑在当时以300元／平方米计总造价需600万元。以50%造价作为基础框架结构，以50%造价做室内外装修及门窗设备等计算，尚可节约300万元。在当时是一项不小的数字。因此作为不同方案的功能与造价分析是极关重要的。只有精心设计仔细分析才能选择出最优方案。

摘自《北京市建筑设计研究院有限公司——纪念集 七十年纪事与述往》

李光耀（1920—2006 年）

李光耀，哈尔滨工业大学建筑师，曾任哈尔滨市城市规划建筑设计院院长、总工程师，哈尔滨市规划局副局长兼总工程师，哈尔滨市科协副主席。他于 1946 年毕业于哈尔滨工业大学城市建筑专业，同年参加工作，1985 年离退休。

第二批中国 20 世纪建筑遗产推介项目：哈尔滨防洪纪念塔

1952 年，李光耀主持设计哈尔滨军事工程学院。该工程由哈尔滨市建筑设计院承担，时任总工程师的李光耀负责领导设计工作。在设计之初，面对学院院长下达的各主楼的屋顶采用"民族形式"的设计要求，李光耀很为难，但负责校园建筑规划的殷之书鼓励他可以通过资料学习，并将从清华带来的一本书借给他作参考。这是一本中国古建筑图集，内容丰富，集中了国内一些著名古建筑的照片和实测图，包括建筑物的正、侧、背立面，墙柱门窗，梁架檩椽，屋面屋檐，斗拱脊兽等的照片和典型的实测细部图，尺寸齐全，十分具体。原来，这是梁思成在中国营造学社工作时，带领莫宗江、罗哲文等学生到国内各古建筑处照相和测量所得的资料。他们援柱骑梁、爬墙上屋，详细测算后才画出各种细部草图来。当时建筑系教师人手一本，而由于殷之书在清华大学土木系教结构计算，也常去营建系蹭课，还被梁思成邀请一起研究古建筑大屋顶的结构改革，使大屋顶的结构更为经济合理，故而有幸得到这本宝书。根据这本书，李光耀了解到了古建筑的特点和构造，后又去北京参观太和殿等古建筑，把握了尺度和设计要点，最终出色完成了这项设计。

20 世纪 50 年代末，随着我国对外关系的发展，黑龙江省召开大型会议和接待外宾机会增多，建设大型宾馆和会议场所迫在眉睫。1959 年初，经省委请示国务院批准，决定建设北方大厦（含花园邨宾馆）。黑龙江省政府成立了建设委员会办公室，省、市主要领导欧阳钦、吕其恩等亲自过问选址、设计方案等问题，要求国庆节前竣工，向新中国十周年献礼。总设计师由哈尔滨建筑设计院李光耀担任，1959 年 3 月 1 日，北方大厦开工，同年 9 月 30 日竣工。哈尔滨的标志性建筑防洪纪念塔及新中国早期的一批建筑如友谊宫等，都是由他主持设计完成的。

熊明（1931年—）

熊明，籍贯江西省丰城县，1957年清华大学建筑系研究生毕业，毕业后到城建部科学研究所工作，1957年进入北京市建筑设计院工作至今。

第二批中国20世纪建筑遗产推介项目：北京工人体育馆

熊明1987年任北京市建筑设计院总建筑师，1990—1994年任北京市建筑设计研究院院长，1990年被授予首批全国工程勘察设计大师称号。主要设计工程项目有：首都体育馆、北京工人体育馆、北京外贸谈判楼、中国银行总部大楼、北京昆仑饭店、北京方庄方古园小区的规划（合作设计），参与北京天安门广场和长安街规划设计等工程。发表《论建筑的本质及美学特性》《建筑创作与文脉》《面向生活、刻意创新》《中低档旅馆标准及功能构成的研究》《大型体育比赛馆设计研究（专著）》学术论文及专著。

熊明是新中国培养的建筑设计专家，工作几十年来负责设计、指导的工程和科研近百项，其中首都体育馆获全国科学大会奖，北京工人体育馆、北京昆仑饭店获中国建筑学会优秀建筑创作奖，另有多项作品获得部、市级奖励。从他的作品中可以看出他勇于探索的精神，创作中注重建筑技术与艺术的有机结合。北京工人体育馆是国内第一个大型圆形比赛场馆，首次采用悬索结构。首都体育馆是国内第一个有室内冰场的体育馆，第一次采用平板斜交大型钢网架，第一次设计大型可拖动木地板、活动折叠看台，北京外贸谈判楼是国内第一个采用核心设计的建筑，中国银行总部大楼是国内第一个采用筒中筒结构设计的高层建筑，并第一次设计斜坡停车库。他总是力求将国外最先进的技术应用于工程设计中，并在现代技术与传统文化结合方面做了种种探索、研究、创造。他在体育建筑领域研究和实践的成果，对国内体育建筑的发展

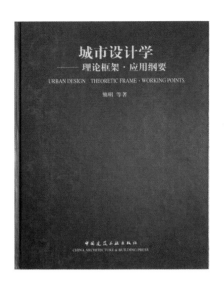

有重大影响,并被国外两所大学编入教材。其中,有关体育馆视觉质量分析和疏散计算方面的研究成果,已纳入《体育建筑设计》(中国建工出版社),成为设计工作的指导性文献。他对"中低档旅馆标准及功能构成"的研究,为扭转盲目兴建高档旅馆做出理论分析和设计依据,首次提出"开放性"概念,论述旅馆与城市结合效益,提供可靠数据和最佳模数。建设部设计管理司确认"有一定指导意义"并向全国推荐。

熊明对建筑创作的理论有许多比较深入的探讨,特别是在建筑与艺术及美学、建筑矛盾的特殊性、建筑创作中的内容与形式等方面有自己的理解和认识,他认为,"建筑创作不能脱离建筑的美学特性。一方面,建筑艺术不可再现人物、再现生活,不可能直接明确地表达具体思想(纯纪念物不如归入雕塑类),另一方面作为生活的空间,建筑可能给人们的感染力远非其他艺术所能比拟,所以,就建筑艺术而言,生活空间是艺术形象存在的依据,艺术形象是生活空间存在的表现"。"建筑不同于纯艺术的第二个重要方面是:建筑形象的实现要依靠相当的技术并耗费大量材料。因而驾驭材料的技术在建筑艺术形象的创造上占有特殊的地位。用材料构成空间和形象的技术证明人的智慧和力量,使人感到自豪,这是美感的一个重要来源"。"建筑不同于其他艺术的第三个重要方面:建筑一经建成就长期固定在环境中(自然环境、城市环境、历史环境),成为更大范围空间和艺术形象的组成部分。和环境的关系不仅影响个体本身的美,而且关系到

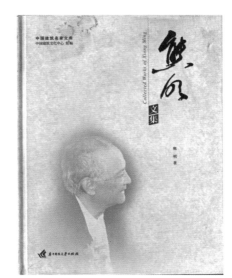

群体的美,需要更加强调"。把建筑美学特性从功能、技术、环境三方面进行理性的分析,从而达到突出形象个性的效果,这正是我国建筑师应当追求的建筑作品要体现"新而中",反映中国特色社会主义现代化的时代精神。

"生活空间和艺术形象这对矛盾标志出建筑区别于其他事物的本质特征"。"生活空间一般总是矛盾的主要方面。可在不同的条件下,矛盾双方力量的对比会发生变化,有时是根本的变化,艺术形象转化为矛盾的主要方面。这就是建筑种类非常多,由类似机器到接近纯艺术的内在原因,也是人们可以对各种建筑提出不同程度的审美要求的内在根据。"这种对建筑矛盾特殊性的具有独创性的表述,抓住了建筑创作问题的根本。

"生活方式和审美观念是建筑创作的内容,生活空间和艺术形象是建筑创作的形式。但因为建筑是生活空间和艺术形象的矛盾统一体,所以建筑形象不是简单地取决于审美观念,还受生活空间的制约。艺术形象的物质形态不能与生活空间的组织背离,艺术形象的精神性格不能和生活空间的活动性质脱节。同时,生活空间对艺术形象的制约作用不仅表现为它限制建筑的艺术形象不能任意塑造,而且表现为它使得艺术形象的物质形态多样化,精神性格丰富化。这正是矛盾双方相辅相成具有同一性的表现。"这些论述透彻地说明了建筑创作中内容与形式的内涵及其相互关系,是建筑创作要求百花齐放,要求不断创新的内在原因。

彭一刚（1932年—）

彭一刚，出生于安徽合肥，1953年于天津大学毕业后留校任教，教授、博士生导师；1995年当选为中国科学院院士；2003年获得第二届梁思成建筑奖。

第二批中国20世纪建筑遗产推介项目：甲午海战馆

彭一刚，1932年出生于安徽，1953年于天津大学土木建筑系毕业。多年来他潜心于建筑理论研究工作，并积极从事建筑创作实践活动，主要研究方向为建筑美学及空间构图理论、建筑设计方法论、传统建筑文化与当代建筑创新。彭一刚曾发表论文几十篇，出版学术著作多部，先后完成甲午海战馆、天津大学建筑系馆、天津水上公园熊猫馆、山东平度公园规划、福建漳浦西湖公园规划、福建南安公园规划、厦门杏林开发区公园规划、郑州高新技术开发区高新技术孵化器大楼、郑州高新技术开发区管委会大楼、舟山市沈家门小学等项目的设计，甲午海战馆等多项工程获建设部、教育部、中国建筑学会及国家级、省部级多项奖励。

在建筑创作中，彭一刚以严肃认真的态度探索传统与创新、形象与意蕴、环境与文脉和建筑个性表现等深层次问题，为创作具有时代性、民族性、和科学性的建筑新风格做出贡献。天津大学建筑系馆的设计，为了求得与所处三角形地形的契合，最终确定为"凸"字形平面，这样可以保证绝大部分房间呈南北向。建筑门厅位于东部，庭院居中既利于通风，又可使走廊获得充足的自然采光。展厅设于二楼，全部敞开。建筑结构柱网极其规整，除大绘图教室采用井字梁结构体系外，其他部分均为7.2米。为突出建筑系馆的个性特征，并赋予其文化内涵，于入口两侧墙面上引用了老子《道德经》中既富于哲理又与建筑有关的一段语录。左侧为中文，与斗拱雕像相呼应，右侧为英文，与爱奥尼克柱头比邻。入口上方为馆徽，也是以斗拱形象为标志。所有这些都是借助符号化的语言来深化建筑的个性，丰富了学校建筑特有的文化气息。

在接受媒体采访时，彭一刚先生说：

传统与创新是一个永恒的话题，又是当前摆在建筑界的一个热点话题。随着全球化进程的加速，特别是加入WTO之后，西方文化对我们的强势冲击势不可挡，如果不做好应对，我国传统建筑文化有可能荡然无存。为此，有不少同志忧心忡忡，但是经过反复思考之后，我反而认为不必把这个问题看得过于严重。只有强势文化的剧烈冲击，才能促进弱势文化发展，甚至是起死回生的发展。回顾我国近代历史，受西方强势文化冲击最剧烈的可以说有两次，一次是鸦片战争之后，另一次是改革开放之后。前一次是被迫的，后一次则是主动引进的。但无论是被迫还是主动引进，客观上都促进了我们的发展，使我们摆脱了传统，朝着现代文明迈进了一步，尽管这之中不免会带来阵痛，甚至是辛酸和苦难。

传统作为一种文化现象，是人类进步的标志，它至少具有两个特征：其一，是从高处向低处流；其二，是具有极强的渗透力。历史上我们习惯于用"同化"这个词来表述文化的强大功能，例如元、清两代入主中原，由于当时文化层次的差异，终于被汉文化所同化，致使汉人产生了无比的优越感。

到了近代，西方文化处于强势地位，再想重演历史上的辉煌，无异于白日做梦。记得鲁迅先生曾在一篇杂文中发过议论，目的无非是提醒国人，切不可妄自尊大。

再从世界历史看，哪一个国家、民族、地区，如果死抱着传统不放，那么他的发展必然滞缓。例如被称为四大文明古国的古埃及、古巴比伦、古印度、古中国，由于过分地迷恋往昔的辉煌而不思进取，到了近代，都相继地沦为发展中国家。幸好，我们觉醒的比较早，提出了改革开放的政策，从学习西方先进文化中，极大、极快地发展、壮大了自己。"三个代表"中提出的代表先进文化的发展方向，我认为就包含有这方面的意思。

那么，是不是就不要传统了？当然不是。

文化，虽然有从高处向低处渗透力极强的特点，但是它毕竟不能以一种文化来取代另一种文化。因为每一种文化的产生都有着和它密不可分的土壤。由此，担心传统的彻底沦丧，我认为是一种过虑。尽管在强势文化的冲击下，暂时的消弭也许是难以避免的，不过从长远看，在吸收了强势文化的营养后，必将由弱变强，在更高的层次上，使外来文化本土化，从而呈现一种崭新的面貌。

程泰宁（1935年—）

出生于江苏，1956年毕业于南京工学院（现东南大学）建筑系。中国工程院院士，全国工程勘察设计大师，教授，博士生导师，东南大学建筑设计与理论研究中心主任，筑境设计主持人。

第二批中国二十世纪建筑遗产推介项目：杭州黄龙饭店

程泰宁曾参加人民大会堂、南京长江大桥桥头建筑方案设计工作。从业以来，他主持设计国内外工程150余项，多项作品获得国家、省部级奖励。其中，杭州黄龙饭店、杭州铁路新客站入选"中华百年建筑经典"；加纳国家剧院、马里共和国议会大厦入选国际建协(UIA)《廿世纪世界建筑精品选》（该选集选出了全球100年中的1000件优秀作品）。2004年，程泰宁获中国建筑师最高奖"梁思成建筑奖"。他一直关注中国特色的建筑设计理论体系建构，曾出版《程泰宁——建筑院士访谈录》《程泰宁文集》《语言与境界》等著作；发表论文多篇，在国内外影响广泛。最近澳大利亚视觉出版集团推出的"大师系列"丛书"程泰宁作品选"中，他是被收入该出版社世界建筑大师系列的第一位中国建筑师。

近年来，程泰宁领衔完成的中国工程院咨询研究课题"当代中国建筑设计现状与发展"及住建部课题"关于提升建筑设计水平的政策措施研究"，是对中国建筑设计领域与行业发展具有指导意义的研究成果。

1990年，程泰宁被授予国家级有突出贡献中青年专家称号，2000年被评为全国工程勘察设计大师，2004年获第三届梁思成建筑奖，2005年当选中国工程院院士。

语言·意境·境界——东方智慧在建筑创作中的运用（节选）

近百年来，中国现代建筑一直处在西方建筑文化的强势影响之下。从好处说，西方现代建筑的引入，推动了中国建筑的发展；从负面来讲，我们的建筑理念一直为西方所裹挟，在跨文化对话中"失语"，是一个不争的客观事实。虽然在这个过程中有不少学者、建筑师以及政府官员，在反思的基础上，倡导过"民族形式""中国风格"等，但由于缺乏有力的理论体系作支撑，只是以形式语言反形式语言，以民粹主义反外来文化，其结果，只能停留在表面上，最后无疾而终。因此，建构自己的哲学和美学思想体系、以支撑中国现代建筑的发展，是一个值得我们重视并加以研究的重要问题。那如何来建构这样一个理论体系？我同意这样的观点："中国文化"更新的希望，就在于深入理解西方思想的来龙去脉，并在此基础上重新理解自己。

……

我认为，建筑创作的哲理——亦即"最高智慧"，是"境界"。何谓"境界"？王国维在《人间词话》的手稿中说"不期工而自工"是文艺创作的理性境界。有学者进一步解释说"妙手造文，能使其纷踏之情思，为极自然之表现即为"境界"。结合建筑创作，我认为这里包含着两方面的含义。

其一，从"天人合一"、万物归于"道"的哲学认知出发，要看到，身处大千世界，建筑从来不是一个孤立的单体，而是"万事万物"的一个组成部分。在创作中，摆正建筑的位置，特别注意把建筑放在包括物质环境和精神环境这样一个大环境、大背景下进行考量，既重分析，更重综合，追求自然和

谐为重中之重；既讲个体，更重整体，追求有机统一，使建筑、人与环境呈现一种"不期工而自工"的整体契合、浑然天成的状态，是我们所追求的"天人境界"。

其二，"境界"不仅诠释并强调了建筑和外部世界的内在联系，而且还揭示了建筑创作本身的内在机制。以"境界"为本体，我们可以看到，在建筑创作中，功能、形式、建构，甚至以致意义、意象等理性和非理性因素之间，并不遵循"内容决定形式"或"形式包括功能"这类线性的逻辑思维模式，也很难区分哪些是"基本范畴"和"派生范畴"。在创作实践中，建筑师所建构的，应该是一个多种因素为节点的、相互连接的网络。当我们游走在这个网络之中，不同的建筑师可以根据自己理解和创意，选择不同的切入点，如果选择的切入点恰当，我们的作品不但能够解决某一个节点（如形式）的问题，而且能够激活整个网络，使所有其他各种问题和要求相应的得到满足。这种使"纷沓的情思"得到"极自然表现"的"自然生成"，是我们追求的创作"境界"。因此，从语言哲学和线性逻辑思维模式中解放出来，以"境界"这一具有东方智慧的哲学思辨来诠释建筑创作机制，建构一种符合建筑创作内在规律的"现象合一"的方法论，将使建筑创作的魅力和价值能够更加充分的显示出来。

此外，以境界为本体，还可以使我们更好地理解并运用那些充满东方智慧的、具有创造性的思维方式。例如直觉、通感、体悟……这些具有创造性的思维活动（方式），需要在反复实践和思考中获得，它也体现了一种建筑境界。

（原载于《语言与境界》）

李拱辰（1936年—）

李拱辰 1959 年毕业于天津大学建筑系建筑专业，现任河北建筑设计研究院有限责任公司资深总建筑师，享受国务院特殊津贴，2011 年被授予河北省首批建筑大师称号。

第二批中国 20 世纪建筑遗产推介项目：唐山抗震纪念碑

李拱辰的主要设计作品有：唐山抗震纪念碑、河北艺术中心、泥河湾博物馆、上海黄浦新苑小区、西柏坡纪念馆、河北建设服务中心等。其中，唐山抗震纪念碑设计获 2009 年中国建筑学会建筑创作大奖，2017 年入选第二批中国 20 世纪建筑遗产名录。河北艺术中心设计获国家第十届优秀工程设计铜质奖，2002 年度全国优秀工程勘察设计行业奖二等奖。河北建设服务中心及泥河湾博物馆设计分别获 2009 年度、2013 年度全国优秀工程勘察设计行业奖二等奖。石家庄铁路客站（老火车站）于 2019 年 5 月经石家庄市人民政府公布为石家庄第一批历史建筑。

通过多年来的设计工作，李拱辰指导或带动一批年轻建筑师走向成熟，指导或参与的设计项目，如石家庄人民广场及西柏坡纪念馆改扩建设计等也多次获得全国行业奖、省部级优秀设计奖。

他多年来曾受聘担任石家庄市城乡规划建设专家咨询委员会委员。曾长期担任住建部注册建筑师考试专家组组长，负责全国注册建筑师考试命题及评分工作。

结缘建筑学（节选）

建筑创作和它所处的年代以及社会环境条件，往往有着千丝万缕的联系。诸如，地域环境、社会发展、科技水平、文化倾向、思维方式等。我所经历的年代，这种时代性的特征十

分明显。

几十年的建筑设计生涯，伴随着的是一条无尽的学习之路。首先是要传承，要向老一辈建筑师学习。走在这条路上，我首先想到的是北京院的傅义通老总，他为人和善，学识渊博。我们相识较早，他对我比较赏识，很多机会我都得到他的举荐。我们经常交谈，从为人处世到设计观点，使我受益匪浅。西北院黄克武大师，编制88J时我们相处十年，老先生讲到技术问题，至情至理，深入浅出，分析透彻且耐心细致。我也耳濡目染，收获良多。其次，利用会议等机会虚心向院士、大师们学习。我曾和多位院士、大师借机交流，彼此之间都成了好朋友。再者，利用机会向兄弟院学习，譬如：我曾多次参加全国评优，报送的都是各院作品的精华。熟读这些设计，悟出他们的精华所在，武装自己，借以丰富自己的创作思维。当然，在讲求文化的当下，更要对传统文化的丰富内容常怀敬畏之心，努力学习，丰富自身的文化认知，提高自己的文化修养。特别是在丰富多彩的日常生活面前，更要细致地体验生活，深入生活，发掘出生活中不曾被人重视的细节，提炼、提升为我们设计的关键点，使设计更出众！让我们贴近生活，了解生活，愿今后设计更精彩，生活更美好！

合作建筑师：
郭卫兵（1967 年—）

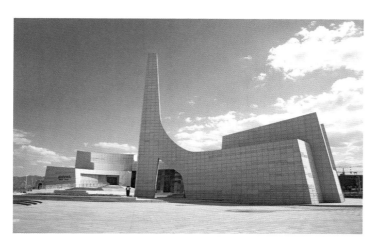

郭卫兵 1989 年毕业于天津大学建筑系，工程硕士学位，现任河北建筑设计研究院有限责任公司党委书记、董事长、总建筑师。

郭卫兵先后荣获河北省建设系统劳动模范、第七届河北青年科技奖、河北省建筑行业新中国 60 周年最具影响力人物、河北省有突出贡献中青年专家、河北省工程勘察设计大师、当代中国百名建筑师、河北省劳动模范等奖项；现任中国建筑学会理事，河北省土木建筑学会建筑师分会理事长，国家行业优秀设计评选专家，国家注册建筑师委员会委员。其主持设计的项目有 20 余项荣获河北省优秀设计一等奖，其中 9 项荣获全国优秀工程设计行业奖，代表作品有河北省博物院、磁州窑博物馆、定州中山博物馆、石家庄大剧院等。郭卫兵有着强烈的社会责任感，先后策划了"建筑的力量""文化遗产与建筑创新""人民的建筑""媒体的力量·让建筑更美好"等大型论坛。

孙国城（1937年—）

孙国城1956年毕业于西安冶金建筑学院（现西安建筑科技大学），全国工程勘察设计大师。先后任职于北京城市建设部民用建筑设计院、新疆维吾尔自治区建工局设计院，现任新疆维吾尔自治区建筑设计研究院有限公司名誉总建筑师。

第一批中国20世纪建筑遗产推介项目：新疆人民会堂

孙国城自述

我出生在苏州，中学毕业后，就读于由柳士英、刘敦桢等前辈于1923年创办的我国第一所建筑院校苏州工专的后续——1947年复校的五年一贯制的苏南工专。学生多半是家境状况不佳的子弟，不勤奋读不下来，别无选择地考进了这所学校的建筑科。1956年毕业那年，正值院系调整，我们被并入西安冶金建筑学院（部分老师并入西安交大），于是我成为该校的最后一批毕业生。母校校址在三元坊，由于苏州美专合并到上海，美专所在地沧浪亭和"罗马大楼"留给建筑系，成了我们末代毕业生最后的记忆。

在校期间给我触动最深的不全是苏州园林的美景，而是一次苏联莫斯科、列宁格勒等地建筑学院的大幅人物与建筑钢笔素描原稿的建筑绘画巡回展。另一次是我到上海的参观实习，亲眼看到当年董大酋建筑师设计的江湾体育馆场以及江湾市政府钢筋混凝土配大屋顶的现代建筑。前者提高了我对建筑绘画的浓厚兴趣，后者领略了前辈建筑师将传统建筑模式赋予现代技术带来的震撼。

1956年毕业后我有幸被分配在北京城市建设部民用建筑设计院，在那里和来自清华大学、同济大学、南京工学院等名校的高材生在一起，做方案参加设计竞赛，并有机会得到董大酋、张家德（重庆市人民大礼堂设计者）、过元熙（留美回国）等院级总建筑师的指点，真是如鱼得水。1957年人员外调、机构合并（大部分并入北京市建筑设计院），我随50名设计人员最先调往位于新疆乌鲁木齐的自治区建工局设计院。然而，这一年成了我一生中最重要的经历和实践，令我回味无穷。

同我一起从北京民用院来疆的孟昭礼建筑师曾在天津基泰事务所张镈大师手下工作，主张现代建筑方向。进疆后，他创作的几个简约风格的设计作品受到当时新疆本地领导和业内人士的赞赏。北京十大建筑问世后，他更加理解张镈大师的作品，他给我认真解读并指导我完成了自治区展览馆等一系列设计，是我一生中最敬重的良师益友。

在60多年的设计生涯中，我从事过少数民族传统民居调研和古建维修，更多的是进行民用住宅和公共建筑设计。20世纪50年代主持设计了乌鲁木齐市人大办公楼；60年代设计了自治区展览馆、军区大楼；70年代设计有乌鲁木齐机场候机楼（T1）；80年代设计有新疆人民会堂、自治区人大办公楼和新疆科技馆；90年代设计有天山百货大楼、新疆国际博览中心、海德酒店和中银广场等一批超高层建筑，还设计了近5万平方米的乌鲁木齐机场T2航站楼。

进入21世纪，和北京市建筑设计研究院合作承接了新疆体育中心设计。2003年主持了新疆迎宾馆9号楼工程设计以及老区改造任务，满足了自治区成立五十年大庆庆典活动。2005年和北京装备部设计院合作设计了新疆民族歌舞学校，为培养少数民族歌舞人才提供了高水准场所。2007年，我们和北京民航设计总院合作设计，负责了10万多平方米的T3

航站楼的建筑设计,并自主完成了喀什、和田、伊犁、库尔勒、喀纳斯、吐鲁番、博乐、阿克苏、库车等一批支线机场航站楼设计任务。2008年以后我还兼任了新疆大学建筑建筑学硕士生导师。1994年以来,历任建设部注册考试中心注册建筑师考试命题专家组长之一,历经24年。

创作理念:形式适应气候,形式顺应环境,形式反映功能,形式诠释文化,形式与时俱进。

主要著述:《建筑与气候——论新疆建筑的地域特色》《穹顶、宣礼塔与传统建筑的降温功能》《探索现代空港设计中的地域性》《并未褪色的地域建筑》。

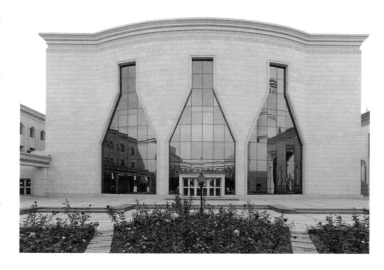

20世纪中国建筑遗产大事要记
（2016年10月—2020年12月）

Highlights of China's Architectual Heritage in the 20th Century (October 2016—December 2020)

2016年

4月19日，以"21世纪的森林建筑"为主题的"2016吸碳建筑研讨会"在北京钓鱼台国宾馆举行。

5月18日—19日，主题为"建筑的春天"2016年中国建筑学会学术年会在北京召开。来自全国各地的建筑科技工作者，会员以及建筑院校师生500余人参加了会议。中国建筑学会理事长修龙在开幕式致辞中指出，中国建筑学会的优势在于能够集中业界众多院士、大师和高水平的建筑科技工作者，畅通民间和政府沟通互动的渠道，搭建国内与国际交流沟通的平台。

6月18日—21日，"敬畏自然 守护遗产 大家眼中的西溪南——重走刘敦桢古建之路徽州行暨第三届建筑师与文学艺术家交流会"在黄山市徽州区的西溪南镇启动。本次活动由中国文物学会、黄山市人民政府主办，中国文物学会20世纪建筑遗产委员会、北京大学建筑与景观设计学院、东南大学建筑学院、《中国建筑文化遗产》《建筑评论》编辑部等联合承办。

9月7日，因国际著名建筑师勒·柯布西耶（Le Corbusier，1887—1965年）跨越7个国家的17座建筑入选《世界遗产名录》这一重大事件，《中国建筑文化遗产》《建筑评论》编辑部与中国建筑技术集团有限公司联合主办"审视与思考：柯布西耶设计思想的当代意义"建筑师茶座。

9月29日，"致敬百年建筑经典：首届中国20世纪建筑遗产项目发布暨中国20世纪建筑思想学术研讨会"在故宫博物院宝蕴楼（建于1914年）隆重召开。来自全国文博界、文化界、城市建筑界、传媒界的专家学者、20世纪建筑遗产入选单位代表、高等院校师生共计200余人出席。会议公布了98项"首批中国20世纪建筑遗产名录"，推出《中国20世纪建筑遗产名录［第一卷］》图书，宣读了《中国20世纪建筑遗产保护与发展建议书（征求意见稿）》。中国20世纪建筑思想学术研讨会同期举行。

10月19日—20日，"世界古代文明保护论坛"在故宫博物院举行，来自国际文物保护与修复研究中心、国际博物馆协会、国际古迹遗址理事会3个国际组织，以及中国、埃及、希腊、印度、伊朗、伊拉克、意大利、墨西哥8个文明古国的近50位考古学家、历史学家和博物馆学者参加了论坛。

2016年6月19日 · 安徽

2016年9月29日 · 北京

2016年10月19日·北京

10月22日—23日，第四届"建筑遗产保护与可持续发展·天津"国际会议召开，本次活动由中国文物学会20世纪建筑遗产委员会、天津大学、天津市历史风貌建筑保护专家委员会联合主办。受大会委托，中国文物学会20世纪遗产委员会承办了主题为"建筑传统营造技艺与建筑文化"的分论坛。

11月19日—21日，"2016中国第七届工业遗产学术研讨会"在同济大学召开。来自中国大陆各省市自治区和台湾地区，以及美国的300多位专家学者参会，会议主题为"工业遗产的科学保护与创新利用"。

12月15日，《万里长城 薪火相传·长城保护条例》颁布10周年纪念展于北京建筑大学开展。

12月17日—19日，由中国文物学会20世纪建筑遗产委员会、中国"三线"建设研究会等单位联合主办的"致敬中国'三线'建设的'符号'816暨20世纪工业建筑遗产传承与发展研讨会"于重庆市涪陵区举行。

10月25日，2016 ENR《建筑时报》"中国承包商80强和工程设计企业60强"颁奖典礼在福州举行。

《建筑时报》评选2016工程设计行业十件大事，"新八字建筑方针出炉"及"梁思成建筑奖首颁给外国建筑师"等事件名列其中。

2016年12月17日·重庆

2017年2月23日·北京

2017年4月21日·江西

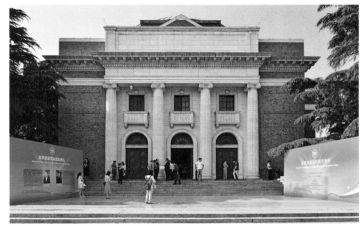
2017年5月17日·北京

2017 年

1月15日，故宫博物院院长单霁翔新著"新视野·文化遗产保护论丛"（第二辑 十卷本）正式出版。该书由《中国建筑文化遗产》编辑部与天津大学出版社推出。

2月4日，中国著名古建筑保护专家、城市规划专家郑孝燮先生追悼会在北京八宝山革命公墓举行。2月23日，郑孝燮先生追思会在故宫博物院举行。

4月1日，中共中央、国务院决定设立河北雄安新区。这是党中央作出的一项重大的历史性战略选择，是继深圳经济特区和上海浦东新区之后又一具有全国意义的新区，是千年大计、国家大事。

4月7日，中国建筑学会城乡建成遗产学术委员会在同济大学正式成立，这是我国建成遗产保护与再生领域的第一个学术组织。中国科学院院士、同济大学教授常青任理事长。

4月13日，为迎接第22个"世界读书日"，《中国建筑文化遗产》《建筑评论》编辑部与中国建筑技术集团有限公司联合主办"建筑师茶座"，同时启动《20世纪建筑遗产读本》（暂定名）第一次编撰策划研讨活动。

4月21日至23日，《中国建筑文化遗产》编辑部与陕西省土木建筑学会建筑师分会联合举办了"重走洪青之路婺源行"活动。本次活动旨在追溯洪青成长之路，向洪青大师学习。考察团队先后前往婺源清华镇的彩虹桥，百年老宅汇集的思溪延村、人文盛行的李坑、徽州古商埠汪口、伟人故里江湾等古村落展开田野考察。4月23日晚，在陶溪川陶瓷

2017年9月15日·山东

2017年11月21日·江苏

创意园区举行座谈会。

4月25日，《慈润山河——义县奉国寺》由天津大学出版社出版。该书系2008年出版的专著《义县奉国寺》的"科普版"，为辽代木构建筑系列"申遗"提供了普惠公众的传播模板。

5月17日，主题为"建筑师的职业责任"的2017年度建筑学会年会在北京清华大学举行，会议举行了2016梁思成建筑奖颁奖典礼。2016年梁思成建筑奖由马来西亚汉沙杨有限公司总裁杨经文、天津华汇工程建筑设计有限公司总建筑师周恺摘得。

5月25日，"何镜堂建筑设计作品展"在北京大学百年纪念讲堂纪念大厅举行。

9月12日，中国建筑师张轲荣获阿尔瓦·阿尔托奖，成为获颁这一国际建筑界重要奖项的首位中国人。

9月15日，由中国文物学会20世纪建筑遗产委员会策划的"致敬中国建筑经典——中国20世纪建筑遗产的事件·作品·人物·思想展览"亮相威海国际人居节并受到业内外赞誉。

10月，天津大学建筑学院迎来80华诞，由《中国建筑文化遗产》《建筑评论》编辑部策划编撰的《建筑师的大学》、布正伟著《建筑美学思维与创作智谋》等图书及其系列活动助兴80华诞。

11月，东南大学建筑学院喜迎90华诞。

2017年12月2日·安徽

11月21日—12月21日，东南大学建筑学院90周年纪念展览：基石——毕业于宾夕法尼亚大学的中国第一代建筑师，在江苏省美术馆老馆举行。

12月2日，"第二批中国20世纪建筑遗产"项目于安徽省池州市发布。本次活动由中国文物学会、中国建筑学会、池州市人民政府、中国建设科技集团股份有限公司联合主办，中国文物学会20世纪建筑遗产委员会、中国城市建设研究院有限公司、池州市城乡规划局等单位承办，共计100项"第二批中国20世纪建筑遗产"向社会公布。

12月3日，系平遥古城"申遗"成功20周年的日子。无论"申遗"前后，各方不断探索对这座"明清华尔街"的保护方式。

《文汇学人》2017年12月1日发表记者署名文章《城市保护是一场接力赛》。

12月12日，《建筑时报》整版刊登"向上世纪承载着历史与文化的建筑致敬"，以对中国文物学会20世纪建筑遗产委员会副会长、秘书长金磊的访谈形式，对"第二批中国20世纪建筑遗产项目"公布的相关情况、20世纪建筑遗产的重要意义进行了深入报道。

12月16日，中国建筑学会建筑评论学术委员会成立大会在同济大学举行，郑时龄院士当选为建筑评论学术委员会主任委员，李翔宁、刘克成、金磊等当选为副主任委员，李翔宁当选为秘书长。

2018年1月27日·北京

2018年

1月，中国文物学会20世纪建筑遗产委员会在《建筑》杂志开办"建筑评论"栏目，首期刊登了《中国需要建筑评论和建筑评论家》专文。

1月27日，"天津大学北京校友会建筑与艺术分会2018年新春联谊会暨大国工匠——校友成果展/北京站、布正伟新著出版座谈会"在中国建筑设计研究院举行。展览及联谊活动由中国工程院院士崔愷主持，布正伟著《建筑美学思维与创作智谋》（天津大学出版社2017年，《中国建筑文化遗产》编辑部承编）北京首发式由金磊总编辑主持。马国馨院士、布正伟总建筑师、李兴钢大师等30余位嘉宾出席。

1月28日，《世界建筑》原主编曾昭奋先生著《建筑论谈》一书正式出版，该书系《中国建筑文化遗产》《建筑评论》编辑部继2015年承编《国·家·大剧院》后，为曾先生推出的又一"建筑评论"著作。全书收录了曾昭奋教授精心挑选的88篇作品，近75万字。

2月25日，中国工程院院士、同济大学建筑与城规学院教授、曾任同济规划建筑设计研究总院总建筑师、中国建筑学会常务理事戴复东（1928-2018）因病逝世，享年90岁。

3月14日，中国事务所MAD设计的卢卡斯叙事艺术博物馆举办动工仪式，为中国建筑师首次赢得海外文化地标的设计权。

3月15日，"文化池州——工业遗产创意设计项目专家研讨会"在故宫博物院宝蕴楼举行，中国文物学会会长、故宫博物院院长单霁翔，中国工程院院士马国馨，池州市人民政

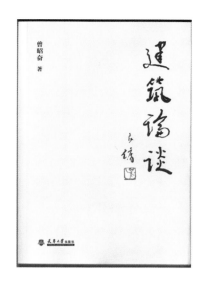

2018年1月28日·北京

2018年3月15日·北京

2018年5月17日·福建

府市长雍成瀚，副市长贾瑄，中元国际资深总建筑师费麟，全国工程勘察设计大师黄星元、周恺、张宇、李兴钢及多位设计机构总建筑师们悉数到会，以建设国润祁红科技文化创意小镇为主题，对文化池州的创意设计深入剖析与研讨。

3月29日，"笃实践履 改革图新 以建筑与文博的名义省思改革：我们与城市建设的四十年北京论坛"举行，标志着"改革开放四十年城市系列论坛"正式启动。随后的4月12日—13日，石家庄举行了"第二届媒体的力量暨以建筑设计的名义纪念改革四十周年"学术活动。

6月26日，深圳南海酒店举行了"以建筑设计的名义纪念改革开放：我们与城市建设的四十年·深圳广州双城论坛"。

4月14日，正值"汶川地震十周年"前夕，中国文物学会20世纪建筑遗产委员会副会长、秘书长金磊赴绵阳参加了由中国文物学会、中国灾害防御协会、绵阳市委市政府主办的"地震遗址保护与利用学术座谈会"，并做了"灾害类纪念建筑品鉴"的学术报告。

5月17日，中国建筑学会2018年学术年会召开前夕，在泉州文化地标威远楼古城广场，由中国文物学会20世纪建筑遗产委员会策划推出了第一、第二批中国20世纪建筑遗产名录展。

5月19日—20日，中国建筑学会学术年会在福建省泉州市开幕，年会以"新时代本土建筑文化的融合与创新"为主题。

5月20日，中国建筑学会学术年会分论坛之九"新中国20世纪建筑遗产的人和事学术研讨会"在泉州海外交通史博物馆隆重召开。本次活动由中国文物学会20世纪建筑遗产委员会主办承办，泉州古城保护发展工作协调组办公室协办。

2018年3月29日·北京

2017年6月27日·天津

5月28日，中国台湾著名现代主义建筑师王大闳先生（1917—2018）去世，享年101岁。台北孙中山纪念馆、林语堂宅是其设计的代表作。

6月1日，俞挺和闵而尼于2013年在上海成立的Wutopia Lab获得Architectural Record评选的2018年全球十佳"设计先锋"，是2018年度唯一入选的中国事务所。

6月，中国建筑学会、《建筑学报》编辑部编著《中国建筑设计作品选2013—2017》出版。

7月2日，中国政府采购网发布《河北雄安新区启动区城市设计方案征集资格预审公告》，雄安新区面向全球征集启动区城市设计方案。

7月15日，历时近一年编撰，由胡明、金磊主编，《中国建筑文化遗产》《建筑评论》编辑部承编的《厚德载物的学者人生——纪念中国结构工程设计大师胡庆昌》，由天津大学出版社正式出版。

7月18日，天津大学建筑设计规划研究总院原院长、执行总规划师，天津大学教授、博士生导师，日本神户大学工学博士洪再生先生（1962—2018年），因病去世，享年56岁。天津市科学技术协会刊发纪念文章，将他誉为"现代城市的规划大师"。对我国建筑评论类学刊的发展，洪再生先生有历史性贡献，2012年10月，与《中国建筑文化遗产》编辑部共同创办国内第一本《建筑评论》学刊。

8月18日，"觉醒的现代性——毕业于宾大的中国第一代建筑师"展在上海当代艺术博物馆展出。

8月29日，中国文物学会20世纪建筑遗产委员会组织了"北京香厂地区近现代建筑考察"活动，该活动由《中国建筑文化遗产》与"建筑文化考察组"主办，对珠市口基督

2018年11月17日·北京

2018年11月20日·北京

教堂、留学路、鹞儿胡同、万明路、香厂华康里与平安里弄堂、香厂仁德医院、东方饭店等近现代历史建筑展开考察，这是对朱启钤20世纪20年代北京城市建设贡献的再回望。

8月30日，"地域、文化、时代——为激变的中国而设计"何镜堂院士美国首展在加州伯克利大学正式开幕。

8月31日，由封面新闻、华西都市报主办"天府十大文化地标"颁奖典礼在历史文化名城南充举行。峨眉山、青城山—都江堰、三星堆遗址、广安邓小平故居、成都武侯祠博物馆、成都杜甫草堂博物馆、阆中古城及"春节源"纪念地、眉山三苏祠、凉山冕宁彝海结盟遗址和五粮液老窖获"天府十大文化地标"称号。

10月，中国文物学会20世纪建筑遗产委员会与中国建筑设计研究院有限公司所辖的亚太建设科技信息研究院有限公司正式合作，就"新中国七十年城市住宅发展的典型命题"联合开展研究，成果刊载于《城市住宅》杂志专版。12月推出首次专版，主题为"新中国城市住宅的70年（1949—2019）：北京"。中国文物学会20世纪建筑遗产委员会副会长、秘书长金磊为本专版撰写了"开篇的话：为了新中国城市住宅的设计记忆"。2019年2月，推出由天津市规划和自然资源局副局长路红、天津大学建筑学院博士王月撰写的《新中国城市住宅70年(1949—2019)之天津》专版文章。

10月14日，《人民日报》第七版刊发中国文物学会20世纪建筑遗产委员会副会长、秘书长金磊专文《认同与保护20世纪建筑遗产》。

10月24日，港珠澳大桥正式通车。港珠澳大桥连接香港、澳门和广东珠海市，是2018年中国基础建设成就最值得纪念的事。作为世界级的超级工程，港珠澳大桥实现"六个最"：世界总体跨度最长、钢结构桥体最长、海底沉管隧道最长跨海大桥，也是世界公路建设史上技术最复杂、施工难度最高、

2018年11月24日·江苏

工程规模最庞大桥梁。

11月1日,"深圳体育馆的保护"专题研讨会在建筑设计博览会期间举行,在崔愷院士主持下,来自建筑设计界、遗产保护界多位专家学者就深圳体育馆拆除引发的近现代建筑保护话题展开交流,中国文物学会20世纪建筑遗产委员会副秘书长李沉代表委员会发言。

11月2日,中央美术学院举办"挑战:反观建筑思想、教育与实践"建筑论坛。

11月20日—12月9日,"意匠清华七十年 关肇邺院士校园营建哲思"展在清华大学举行,本次展览萃集了关肇邺院士数十年来参与清华校园营建工作的全部成果,通过建筑模型、设计图纸、实景照片与视频的形式综合展出。

11月22日,"都·城——我们与这座城市"专题展览及学术研讨活动在中国国家博物馆隆重开幕。本次活动由北京市人民政府国有资产监督管理委员会与北京市建筑设计研究院有限公司联合主办,共分五个版块:"都·城——我们与这座城市"专题展览开幕式;都·城——我们与这座城市展览主题论坛;北京·伦敦城市发展论坛马国馨院士新书《南礼士路62号:半个世纪建院情》、北京建院文创产品发布会。

11月22日,2018年度英国皇家风景园林学会奖(LI Award)获奖作品宣布,全球有36个项目,中国斩获3项,北京清华同衡城市规划设计研究院设计的上海辰山植物园矿坑花园、同济大学园林系设计的米囿(Rice Garden)、易兰与世界著名建筑大师扎哈·哈迪德(ZAHA HADID)合作设计的望京SOHO公园。

11月24日—25日,"第三批中国20世纪建筑遗产"项目于曾荣获"首批中国20世纪建筑遗产"的中央大学旧址——东南大学四牌楼校区大礼堂隆重举行。在中国建筑学会理事长修龙、东南大学校长张广军院士、中国工程院钟训正院士、全国工程勘察设计大师刘景樑等的见证下,推介公

2018年12月18日·北京

布100项"第三批中国20世纪建筑遗产"。主旨演讲环节，金磊副会长受中国文物学会会长单霁翔的委托，代表中国文物学会20世纪建筑遗产委员会向与会专家、东南大学师生做了主题为"20世纪建筑遗产的保护与创新"的报告。11月24日晚，举办"第三批中国20世纪建筑遗产项目公布学术研讨会"，25日代表团考察了南京20世纪代表性建筑遗产项目。

11月30日，"建筑界的奥斯卡"世界建筑节（WAF）公布了2018年建筑制图奖获奖名单公布，李涵获总冠军。

12月7日，日建设计（ArchDaily）2018年度建筑大奖公布，中国建筑师在各类奖项的15席中占据了3席，中国建筑师马岩松、李兴钢、韩文强获奖。

12月18日，作为"都·城 我们与这座城市"展览系列活动之一，"《中国20世纪建筑遗产大典（北京卷）》首发暨学术报告研讨会"在故宫博物院报告厅举行。单霁翔院长以"20世纪遗产与北京这座城"为题发表主旨演讲。单霁翔院长、马国馨院士、赵知敬主任、徐全胜董事长为"BIAD建筑与文化遗产设计研究中心"成立揭牌。在北京城市建筑遗产沙龙中，马国馨院士、邱跃主任等八位专家展开交流。

12月21日，中国文物学会第八次代表大会在北京召开。大会选举出69名理事组成的中国文物学会第八届理事会，单霁翔当选为会长。

12月末，《周治良先生纪念文集》正式出版，该书由《中国建筑文化遗产》《建筑评论》编辑部承编，团队历经近2年时间，多次走访周治良先生亲朋好友及工作伙伴，采集了大量一手文献资料。

2019年4月3日·北京

2019年

1月21日，国家住建部、文物局联合发布第七批中国历史文化名镇名村名单。

2月19日，北京冬奥组委会发布《2022年北京冬奥会和冬残奥会遗产战略计划》，共涉及7个方面，35个重点领域。

4月3日，"感悟润思祁红·体验文化池州——《悠远的祁红——文化池州的"茶"故事》首发式"在故宫博物院建福宫花园举行。本次活动由中国文物学会20世纪建筑遗产委员会、池州市人民政府联合主办，北京市建筑设计研究院有限公司建筑与文化遗产设计研究中心等单位联合承办。在现场百余位来自全国文博界、建筑界、出版界、茶行业百余位专家领导的见证下，由《中国建筑文化遗产》编辑部联合众专家编撰完成的《悠远的祁红——文化池州的"茶"故事》一书隆重面世。

4月12—13日，第三届"媒体的力量——让建筑更美好"学术沙龙暨以建筑设计的名义纪念新中国70周年学术沙龙活动在河北建筑设计研究院有限责任公司举办。金磊总编辑以《40周年——真实的建筑时代记忆》为题，介绍了即将出版的《中国建筑历程1978—2018》一书。

4月16日，"品质设计：以文化建筑的名义纪念新中国70年暨《天津滨海文化中心》首发座谈会"在天津滨海文化中心举行。

5月11日，"境"——程泰宁建筑作品展，在北京中国国家博物馆举行。5月18日"建筑师的文化自觉和建筑创作的社会责任"双城论坛于5月18日在中国国家博物馆学术报告厅成功举办。论坛由北京市建筑设计研究院有限公司、

2019年4月3日·北京

2019年4月16日·天津

筑境设计、《建筑评论》编辑部等联合主办。

5月16日，华裔建筑大师贝聿铭先生（1917—2019）于美国逝世，享年102岁。法国卢浮宫博物馆发表公告称"贝聿铭为一位富有远见的建筑师，拥有绵长而灿烂的职业生涯，他的设计不仅延续了12世纪以来卢浮宫的建造精髓，更以果敢的精神带领卢浮宫走向现代。"《中国建筑文化遗产》23辑、24辑推出纪念专文。

5月，受国际古迹遗址理事会20世纪遗产科学委员会、澳大利亚20世纪建筑保护委员会、南澳大学西校区艺术建筑和设计学院、阿德莱德大学等组织邀请，由中国文物学会20世纪建筑遗产委员会秘书处策划组织"中国20世纪建筑遗产考察团"赴新西兰、澳大利亚考察20世纪建筑遗产经典项目并在悉尼举办"20世纪建筑遗产在中国"学术研讨。2020年8月，《田野新考察报告（第七卷）》出版，内容包括中国、新西兰、澳大利亚20世纪建筑遗产考察交流研讨报告等。

5月22日，主题为"新时代本土建筑文化和技艺的融合与创新"2019中国建筑学会学术年会在苏州市太湖国际会议中心隆重开幕。

6月21日，北京市第一批历史建筑公示，共计收录429处建筑物。6月27中国文物学会20世纪建筑遗产委员会秘书处提交了《北京首批历史建筑公示意见反馈》。10月22日，北京市第二批历史建筑公示，共计315处建筑物被收录。

7月3日，全国注册建筑师管理委员会换届会议在北京召开，成立了新一届全国注册建筑师管理委员会。

7月23日—10月13日，由清华大学艺术博物馆与清华大学建筑学院联合主办的展览"归成——毕业于美国宾夕法尼亚大学的第一代中国建筑师"在清华大学艺术博物馆举行。

2019年8月15日·北京

8月4日，"参合中西—吕彦直先生（1894—1929）逝世90周年纪念座谈会暨《吕彦直集传》首发式"在北京举行。

8月15日，在中国建筑学会支持下，中国建筑学会建筑师分会、中国文物学会20世纪建筑遗产委员会、北京市建筑设计研究院有限公司、《建筑评论》编辑部共同组织，历时一年多编撰出版《中国建筑历程 1978—2018》书（《建筑评论》编，2019年6月第一版，天津大学出版社）。首发座谈会在北京市建筑设计研究院有限公司隆重举行，共有来自全国四十余位设计大师、总建筑师及建筑出版媒体人参会。

8月31日，《中国文化报》在"古韵新妍——我们的中轴线"主题下，发表了单霁翔《保护好、利用好、传承好北京中轴线文化遗产》的署名文章。该文用全新思想梳理了属于古都，更面向未来的北京中轴线，从"四方面"作了新定位，北京中轴线系思想文化轴线、文化艺术轴线、民族融合轴线、发展创新轴线。文章全文见《中国建筑文化遗产》总第25辑。

9月1日—9月5日，第27届国际文化遗产记录科学委员会（CIPA 2019）全球双年会在西班牙中西部世界文化遗产地阿维拉古城召开。本届年会的主题是"为了更好的未来而记录过去（Documenting the Past for Better Future）"。在闭幕式上，CIPA主席Stratos先生揭晓了CIPA2021的举办城市：中国北京，并与全场会议代表一起向北京组委会表示了祝贺。这将是CIPA自1968年创建50周年以来，第一次在中国大陆举办双年会。CIPA2021将于2021年8月28日至9月1日在清华大学召开，由清华大学、清城睿现数字科技研究院有限公司、北京建筑大学三家联合举办。

9月9日，作为北京国际设计周设计之旅的特别活动，"新中国大工匠智慧——人民大会堂"特展在北京市规划展览馆开幕。展览以设计者的视角回溯了人民大会堂设计建设中的

2019年9月9日·北京

2019年9月19日·北京

2019年10月25日·北京

经典瞬间，还原了这一伟大建筑的缔造者们谱写的传奇故事。

9月16日，《人民日报》（海外版）第11版刊发了对中国文物学会20世纪建筑遗产委员会题为《20世纪能留下多少建筑遗产》的采访文章。

9月19日，在中国文物学会、中国建筑学会大力支持下，"第四批中国20世纪建筑遗产项目终评"活动在北京市建筑设计研究院有限公司A座举行。在公证处公证下，共计诞生98项"第四批中国20世纪建筑遗产项目"。随后，中国文物学会会长单霁翔、中国建筑学会会长修龙、中国工程院院士马国馨共同为中国文物学会20世纪建筑遗产委员会、马国馨院士学术研究室、BIAD建筑与文化遗产设计研究中心三机构揭牌。

9月24日，第九届梁思成建筑奖颁奖典礼在北京中国科技会堂举行。经过评审，中国建筑师庄惟敏、德国建筑师冯·格康（Meinhard von Gerkan）获第九届梁思成建筑奖。

10月24日，亚投行总部大楼暨亚洲金融大厦竣工仪式在北京奥林匹克公园中心区举行。亚投行总部大楼占地6.1公顷，具有浓厚的中国特色和国际风尚，体现出简洁优雅、隽永经典的建筑风格。

10月，中国文物学会20世纪建筑遗产委员会秘书处完成《庆祝共和国七十华诞 绘制城乡建设大蓝图——中国建筑设计70年简述》一文，刊于中国勘察设计协会主办《中国勘察设计》杂志2019年10期

10月25日，由北京市建筑设计研究院有限公司主办，北京建院文化传播有限公司承办的"北京建院成立70周年主旨论坛——高质量发展的建筑与城市"在北京城奥大厦隆重举行。活动过程中，多部"BIAD70周年系列丛书"进行了预发布。

2019年12月3日·北京

2019年12月13日·海南

10月26—28日，中国文物学会工业遗产委员会与20世纪建筑遗产委员会共同举办第十届中国工业遗产学术研讨会。会议在具有悠久历史和深厚工业文化底蕴的历史文化名城郑州召开，围绕"砥砺奋进、铸就辉煌——新中国工业建设的发展历程、伟大成就、记忆及遗产"这一主题进行为期三天的学术研讨与考察。

10月27日，2019全球城市论坛在上海交通大学举行。本次论坛系"世界城市日"主题活动之一，主题为"城市转型·创新发展"。近400名嘉宾围绕"全球城市·协同治理""全球海洋·中心城市""全球城市·区域规划""城市创新·智库实践"等展开研讨与交流

10月28日—29日，由联合国人居署、国际地下空间联合研究中心、住建部建筑杂志社、上海市土木工程学会、同济大学和深圳大学联合主办的"全球城市地下空间开发利用上海峰会暨2019第七届中国（上海）地下空间开发大会"在沪召开。这也是继1991年的《东京宣言》后，人们有关全球地下空间领域利用所达成的又一次共识，《上海宣言》是《东京宣言》的继承与提升。

12月3日，由中国文物学会、中国建筑学会等单位联合主办的"致敬百年建筑经典——第四批中国20世纪建筑遗产项目公布暨新中国70年建筑遗产传承创新研讨会"在与新中国同龄的"中国第一院"——北京市建筑设计研究院有限公司举行，共计98项"第四批中国20世纪建筑遗产"问世。会议通过了由中国文物学会20世纪建筑遗产委员会发出的《聚共识·续文脉·谋新篇 中国20世纪建筑遗产传承创新发展倡言》。

2019年12月25日·北京

12月12日，第十五届(2019)光华龙腾奖颁奖典礼在人民大会堂小礼堂隆重举行。2019年特别设立"第十五届(2019)光华龙腾奖特别奖·中国设计贡献奖金质奖章 新中国成立70周年中国设计70人"，他们中有中国现代建筑开拓者和奠基者梁思成、天安门观礼台总设计师张开济等。

12月13—14日，以"谱写人居环境新篇章"为主题的第三届全国建筑评论研讨会，在海南省海口市召开。会议由第三届组委会与《建筑评论》编辑部、海南省土木建筑学会主办。1987年，第一届建筑评论研讨会在浙江东阳举行；1991年，第二届建筑评论研讨会在四川德阳举办。

12月21日，"第八届深港城市/建筑双城双年展（深圳）"开幕，主题为"城市交互"，展览包含两个板块，即"城市之眼"和"城市升维"。

12月25日，由中国建筑学会、《建筑学报》杂志社主办的"中国营造学社九十周年纪念展"在中国建筑设计研究院举行。

2020 年

1月13日，住房和城乡建设部发布《关于公布第九批全国工程勘察设计大师名单》的公告，共计60位工程勘察设计行业优秀代表入选。

3月19日，中国工程院院士、我国著名建筑学家、建筑教育家、清华大学建筑学院教授李道增先生（1930—2020年），因病在北京逝世，享年90岁。

4月27日，《住房和城乡建设部、国家发展改革委关于进一步加强城市与建筑风貌管理的通知》发布，"通知"旨在贯彻落实"适用、经济、绿色、美观"新时期建筑方针，治理"贪大、媚洋、求怪"等建筑乱象，进一步加强城市与建筑风貌管理。

4月，中国出版协会公布了第七届中华优秀出版物奖获奖名单。由中国文物学会、中国建筑学会主编、中国文物学会20世纪建筑遗产委员会承编，天津大学出版社出版的《中国20世纪建筑遗产名录[第一卷]》继2019年荣获天津市委宣传部等单位颁发的"天津市优秀图书奖"外，获得第七届中华优秀出版物奖提名奖。

5月中旬，由中国科学院院士、著名建筑评论家、同济大学教授郑时龄先生著，同济大学出版社出版的《上海近代建筑风格（新版）》问世，全书143.5万字。

5月20日，继"BIAD70周年系列丛书"之马国馨院士编《都·城 我们与这座城市——北京建院首都建筑作品展》推出后，由《中国建筑文化遗产》《建筑评论》编辑部策划承编、北京市建筑设计研究院有限公司主编《北京市建筑设计研究院有限公司五十年代"八大总"》《北京市建筑设

2020年8月10日·河北

研究院有限公司纪念集七十年纪事与述往》正式出版。

6月，由河北省建筑设计研究院有限责任公司资深总建筑师李拱辰著，《中国建筑文化遗产》《建筑评论》编辑部承编的《时光筑梦——六十载从业建筑札记》一书由天津大学出版社出版，该书展示了李拱辰总建筑师在燕赵大地60年奋斗中积累的宝贵设计理念及所设计的经典建筑作品。

6月6日，中国文物学会20世纪建筑遗产委员会作为主组稿方，同《当代建筑》杂志社联合推出《20世纪建筑遗产传承与创新》专刊。中国文物学会20世纪建筑遗产委员会副会长、秘书长金磊为本专版撰写了《卷首语："中国20世纪建筑经典"乃遗产新类型》。专刊共计收录包括单霁翔会长在内的十余位国内建筑与文博界专家学者学术专文。

6月8日，我国著名建筑学家、建筑教育家，同济大学建筑与城市规划学院教授、博士生导师，上海市建筑学会名誉理事长，罗小未先生（1925—2020年），因病在上海逝世，享年95岁。

8月5日，由北京大学文化产业研究院，新华网等联合发布"2019中国城市文化创意指数排行榜"，分别从城市创意生态、赋能能力、审美驱动力、创新驱动力等予以评分。排名前十名的城市：北京、上海、深圳、广州、杭州、重庆、西安、成都、东莞、南京。

8月10日，河北雄安新区勘察设计协会成立大会在雄安设计活动中心举行，中国文物学会会长、故宫博物院原院长单霁翔在"雄安设计讲坛"中做了题为《从建筑、城市规划、文遗保护和博物馆多维视角看建筑师的文化传承与发展》的主旨演讲。

9月21日，2020北京国际设计周产业合作单元暨北京城市建筑双年展2020先导展开幕式在北京城市副中心张家湾设计小镇举办，展览共包含十个主题展和十个专题活动，北京大兴国际机场代表了北京最新的重大建设项目成就，其展陈

2020年10月3日·辽宁

方式极其耀眼。中国文物学会20世纪建筑遗产委员会与北京建院合办的"致敬中国百年建筑经典——北京20世纪建筑遗产"特展在活动中推出。本展览涵盖了88项"中国20世纪建筑遗产"的北京项目，充分反映了新中国建筑师对北京城市建设发展的贡献，从国际20世纪建筑遗产发展的语境与《世界遗产名录》的最新趋势出发，解读国内城市间的比较，同时说明20世纪遗产属已必要关注且保护的新类型。

9月24日，《人民日报》刊登由中国文物学会20世纪建筑遗产委员会署名《珍视二十世纪建筑遗产》文章。

9月，在年逾90周岁的张钦楠老局长的主持下，生活·读书·新知三联书店出版了《20世纪世界建筑精品1000件（十卷本）》（总主编：K.弗兰姆普敦，副总主编：张钦楠），该书为1999年在北京召开的第20届世界建筑师大会中推出的《20世纪世界建筑精品集（十卷本）》（中国建筑工业出版社，1999年，第一版）中文普及版。本丛书在展现全球20世纪建筑经典风采的同时，也成为向社会传播世界建筑文化的"普及教材"。

9月8日，为纪念紫禁城建成600周年，由故宫学院（上海）指导，同济大学、天津大学等主办"延续二百多年的建筑世家——样式雷建筑图档同济大学特展"在同济大学博物馆举办，展览以视频、展板、图纸和宣传册四种形式，表现了中

2020年10月3日·辽宁

2020年10月28日·广东

国建筑"千古一系"的伟大传承。

9月9日，武汉市自然资源和规划局公布《武汉市疫后重振规划（三年行动规划）》。

9月，华建集团发布用设计延续传承、开拓创新，助力上海南京路步行街东拓工程。南京路步行街是上海开埠后最早的商业街，1865年正式定名，20世纪30年代号称"中华商业第一街"，在坚持"三大"特色分区的多元化国际布局下，镌刻历史记忆尤为凸显，如在江西中路西北转角新增一处"1882广场"，一盏灯柱站立，它是1882年上海电气公司在南京路点亮的第一盏电灯。

10月3日，"守望千年奉国寺·辽代建筑遗产保护研讨暨第五批中国20世纪建筑遗产项目公布推介学术活动"于辽宁省锦州市义县奉国寺隆重举行，共计500余人参会。中国文物学会、中国建筑学会联合推介了101项"第五批中国20世纪建筑遗产"项目，自2016年至此，共推介五批497项中国20世纪建筑遗产。中国文物学会会长、故宫博物院原院长单霁翔做了"让文化遗产活起来"主题演讲。会议通过了《中国建筑文化遗产传承发展·奉国寺倡议》，同时举办"纪念中国营造学社成立九十周年 千年奉国寺·辽代建筑遗产研究与保护传承"学术研讨会，并对锦州义县文化遗产做了深度考察。

10月5日—6日，2020年"世界人居日"全球主场活动在印度尼西亚海滨城市泗水市以在线方式举办。联合国副秘书长、人居署执行主任谢里夫说"2020年世界人居日提供了一个重要机会，让我们反思新冠肺炎危机的影响，以及我们如何使未来的人类住区更具韧性。"联合国可持续发展目标 11（SDG 11）旨在到2020年使城市有韧性、包容、安全、多元化。

10月14日，深圳经济特区建立40周年庆祝大会在广东深圳市举行。习近平发表重要讲话，提出"五大历史跨越""十条宝贵经验""六大重要部署"。其中建筑与城市界瞩目的集聚粤港澳大湾区的综合设计力量，促进湾区社会发展和城市建设，成为深圳先行示范区建设的中国机遇，这里展现了新城市发展基因的生成与绽放的"样板"。

10月19日，原《世界建筑》主编、清华大学建筑学院教授、建筑评论家、学者曾昭奋先生（1935—2020年）因病于北京逝世，享年85岁。他是那种不顾建筑师的名衔、资历而阐发自己的批评见解，并热情肯定中青年建筑师的探索方向的建筑评论家。

10月26日—29日，十九届五中全会在北京召开，提出到2035年建成文化强国的战略目标，并对如何实现这一战略目标作出新的谋划和部署。会议还通过了《中共中央关于制定国民经济和社会发展第十四个五年规划和二〇三五年远景目标的建议》明确提出实施城市更新行动。

10月28日，北京市十五届人大常委会第二十五次会议，对《北京历史文化名城保护条例（修订草案）》进行第一次审议，修订草案提出，北京历史文化名城的范围涵盖本市全部行政区域，主要包括老城、三山五园地区以及大运河文化带、长城文化带、西山永定河文化带等。同时，明确保护对象包括世界遗产、文物；历史建筑和革命史迹；历史文化街区、特色地区和地下文物埋藏区；名镇、名村和传统村落等9大类。

10月28日—30日，主题为"好设计·好营造——推动城乡建设高质量发展"的中国建筑学会学术年会在深圳国际会展中心举办，共有10场主旨报告及32场专题论坛，近2000人与会。年会的"云开幕"和累计126小时的全程线上直播，使更多学人共享学术盛宴。同时，以程泰宁院士为主任委员的中国建筑学会建筑文化学术委员会宣告成立。

11月9日，在中国文物学会单霁翔会长的委派下，中国文物学会20世纪建筑遗产委员会与《中国建筑文化遗产》编辑部组成专家组，赴东四八条111号朱启钤旧居调研，与朱启钤先生（1872—1964年）曾孙朱延琦进行了建筑遗产保护相关事件、人物的交流，也作为开启中国营造学社90周年纪念的活动。《中国建筑文化遗产》第27辑推出"纪念中国营造学社成立90周年"专辑，收录有十余篇各界专家纪念朱启钤及营造学社的文章。

11月17日，科技部发布《全球生态环境遥感监测2020年度报告》，指出全球城市面积20年间扩展了一倍多，即土地面积由2000年的23.9万平方公里增加到2020年的51.98万平方公里。亚洲和北美洲是全球城市扩展的重点区域。

11月27日，在江苏南通由美国《工程新闻记录》（ENR）和中国《建筑时报》共同发布2020年度"中国承包商80强和工程设计企业60强"。以2019年设计营业收入为依据前十名的建筑设计机构：中国建设科技集团股份有限公司、中国建筑股份有限公司、华东建筑集团股份有限公司、同济大学建筑设计研究院（集团）有限公司、华设设计集团股份有限公司、中国联合工程有限公司、CCDI悉地国际、上海天华建筑设计有限公司、北京市建筑设计研究院有限公司、广东省建筑设计研究院有限公司。

11月29日，由英国皇家建筑师学会（RIBA）主办的"中国88城市建筑与空间摄影大赛"在沪揭晓，《浙水村书屋》《景德镇御窑博物馆》等作品获奖。1834年成立的RIBA，其宗旨是为全球行业社群和社会服务，以提供更好的建筑和场所、更强大的专业社群和可持续的发展环境。2019年12月，RIBA在上海成立办事处。

12月3日，北京市规划和自然资源委员会发布《北京市建筑师负责制试点指导意见（征求意见稿）》，并向社会公开征求意见，《指导意见》旨在充分发挥建筑师的专业优势和技术主导作用，传承和发展优秀建筑文化，鼓励设计创新，提升首都工程建设质量和建筑品质等。

12月3日—4日，中国文物学会20世纪建筑遗产委员会派建筑摄影师，在秦皇岛市文保所和秦皇岛市建筑设计研究院的协助下，对朱启钤保护的北戴河近现代建筑及别墅区进行了拍摄及资料收集。

12月7日，继《北京市建筑设计研究院有限公司五十年代"八大总"》书出版之后，为传承与新中国同龄的设计大院的历史，北京建院筹备编撰全国首批勘察设计大师熊明老院长的作品与创作思想文集。

12月8日，《中国20世纪建筑遗产名录[第二卷]》完成，修订稿进入出版阶段。

中国 20 世纪建筑遗产认定标准 2014

The Identification Standard for 20th-Century Chinese Architectural Heritage, 2014

CSCR-C20C

中国文物学会 20 世纪建筑遗产委员会

Chinese Society of Cultural Relics, Committee on Twentieth-Century Architectural Heritage，CSCR-C20C

中国 20 世纪建筑遗产认定标准 2014

北京，2014 年 8 月

一、编制说明

1.1 依据

《中国 20 世纪建筑遗产认定标准（试行稿）》（以下简称《认定标准》）是根据联合国教科文组织《实施保护世界文化与自然遗产公约的操作指南（12-28-2007）》、国际古迹遗址理事会 20 世纪遗产科学委员会《关于 20 世纪建筑遗产保护办法的马德里文件 2011》、《中华人民共和国文物保护法（2007）》、国家文物局《关于加强 20 世纪建筑遗产保护工作的通知（2008）》等文献完成的。

本认定标准由中国文物学会 20 世纪建筑遗产委员会（Chinese Society of Cultural Relics, Committee on Twentieth-Century Architectural Heritage，简称 CSCR-C20C）编制完成，最终解释权归 CSCR-C20C 秘书处。

1.2 原则

中国 20 世纪建筑遗产是按时间段划分的遗产集合，包括 20 世纪历史进程中产生的不同种类的建筑遗产。20 世纪建筑遗产主要分成两部分：中国近代建筑（1840-1949 年）；中国现当代建筑（1949 年 -21 世纪初中叶）。世界文化与遗产的多样性对中国 20 世纪建筑遗产的认定是一项丰富的精神源泉，具有重要的指导与借鉴作用。中国 20 世纪建筑遗产项目不同于传统文物建筑，其可持续利用是对文化遗产"活"态的尊重及最本质的传承，保护并兼顾发展与利用是中国 20 世纪建筑遗产的主要特点。研究并认定 20 世纪建筑遗产重在传承中国建筑设计思想、梳理中国建筑作品，不仅是敬畏历史、更是为繁荣当今的建筑创作服务。

1.3 价值

城市和建筑的发展历程是人类文明的重要组成部分。建筑设计并非创造世界上没有的东西，而是对所提出的问题作出好的解决方案。坚持建筑本质、树立文化自觉与自信是开展 20 世纪建筑遗产认定的使命之一。无论是历史研究评估，还是通过评估认定建筑群的历史价值、艺术价值、科学价值等，都不限定建筑的建成年代，而要充分关注它们是否对社会及城市曾产生或正在产生深远影响，如"开中国之先河"等。此外还要特别研究、评估那些有潜在价值的建筑是否具备提升或拓展的价值（诸如建筑美学价值、规划设计特点、体量造型、材料质感、色彩运用、细部节点、科技与工艺创新等）。

二、认定标准

世界遗产语境下的遗产名录是基于"突出的普遍价值"而设立的，其含义是纪念性建筑物或建筑群应从历史、艺术以及科技的视角发现其具有怎样的普遍价值。完成中国20世纪建筑遗产项目的认定并登录，对中国新型城镇化建设中20世纪建筑遗产项目的保护、修缮、再利用乃至公众参与及明晰产权关系、政府与企业资助等方面均有重要意义。

在中国20世纪建筑遗产认定中坚持文化遗产的普遍性、多样性、真实性、完整性原则要注意如下文化遗产特征的应用。

该建筑应是创造性的天才杰作；具有突出的影响力；文明或文化传统的特殊见证；人类历史阶段的标志性作品；具有历史文化特征的居住建筑；与传统或信仰相关联的建筑（就中国20世纪建筑而言，中山纪念建筑、抗战纪念建筑、辛亥革命纪念建筑等均属该类项目）。

凡符合下列条件之一者即具备申报"中国20世纪建筑遗产"项目的资格。

2.1 在近现代中国城市建设史上有重要地位，是重大历史事件的见证，是体现中国城市精神的代表性作品；

2.2 能反映近现代中国历史且与重要事件相对应的建筑遗迹与纪念建筑，是城市空间历史性文化景观的记忆载体，同时，也要兼顾不太重要时期的历史见证作品，以体现建筑遗产的完整性；

2.3 反映城市历史文脉，具有时代特征、地域文化综合价值的创新型设计作品；

2.4 对城市规划与景观设计诸方面产生过重大影响，是技术进步与设计精湛的代表作，具有建筑类型、建筑样式、建筑材料、建筑环境、建筑人文乃至施工工艺等方面的特色及研究价值的建筑物或构筑物；

2.5 在中国产业发展史上有重要地位的作坊、商铺、厂房、港口及仓库等；

2.6 中国著名建筑师的代表性作品、国外著名建筑师在华的代表性作品，包括20世纪建筑设计思想与方法在中国的创作实践的杰作，或有异国建筑风格特点的优秀项目；

2.7 体现"人民的建筑"设计理念的优秀住宅和居住区设计，完整的建筑群，尤其应保护新中国经典居住区的建筑作品；

2.8 为体现20世纪建筑遗产概念的广泛性，认定项目不仅包括单体建筑，也包括公共空间规划、综合体及各类园区，20世纪建筑遗产认定除了建筑外部与内部装饰外，还包括与建筑同时产生并共同支撑创作文化内涵的有时代特色的室内陈设、家具设计等；

2.9 为鼓励建筑创作，凡获得国家设计与科研优秀奖，并具备上述条款中至少一项的作品。

三、认定理由说明

3.1 20世纪建筑遗产认定是一项权威性、科学性、文化历史性极强的工作，因此要考虑所选项目具有：

· 可达性：建筑与城市道路、交通枢纽的位置关系，反映人们到达建筑的便捷性；

· 可视性：建筑要有风貌特点及环境景观价值（质量、风貌、色彩）；

·可用性：建筑的功能使用安全状况及现在的完好程度；

·关联性：建筑融于城市设计并与周边项目的整体协调性。

3.2 20世纪建筑遗产类型十分丰富，至少涉及文教、办公、博览、体育、居住、医疗、商业、科技、纪念、工业、交通等建筑物或构筑物。

3.3 认定理由示例。

以进入联合国教科文组织"世界文化遗产名录"的部分当代建筑（属中国20世纪建筑遗产可借鉴的项目及认定方式）为例：

·1930-1950年建成的特拉维夫项目（2003年入选），在空地建起的白色建筑群"白城"体现了现代城市规划的准则，成为欧洲现代主义艺术运动理念传播到的最远地域；

·1948年建成的墨西哥路易斯·巴拉干故居及工作室（2004年入选），属"二战"后建筑创意工作的杰出代表，将现代艺术与传统艺术、本国风格与流行风格相结合，形成一种全新的特色，其影响力促进了风景园林设计的当代发展与品质；

·1956年建成的巴西利亚项目（1987年入选），其城市格局充满现代理念，建筑构思新颖别致，雕塑寓意丰富；

·1958年建成的广岛和平纪念公园（1996年入选），伴随着"二战"的意义，广岛和平纪念公园成为人类半个世纪以来为争取世界和平所作努力的象征地；

·1973年建成的悉尼歌剧院（2007年入选），该项目是全球公认的20世纪世界十大奇迹之一，作为20世纪悉尼的地标性建筑，它已成为澳大利亚的象征建筑；

·1999年建成的墨西哥大学城（2007年入选），该建筑群别具特色，将现代工程、园林景观及艺术元素有机融于一体，成为20世纪现代综合艺术体的世界独特范例。

四、使用说明

鉴于"中国20世纪建筑遗产认定标准"是彰显中国20世纪优秀建筑作品、传承建筑师设计思想这一宏大工程的标志性文件，而认定实践尚缺少可借鉴性，所以本标准的使用原则是，在按照标准认定的同时，要做到边使用、边完善、边实践。

4.1 严格按照认定标准的条目进行，通过调研、分析、发现、判断、甄别认定那些有价值（含潜在价值）且必须保护和传承的20世纪建筑遗产项目，旨在形成不同建筑类型、不同事件背景下的预备登录名单；

4.2 认定过程要视不同类型的20世纪建筑遗产做必要的工作流程设计，即每届评选认定都需编制与主题相适合的工作计划及"评选认定手册"；

4.3 专家评选认定工作需要提交的申报材料清单（略）。

第二批中国 20 世纪建筑遗产名录

The Second List of the 20th-Century Chinese Architectural Heritage

序号	项目名称	项目地点
1	国立中央博物院（旧址）	江苏省南京市
2	鼓浪屿近现代建筑群	福建省厦门市
3	开平碉楼	广东省江门市
4	黄埔军校旧址	广东省广州市
5	中国人民革命军事博物馆	北京市
6	故宫博物院宝蕴楼	北京市
7	金陵女子大学旧址	江苏省南京市
8	全国农业展览馆	北京市
9	北平图书馆旧址	北京市
10	青岛八大关近代建筑	山东省青岛市
11	云南陆军讲武堂旧址	云南省昆明市
12	民族饭店	北京市
13	国立中央研究院旧址	江苏省南京市
14	国民大会堂旧址	江苏省南京市
15	三坊七巷和朱紫坊建筑群	福建省福州市
16	中东铁路附属建筑群	内蒙古满洲里、黑龙江省哈尔滨市、吉林省长春市、辽宁省沈阳市等
17	广州沙面建筑群	广东省广州市
18	马尾船政	福建省福州市
19	南京中山陵音乐台	江苏省南京市
20	重庆大学近代建筑群	重庆市
21	北京工人体育馆	北京市
22	梁启超故居和梁启超纪念馆（饮冰室）	天津市
23	石景山钢铁厂	北京市
24	中国银行南京分行旧址	江苏省南京市
25	中央体育场旧址	江苏省南京市
26	西安事变旧址	陕西省西安市
27	保定陆军军官学校	河北省保定市
28	大邑刘氏庄园	四川省成都市
29	湖南大学早期建筑群	湖南省长沙市
30	北戴河近现代建筑群	河北省秦皇岛市
31	国民党"一大"旧址（包括革命广场）	广东省广州市
32	北京国会旧址	北京市

序号	项目名称	项目地点
33	京张铁路，京张铁路南段至八达岭段	河北省张家口市、北京市
34	齐鲁大学近现代建筑群	山东省济南市
35	东交民巷使馆建筑群	北京市
36	庐山会议旧址及庐山别墅建筑群	江西省九江市
37	马迭尔宾馆	黑龙江哈尔滨市
38	上海邮政总局	上海市
39	四行仓库	上海市
40	望海楼教堂	天津市
41	长春电影制片厂早期建筑	吉林省长春市
42	798近现代建筑群	北京市
43	大庆油田工业建筑群	黑龙江省大庆市
44	国殇墓园	云南省保山市
45	蒋氏故居	浙江省宁波市
46	天津利顺德饭店旧址	天津市
47	京师女子师范学堂	北京市
48	南开学校旧址	天津市
49	广州白云宾馆	广东省广州市
50	旅顺火车站	辽宁省大连市
51	上海中山故居	上海市
52	西柏坡中共中央旧址	河北省石家庄市
53	百万庄住宅区	北京市
54	马勒住宅	上海市
55	盛宣怀住宅	上海市
56	四川大学早期建筑群	四川省成都市
57	中央银行、农民银行暨美丰银行旧址	重庆市
58	井冈山革命遗址	江西省吉安市
59	青岛火车站	山东省青岛市
60	伪满皇宫及日伪军政机构旧址	吉林省长春市
61	中国共产党代表团办事处旧址（梅园新村）	江苏省南京市
62	百乐门舞厅	上海市
63	哈尔滨颐园街一号欧式建筑	黑龙江省哈尔滨市
64	宣武门天主堂	北京市
65	郑州二七罢工纪念塔和纪念堂	河南省郑州市
66	北海近代建筑	广西壮族自治区北海市

序号	项目名称	项目地点
67	首都国际机场航站楼群	北京市
68	天津市解放北路近代建筑群	天津市
69	西安易俗社	陕西省西安市
70	816 工程遗址	重庆市
71	大雁塔风景区三唐工程	陕西省西安市
72	茅台酒酿酒工业遗产群	贵州省遵义市
73	茂新面粉厂旧址	江苏省无锡市
74	于田艾提卡清真寺	新疆维吾尔自治区于田县
75	大连中山广场近代建筑群	辽宁省大连市
76	旅顺监狱旧址	辽宁省大连市
77	唐山大地震纪念碑	河北省唐山市
78	中苏友谊纪念塔	辽宁省大连市
79	金陵兵工厂	江苏省南京市
80	天津广东会馆	天津市
81	南通大生纱厂	江苏省南通市
82	张学良旧居	辽宁省沈阳市
83	中华民国临时参议院旧址	江苏省南京市
84	汉冶萍煤铁厂矿旧址	湖北省黄石市
85	甲午海战馆	山东省威海市
86	国润茶业祁门红茶旧厂房	安徽省池州市
87	本溪湖工业遗产群	辽宁省本溪市
88	哈尔滨防洪纪念塔	黑龙江省哈尔滨市
89	哈尔滨犹太人活动旧址群	黑龙江省哈尔滨市
90	武汉金城银行（现市少年儿童图书馆）	湖北省武汉市
91	民国中央陆军军官学校（南京）	江苏省南京市
92	沈阳中山广场建筑群	辽宁省沈阳市
93	抗日胜利芷江洽降旧址	湖南省怀化市
94	山西大学堂旧址	山西省太原市
95	北京大学地质学馆旧址	北京市
96	杭州西湖国宾馆	浙江省杭州市
97	杭州黄龙饭店	浙江省杭州市
98	淮海战役烈士纪念塔	江苏省徐州市
99	罗斯福图书馆暨中央图书馆旧址	重庆市
100	洛阳拖拉机场早期建筑	河南省洛阳市

中国文物学会 20 世纪建筑遗产委员会顾问及专家委员名单

The List of Consultants and Experts of CSCR Committee on Twentieth-Century Architectural Heritage

顾问（20人）	吴良镛	中国科学院院士、中国工程院院士、清华大学教授
	谢辰生	国家文物局原顾问、中国文物学会名誉会长
	关肇邺	中国工程院院士、清华大学建筑学院教授
	傅熹年	中国工程院院士、中国建筑设计研究院建筑历史研究所研究员
	彭一刚	中国科学院院士、天津大学建筑学院教授
	陈志华	清华大学建筑学院教授
	张锦秋	中国工程院院士、全国工程勘察设计大师
	程泰宁	中国工程院院士、全国工程勘察设计大师
	何镜堂	中国工程院院士、华南理工大学建筑学院院长
	邹德侬	天津大学建筑学院教授
	郑时龄	中国科学院院士、原同济大学副校长
	费 麟	中元国际公司资深总建筑师
	刘景樑	全国工程勘察设计大师、天津市建筑设计院名誉院长
	钟训正	中国工程院院士、东南大学教授
	魏敦山	中国工程院院士、上海现代建筑集团有限公司顾问总建筑师
	王小东	中国工程院院士、西安建筑科技大学建筑系博士生导师
	王瑞珠	中国工程院院士、中国城市规划设计研究院研究员
	刘叙杰	东南大学古建筑研究所教授
	郭 旃	国际古迹遗址理事会副主席
	黄星元	全国工程勘察设计大师、中国电子工程设计院有限公司顾问总建筑师
委员（91人）	马国馨	中国工程院院士、全国工程勘察设计大师
	王时伟	故宫博物院古建部原副主任
	付清远	中国文化遗产研究院原总工程师、中国文物学会传统建筑园林委员会主任委员
	孙宗列	中元国际工程设计研究院原执行总建筑师
	伍 江	同济大学原常务副校长、教授
	刘伯英	清华大学建筑学院教授
	刘克成	西安建筑科技大学建筑学院原院长、教授
	刘若梅	中国文物学会副会长、中国文物学会传统建筑园林委员会副主任委员、秘书长
	刘临安	北京建筑大学建筑与城市规划学院原院长、教授
	刘 谞	新疆城乡规划设计研究院有限公司董事长
	刘燕辉	中国建筑设计院建筑设计总院原党委书记、顾问总建筑师
	吕 舟	中国古遗址保护协会副主席、清华大学国家遗产中心主任
	庄惟敏	中国工程院院士、全国工程勘察设计大师、清华大学建筑设计研究院有限公司院长
	朱光亚	东南大学建筑学院教授
	汤羽扬	北京建筑大学教授
	邵韦平	全国工程勘察设计大师、北京市建筑设计研究院有限公司执行总建筑师
	邱 跃	北京城市规划学会理事长、教授级高级工程师

何智亚	重庆历史文化名城保护专委会主任委员
张之平	中国文化遗产研究院研究员
张玉坤	天津大学建筑学院教授
张立方	河北省文物局局长、中国文物学会副会长
张　宇	全国工程勘察设计大师、北京市建筑设计研究院有限公司副董事长、总经理、总建筑师
张　兵	自然资源部国土空间规划局局长、教授级高级城市规划师
张　杰	全国工程勘察设计大师、北京建筑大学建筑与城市规划学院院长
张　松	同济大学城市规划设计研究院总规划师、建筑与城市规划学院教授
张　谨	中国文化遗产研究院保护工程与规划所副所长
李华东	北京工业大学建筑与城市规划学院教授
李　季	故宫博物院原副院长、中国文物学会副会长
李秉奇	重庆市设计院有限公司原院长、教授级高级建筑师
杨昌鸣	北京工业大学建筑与城市规划学院教授
吴　晓	湖北省文化厅古建筑保护中心副主任、总工程师、研究员
单霁翔	故宫博物院前院长、中国文物学会会长
陈同滨	中国建筑设计研究院建筑历史研究所原所长、研究员
陈伯超	沈阳建筑大学教授、博士生导师
陈爱兰	河南省文物局原局长、中国文物学会副会长
陈　薇	东南大学建筑学教授、博士生导师
周　岚	江苏省住房和城乡建设厅厅长
周　恺	全国工程勘察设计大师、华汇工程建筑设计有限公司董事长、总建筑师
金　磊	中国文物学会20世纪建筑遗产委员会副主任委员、秘书长、中国建筑学会建筑评论学术委员会副理事长、《中国建筑文化遗产》《建筑评论》总编辑
孟建民	中国工程院院士、全国工程勘察设计大师、深圳市建筑设计研究总院有限公司总建筑师
罗　隽	四川大学城镇化战略与建筑研究所所长、教授
侯卫东	中国文化遗产研究员原副院长、总工程师
胡　越	全国工程勘察设计大师、北京市建筑设计研究院有限公司总建筑师
赵万民	重庆大学建筑城规学院原院长、教授、博士生导师
赵元超	全国工程勘察设计大师、中国建筑西北设计研究院执行总建筑师
赵燕菁	厦门市规划局原局长、教授级高级规划师
唐玉恩	全国工程勘察设计大师、上海现代建筑集团有限公司资深总建筑师
唐　凯	住房和城乡建设部原总规划师、高级城市规划师
贾　珺	清华大学建筑学院教授、博士生导师
奚江琳	中国人民解放军陆军工程大学教授
徐苏斌	天津大学中国文化遗产保护国际研究中心副主任、教授
徐　锋	云南省设计院集团有限公司原总建筑师、教授级高级建筑师
殷力欣	《中国建筑文化遗产》副主编、中国艺术研究院研究员
郭卫兵	河北建筑设计研究院有限公司董事长、总建筑师
郭　玲	《中国建筑文化遗产》编委
崔　彤	全国工程勘察设计大师、中科院建筑设计研究院副院长、总建筑师

崔　愷	中国工程院院士、全国工程勘察设计大师、中国建筑设计研究院有限公司总建筑师
崔　勇	中国艺术研究院研究员、《中国建筑文化遗产》副主编
梅洪元	全国工程勘察设计大师、哈尔滨工业大学建筑设计研究院院长
龚　良	南京博物院院长
彭长歆	华南理工大学建筑学院副院长、教授
韩林飞	北京交通大学建筑与艺术学院教授
韩振平	天津大学出版社原副社长、《中国建筑文化遗产》副主编
赖德霖	美国路易维尔大学美术系亚洲美术与建筑助教授、《中国建筑文化遗产》副主编
路　红	天津市规划和自然资源局原副院长、教授级高级工程师
谭玉峰	上海市文物局原副总工程师、研究员
薛　明	中国建筑科学研究院建筑设计院总建筑师
戴　俭	北京工业大学建筑与城市规划学院院长、教授
杨秉德	浙江大学建筑系教授、博士生导师
宋　昆	天津大学建筑学院党委书记、教授
青木信夫	天津大学中国文化遗产保护国际研究中心主任、教授
王建国	中国工程院院士、东南大学城市设计研究中心主任、教授
常　青	中国科学院院士、同济大学城乡历史环境再生研究中心主任、教授
刘加平	中国工程院院士、西安建筑科技大学建筑学院院长、教授
徐千里	重庆市设计院有限公司院长
陈　飞	湖北省文化和旅游厅文物保护与考古处处长
韩冬青	东南大学建筑设计研究院有限公司院长、教授
夏　青	天津大学建筑学院教授
刘丛红	天津大学建筑学院教授
戴　路	天津大学建筑学院教授
姜书明	天津土地利用事务中心正高级工程师
孙兆杰	中国兵器工业集团北方工程设计研究院有限公司总经理、首席总建筑师
徐俊辉	武汉理工大学艺术与设计学院副教授
陈　勇	北京建筑大学建筑与城市规划学院副教授
柳　肃	湖南大学建筑学院教授、中国科学技术史学会建筑史专业委员会副主任委员兼秘书长
周学鹰	南京大学历史学系教授
舒　莺	四川美术学院公共艺术学院副教授
陈　雄	全国工程勘察设计大师、广东省建筑设计研究院有限公司副院长、总建筑师
覃　力	深圳大学建筑设计研究院有限公司总建筑师
李　琦	中建三局集团有限公司总经理
桂学文	全国工程勘察设计大师、中南建筑设计院股份有限公司总建筑师

参考文献

References

1　单霁翔. 20世纪遗产保护 [M]. 天津：天津大学出版社，2015.

2　刘先觉. 现代建筑理论——建筑界和人文科学自然科学与技术科学的新成就 [M]. 北京：中国建筑工业出版社，1999.

3　陈志华. 建筑急需理论 [N]. 人民日报，1988-8-20.

4　金磊. 珍视20世纪建筑遗产 [N]. 人民日报，2020-9-24.

5　张复合，刘亦师. 中国近代建筑研究与保护（十）[M]. 北京：清华大学出版社，2016.

6　彭长歆. 张之洞与清末广东钱局的创建 [J]. 建筑学报，2015，（6）.

7　庞学臣. 中东铁路大画册 [M]. 哈尔滨：黑龙江人民出版社，2013.

8　郑长椿. 中东铁路历史编年（1895—1952）[M]. 哈尔滨：黑龙江人民出版，1987.

9　王尔敏. 五口通商变局 [M]. 桂林：广西师范大学出版社，2006.

10　张松. 历史城市保护学导论 [M]. 上海：同济大学出版社，2002.

11　付雨干，刘松茯. 昂昂溪中东铁路保护建筑群研究 [C]// 浙江省文物考古研究所，宁波市保国寺古建筑博物馆. 2013年保国寺大殿建成1000周年系列学术研讨会论文合集. 北京：科学出版社，2013.

12　顾方哲. 美国波士顿贝肯山历史街区保护模式研究 [M]. 济南：山东大学出版社，2013.

13　刘思铎，陈博. 沈阳近代城市公园的兴起与发展 [M]. 建筑与文化，2012，（9）.

14　朱启钤. 中国营造学社缘起 中国营造学社汇刊（第一册）[M]. 北京：中国营造学社，1930.

15　李海清. 20世纪上半叶中国建筑工程建造模式地区差异之考量 [J]. 建筑学报，2015，（6）.

16　吴焕加. 20世纪西方建筑史 [M]. 开封：河南科学技术出版社，1998.

17　金磊. 传承城市精神的"北京十大建筑" [J]. 百年潮，2019，（6）.

18　金磊. 回望中国20世纪的建筑巨匠 [J]. 百年潮，2020，（9）.

19　刘珊珊. 中国建筑技术发展研究 [M]. 北京：清华大学出版社，2015.

20　钱毅. 从殖民地外廊式到"厦门装饰风格"——鼓浪屿近代外廊建筑演变 [J]. 建筑学报，2011，（5）.

21　建筑文化考察组. 中山纪念建筑 [M]. 天津：天津大学出版社 2009.

22　建筑文化考察组. 抗战纪念建筑 [M]. 天津：天津大学出版社 2010.

23　建筑文化考察组. 辛亥革命纪念建筑 [M]. 天津：天津大学出版社 2011.

24　王斌. 近代铁路技术向中国的转移——以胶济铁路为例（1898-1914）[M]. 济南：山东教育出版社 2012.

25　《建筑评论》编辑部. 中国建筑历程 1978—2018[M]. 天津：天津大学出版社，2019.

26　潘君祥. 中国近代国货运动 [M]. 北京：中国文史出版社，1996.

27　董黎. 中国近代教会大学建筑史研究 [M]. 北京：科学出版社，2010.

28　陈伯超. 沈阳建筑近代化的标志性特征 [J]. 建筑师，2011，（6）.

29　梁思成. 建筑和建筑的艺术 [N]. 人民日报，1961-7-26.

30　伍江. 上海百年建筑史（1840—1949）[M]. 上海：同济大学出版社，2008.

31　周贻白. 中国剧场史 [M]. 长沙：湖南教育出版社，2007.

32　罗小未. 外国近现代建筑史 [M]. 北京：中国建筑工业出版社，2009.

33 上海市城市规划设计研究院. 大上海都市计划 [M]. 上海：同济大学出版社，2014.

34 金磊. 跨越21世纪的《北京宣言》20年——《世界遗产名录》20世纪建筑遗产乃项目研究与借鉴 [M]// 金磊. 中国建筑设计研究历程（三）. 天津：天津大学出版社，2019.

35 金磊. 致敬中国20世纪现代建筑设计的先驱——以传承建筑师沈理源、戴念慈的建筑思想为例 [M]// 金磊. 中国建筑设计研究历程（三）. 天津：天津大学出版社，2020.

36 杨永生. 中国四代建筑师 [M]. 北京：中国建筑工业出版社，2002.

37 刘既漂. 武汉大学建筑之研究 [J]. 前途杂志，1933.

38 薛林平. 建筑遗产保护概论 [M]. 北京：中国建筑工业出版社，2013.

39 国酒茅台. 贵州茅台酒酿酒工业遗产群入选中国20世纪建筑遗产名录 [N/OL]. (2017-12-18) [2021-5-8]. https://www.sohu.com/a/211356564_655363.

40 杨永生. 哲匠录 [M]. 中国建筑工业出版社，2005.

41 杨永生. 建筑百家书信集 [M]. 北京：中国建筑工业出版社，2000.

42 《安全减灾学人写真》编委会. 安全减灾学人写真 [M]. 北京：科学出版社，2000.

43 金磊，李沉，苗淼. "北京建院"援外设计片断的"集体史"[J]. 华建筑，2020，（10）.

44 赖德霖. 近代哲匠录——中国近代重要建筑师、建筑事务所名录 [M]. 中国水利水电出版社，知识产权出版社，2006.

45 黄元炤. 中国近代建筑师系列 柳士英 [M]. 北京：中国建筑工业出版社，2015.

46 胡明，金磊. 厚德载物的学者人生——纪念中国结构工程设计大师胡庆昌 [M]. 天津：天津大学出版社，2018.

47 张复合. 北京近代建筑史 [M]. 北京：清华大学出版社，2004.

48 邹德侬. 中国现代建筑史 [M]. 天津：天津科学技术出版社，2001.

49 北京市规划委员会，北京城市规划学会. 北京十大建筑设计 [M]. 天津：天津大学出版社，2002.

50 杨永生. 建筑百家回忆录 [M]. 北京：中国建筑工业出版社，2000.

51 杨永生. 中国建筑师 [M]. 北京：中国建筑工业出版社，1999.

52 金磊. 中国建筑设计30年（1978—2008）[M]. 天津：天津大学出版社，2009.

53 北京市建筑设计研究院成立50周年纪念丛书编委会. 北京市建筑设计研究院纪念集 [M]. 北京：中国建筑工业出版社，1999.

54 杨永生. 建筑百家轶事 [M]. 北京：中国建筑工业出版社，2000.

55 杨永生，顾孟潮. 20世纪中国建筑 [M]. 天津：天津科学技术出版社，1999.

56 本书编委会. 岁月·情怀——原建工部北京工业建筑设计院同仁回忆 [M]. 上海：同济大学出版社，2015.

57 中国大百科全书·建筑 园林 城市规划 [M]. 北京：中国大百科全书出版社，1988.

58 中国建筑设计研究院建筑历史研究所. 北京近代建筑 [M]. 北京：中国建筑工业出版社，2006.

59 北京古代建筑研究所. 北京古建文化丛书——近代建筑 [M]. 北京：北京美术摄影出版社，2014.

60 首都建筑艺术委员会，北京日报社. 首都新建筑——群众喜爱的具有民族风格的新建筑 [M]. 北京：北京出版社，1995.

61 上海市地方志办公室. 上海名建筑志 [M]. 上海：上海社会科学出版社，2005.

62 天津市历史风貌建筑委员会办公室，天津市国土资源和房屋管理局. 天津历史风貌建筑图志[M]. 天津：天津大学出版社，2013.

63 岭南建筑丛书编辑委员会. 岭南建筑丛书 莫伯治集[M]. 广州：华南理工大学出版社，1994.

64 《中国建筑文化遗产》编辑部. 悠远的祁红——文化池州的"茶"故事[M]. 天津：天津大学出版社，2018.

65 李拱辰. 时光筑梦——从业六十载建筑札记[M]. 天津：天津大学出版社，2020.

66 本书编委会. 新时代 新经典——中国建筑学会建筑创作大奖获奖作品集[M]. 北京：中国城市出版社，2012.

67 卢海鸣，杨新华. 南京民国建筑[M]. 南京：南京大学出版社，2001.

图书在版编目（CIP）数据

中国20世纪建筑遗产名录. 第二卷 / 北京市建筑设计研究院有限公司,中国文物学会20世纪建筑遗产委员会,中国建筑学会建筑师分会编著. -- 天津 : 天津大学出版社,2021.5

（中国20世纪建筑遗产）

ISBN 978-7-5618-6939-0

Ⅰ．①中… Ⅱ．①北…②中…③中… Ⅲ．①建筑－文化遗产－中国－20世纪－名录 Ⅳ．①TU-092

中国版本图书馆CIP数据核字(2021)第095611号

策划编辑　金　磊　韩振平
责任编辑　郭　颖
装帧设计　朱有恒

ZHONGGUO 20 SHIJI JIANZHU YICHAN MINGLU（DI ER JUAN）

出版发行	天津大学出版社
地　　址	天津市卫津路92号天津大学内（邮编：300072）
电　　话	发行部：022-27403647
网　　址	publish.tju.edu.cn
经　　销	全国各地新华书店
制版印刷	北京华联印刷有限公司
开　　本	235mm×305mm
印　　张	36.25
字　　数	540千
版　　次	2021年5月第1版
印　　次	2021年5月第1次
定　　价	368.00元